Cave and Karst Systems of the World

Series Editor

James W. LaMoreaux, P. E. LaMoreaux and Associates, Tuscaloosa, AL, USA

This book series furthers the understanding of cave and karst related processes and facilitates the translation of current discipline-specific research to an interdisciplinary readership by dealing with specific cave or karst systems. Books in this series focus on a specific cave or karst system, on the cave or karst systems of a specific region, on a specific type of cave or karst system, or on any other perspective related to cave and karst systems of the world. The book series addresses a multidisciplinary audience involved in anthropology, archaeology, biology, chemistry, geography, geology, geomorphology, hydrogeology, paleontology, sedimentology, and all other disciplines related to speleology and karst terrains.

Goran Barovic
Editor

Speleology of Montenegro

Springer

Editor
Goran Barovic
Nikšić, Montenegro

ISSN 2364-4591 ISSN 2364-4605 (electronic)
Cave and Karst Systems of the World
ISBN 978-3-031-49374-4 ISBN 978-3-031-49375-1 (eBook)
https://doi.org/10.1007/978-3-031-49375-1

This Springer imprint is published by the registered company Springer Nature Switzerland AG
The registered company address is: Gewerbestrasse 11, 6330 Cham, Switzerland

Paper in this product is recyclable.

Preface

Although it belongs to the group of small European countries in terms of its surface area, Montenegro, when you take into account its physico-geographical and especially speleological characteristics, is an extremely interesting area. Due to its characteristics, it has earned the interest of the most important scientists who have shown and proven with their research that it is a very interesting and unusual area, the results of this research can be found in numerous articles from several centuries ago. It is important to point out two specificities that characterize this type of research. From the sources known so far, until the middle of the last century there are no traces of scientific research in this area by any local researcher, and another important fact is that its territory was explored and important traces were left by the most famous researchers of that time.

Since there was no established order and planned research by the state authorities, the traces of the research in Montenegro are scattered in various foreign journals, which fortunately preserved these valuable works. The turning point in the speleological research of Montenegro is represented by its development within the SFRY (Socialist Federal Republic of Yugoslavia) when numerous organizations and institutions were formed that directly or indirectly dealt with the research of speleological objects on its territory.

Articles describing speleological characteristics of Montenegro were published at various scientific gatherings, congresses, and consultations in numerous scientific fields. It is not possible to find a systematized, precise, and comprehensive presentation of this very interesting scientific field, but rather individual papers from individual research endeavors. It is also important to point out that certain parts of Montenegro have been the target of numerous research teams and that a large number of papers have been published about them. The records available so far tell us that on the territory of Montenegro, about 1,500 objects have been fully or partially explored so far, among which two stand out, which by their dimensions occupy first placements in this part of Europe: the cave in Đalovića gorge with about 20km of channel explored so far and the Iron Pit with a depth of about 1100 m tested so far.

The monograph consists of thirteen chapters in which all the most significant characteristics of its area are given, from an overview of the history of research in this area, its geological, geomorphological, climatological, and hydrological characteristics, i.e. influence of these segments on the creation and development of the speleological object. Special articles deal with biodiversity, soil characteristics, protected speleological objects, the influence of forest cover on the creation and development of speleological objects, objects where archaeological research was carried out, and potential objects with prominent tourist values. Two articles are devoted to characteristic underwater speleological objects as well as important speleological sites of the Montenegrin coast. A separate article is dedicated to the region of Prokletije, which has received the least amount of previous research, because the least amount of speleological research was carried out in that part of Montenegro in general.

This time, in front of the scientific and professional public as well as interested individuals, for the first time everything that is most important and significant, related to speleological research in the area of Montenegro, is gathered in one place. The authors of the articles are eminent experts in the fields in which they covered this extremely interesting topic. We expect that this monograph will attract a lot of attention, but also that it will be only the first in a series of further efforts to disclose the speleological characteristics of Montenegro.

Goran Barovic, Ph.D.
Faculty of Philosophy
University of Montenegro ul.
Danila Bojovica bb.
Niksic, Montenegro

Contents

Historical Overview of Speleological Research on the Territory of Montenegro 1
Goran Barović

Hydrogeological Characteristics of Montenegro 11
Milan M. Radulović

Geology of Montenegro 21
Slobodan Radusinović and Marko Pajović

Hydrology of Surface Water and Climate Characteristics of Montenegro 51
Golub Ćulafić and Jelena Krstajić

Geomorphological Characteristics of Montenegro 67
Gojko Nikolić, Filip Vujović, Goran Grozdanić and Aleksandar Valjarević

Biodiversity of Montenegrin Caves 79
Lidija Polović

Pedological Characteristics of Montenegro 85
Velibor Spalevic

Protected Speleological Objects 99
Duško Vujačić, Đurđica Perović, Goran Grozdanić and Filip Vujović

Speleological Objects as Tourist Motives 109
Đurđica Perović, Duško Vujačić and Goran Grozdanić

Forests of the Karst Region of Montenegro—Main Characteristics and Vulnerabilities 115
Milić Čurović

Archaeological Studies of Caves in Montenegro 123
Dragica Mijanović, Branka Manojlović and Đurđica Perović

Underwater Cave Systems in the Montenegrin Littoral 137
Vladan Dubljević

Marine Caves—Biodiversity and Threats 143
Mačić Vesna, Petović Slavica, Đorđević Nikola, Varda Dušan and Panou Aliki

Caves and Karst of the Prokletije Mountains 157
Ditta Kicińska, Krzysztof Najdek and Adam Łada

Historical Overview of Speleological Research on the Territory of Montenegro

Goran Barović

Abstract

Montenegro represents a very suitable area for the creation of speleological objects, which has been confirmed in its previous research. The geological composition of most of its area, along with the appropriate climatic conditions, caused a very large number of extremely interesting objects to develop in its area. This, in terms of surface, small area, was the area of interest of numerous researchers. It is important to note two elements related to the research of speleological objects in Montenegro. Until the middle of the last century, we didn't have a single domestic researcher who left any trace of this type of research, and on the other hand, almost all the most famous European researchers dealt with its area and left significant results from the conducted research. This paper deals with an overview of the most significant researches, from their beginning to the present day. Today, the past period can be clearly divided into two parts, up to the middle of the last century and after. The first part represents the dominance of foreign researchers, while in the second part we have an organized, systematic and significantly higher quality and advanced research of almost all of Montenegro and almost all areas of speleology conducted by numerous institutions from Montenegro and clubs from the country and abroad.

Keywords

Caves · Karst · Montenegro · Researchers

G. Barović (✉)
Faculty of Philosophy, University of Montenegro,
ul. Danila Bojovica bb., Niksic 81400, Montenegro
e-mail: barovicg@ucg.ac.me

1 Position

Montenegro covers an area between 41°52′ in the south (Vada Island) and 43°22′ in the north (Moćevići in the municipality of Pljevlja), as well as between 18°26′ (Sutorina) in the west to 30°21′ in the east (Jablanica—Rožaje) which positions it in space which provides it with temperate climates in most of the territory. The climate in the whole area is not the same, it changes from coastal in the coastal area through temperate-continental in the central part to continental and mountainous in the mountainous hinterland and the extreme northern part of its territory.

2 General Characteristics

The territory of Montenegro belongs to the southeastern part of the Dinarides, which has a very complex geological-lithological basis on which the influence of the erosion process has left a deep mark and conditioned the formation of a very dynamic relief. Basically, the relief of Montenegro can be divided into an outer and inner part (according to the division of the Dinarides). The outer part is a geospace of deep karst, and the inner geospace is fluvio-karst and fluvio-glacial relief. The main characteristic of the relief of Montenegro is that in a very small geospace, there are very large altitude differences which we see from the following: "Of the total area of Montenegro (13 812 km^2) only 10% of land is up to 200 m above sea level, 35% is between 200 and 1000 m above sea level, 40% is between 1000 and 1500 m above sea level and the remaining 15% of the land is over 1500 m"(Radojičić 1996).

Very favorable climatic conditions with a specific geological and geomorphological basis have caused the development of a large number of speleological objects on the territory of Montenegro, which have become the subject of interest of many scientists who visited it to conduct their research. Researching this topic, we have singled out two

G. Barovic (ed.), *Speleology of Montenegro*, Cave and Karst Systems of the World,
https://doi.org/10.1007/978-3-031-49375-1_1

important characteristics. The first is that, unfortunately, none of the domestic researchers are among the names listed in the research of this area until the second half of the last century, i.e., there is no one from Montenegro who has left an important trace of research procedures. Another important fact is that the area of Montenegro has been researched by a large number of foreign researchers who represent the very top of speleology as a science, but also disciplines that are directly related to speleology and related sciences. This is also a problem due to the collection of traces of research procedures and finding them in general in the fund material in which the results from the performed research are recorded. The first researchers to engage in speleological research were foreigners, French, Czechs, Austrians and Germans (Pretner 1961).

3 Previous Research

Considering the previous research of the area of Montenegro, we can conditionally divide it into two parts. The first period can be attributed to research up to the middle of the last century, which can be said to be mostly accidental, spontaneous and generally not planned, or even done. It was the foreign researchers who traveled to Montenegro through various private channels and visited the then famous facilities based on the stories of residents. Unfortunately, records of this research must be sought in the archives of the countries from which the researchers came, their clubs and their private archives. Fortunately, individuals and groups that have conducted these research expeditions have a long tradition and a high degree of responsibility and seriousness for keeping records and publishing research procedures. The second period of speleological research in Montenegro is significantly more advanced, more successful, more organized and has incomparably better results. In the fifties of the last century, in the then-formed SFR Yugoslavia, numerous organizations and institutions were organized that did not exist before or were only operating in some specific republics. At the initiative of the formed Speleological Association of Yugoslavia, the Speleological Society in Montenegro was formed, which will later turn out to be the backbone of the development of speleological research in Montenegro. After the formation of this society, came the formations of numerous other clubs (mountaineering, archeological,...) which, among other things, dealt with speleological research, which also contributed to the overall results of speleological research in Montenegro. In addition to clubs and societies that directly or indirectly engaged in speleological research in this period, various scientific, professional and economic organizations were formed that, through the nature of their work, needed these types of research. Among the most important state institutions were: the Institute for Geological Research, the Institute for Nature Protection (which was unfortunately closed and merged with the current Environmental Protection Agency), the Institute for Technical Research, and the companies: Electric power industry of Montenegro, Coal Mine in Pljevlja. Bauxite mine in Nikšić and the Yugoslav People's Army, an important instigator of research actions. Due to the implementation of numerous projects, all these institutions and companies have made a huge contribution to the development of speleological research, because, for the first time, actual organized research began, which includes hiring domestic researchers, connecting with fellow researchers from Yugoslavia and abroad. As is the logical sequence of each research, behind every research there are written traces with a serious elaboration of the conducted research procedure.

Clubs from the surrounding area also influenced a big turnaround in intensifying speleological research in Montenegro. Due to the great contribution to the disclosure of research results in Montenegro, the following must be mentioned: Academic Speleological and Mountaineering Club (ASAK)—Belgrade, Speleological Department of Belgrade (SOB)—Belgrade, Society of Researchers "Vladimir Mandic Manda"—Valjevo, Geographical Institute "Jovan Cvijić"—SANU—Belgrade, Speleological group of the Faculty of Geography (SGGF)—Belgrade, HSS (Croatian Speleological Association)—Zagreb, DISKF (Society for Research and Recording of Karst Phenomena)—Zagreb, Cave Association of Slovenia—Ljubljana, Center for Karst and Speleology—Sarajevo, as well as a number of other smaller clubs that were research associates. In addition to the mentioned research groups, researchers from Poland, France, England, the Czech Republic, Slovakia, Hungary and many other countries also conducted research in Montenegro. A major problem here for the speleological science of Montenegro is that, during an extremely advanced and fruitful period, there were no legal regulations that would regulate the procedures for the stay of foreign researchers. As there were no legal norms, the keeping of records on research was not kept in the right way, so researchers from abroad came and performed research actions independently, and the documentation about them ended up in their archives and unfortunately, in an unknown number of cases will remain unavailable for inclusion in the official documentation (a kind of cadastre) kept today by the Environmental Protection Agency. The formation of the list of researched or recorded speleological objects on the territory of Montenegro was organized by the sector of the Environmental Protection Agency in charge of speleology, which sent a request to registered clubs from Montenegro to submit available data on recorded and researched objects. The realization of this important project (in accordance with their abilities) was helped by: Speleological society "Nikšić"—Nikšić, Mountaineering club "Pestingrad"—Kotor, Mountaineering club "Subra"—Herceg Novi, Speleological club Akovo—Bijelo Polje and Biospeleological society of Montenegro—Podgorica. The current records of researched objects are given in Table 7.

The mentioned formation of associations and clubs on the territory of Yugoslavia, and later on Montenegro, increased the number of research actions carried out on the territory of Montenegro, as well as the number of published works in which the results of the mentioned research were presented. Also, a significant contribution was given by the publications of individual professional and scientific institutions that had their regular editions that published the results of speleological research of the entire Yugoslav area and even Montenegro. Of the former Yugoslav (now foreign) publishers who have paid attention to the results of speleological research in Montenegro in their publications, we must single out: Acta Carsologica—Institute for Karst Research ZRC SAZU—Postojna; Naše jama—newsletter of the Cave Research Society of Slovenia—Ljubljana, Speleologist—Speleological Department of the Croatian Mountaineering Association "Željezničar"; Professional bulletin "Our Karst"—published by the Speleological Society of Bosnia and Herzegovina Karst; Proceedings of the Geographical Institute "Jovan Cvijić" SANU; Special editions of the Geographical Institute "Jovan Cvijić" SANU and the Gazette of the Serbian Geographical Society—Belgrade. Of the editions printed in Montenegro we will single out: Special editions of CANU—Podgorica, Gazette of the Department of Natural Sciences of CANU—Podgorica, Gazette of the Republic Institute for Nature Protection and Natural History Museum—Podgorica, Special editions of the Institute for Geological Research—Podgorica, Geological Gazette—geological research—Podgorica.

In the already defined period, a significant contribution to the research of speleological objects of Montenegro was given by numerous economic collectives, which during the implementation of their projects, were often obliged to engage in speleological research. First of all, it can be freely stated that the greatest contribution was given by the Electric power industry of Montenegro, which in the realization of two major projects (construction of Hydrosystem "Gornaj Zeta", later "HPP Perućica" and "HPP Piva") hired speleologists from the region to research speleological facilities in the zones covered by the accumulation lakes. This type of research was of special importance, especially when it came to the construction of reservoirs in the zone of Nikšić field. The first results of the research at the mentioned accumulations gave the result that 330 springs, 880 abysses, 30 estavels and one intermittent spring (mukavica) were found in Nikšić field. In general, the largest number of these objects was found in the zones of construction of reservoirs, so it was necessary to work on their detailed research, so that the continuation of construction would give the planned results. Something similar, but to a lesser extent, was done during the construction of the reservoir in the Piva River Basin, but again with researchers from outside of Montenegro because, back then, there was no qualified research team in Montenegro that would respond to the task in a sufficient way. A significant contribution to the research of this topic was certainly given by the Institute for Geological Research in Podgorica, which initiated the research of a large number of speleological objects through the implementation of a number of projects which were then published in their aforementioned Herald.

Significant contribution was also certainly made by publications from seminars, congresses and other types of professional and scientific gatherings in which papers were published in the form of a collection of papers presenting the results of research from Montenegro. Certainly, most of the papers were published in editions published by the Congress of Speleologists of Yugoslavia.

At the First Congress of Speleologists of Yugoslavia, in a short review, from only one page, "Caves in Montenegro" by B. Pejović, some information about speleological objects of Montenegro was given. Admittedly, the mentioned presentation is quite inaccurate, not systematic and very abbreviated and looks more like a bad newspaper article, but it certainly represents a clue that should be taken into consideration. It is stated that the caves in Montenegro are poorly researched and that the known ones are usually located next to the roads. The paper draws attention to the Lipska cave and explains its closer locality to it, stating that "there are larger extensions, with rich jewelry and lakes" (Pejović 1955). It is noted that some interventions were already made inside the cave at that time, on digging a canal that will drain excess water, which will enable and facilitate movement. It has been announced that the cave canal will be lit by electric lighting and that this cave will serve as a safe place for the water supply to the locals, which will almost never dry out. The author notes that it is assumed that the channel of the Lipska cave is connected with the Strugarska cave, and that one connects to the Obod cave. The paper also mentions the well-known abyss in the Cetinje field and found to be connected with the Crnojević River by Milojević (1935). The existence of several caves that have not been explored (Orjuška, Prolačka, Kozja and Sokolska) was also stated, and caves in some nearby localities (Lisinje, Oštrome, Grbe, Pećki brdo and Građani) are also mentioned. In Nikšić field, a small cave on the hill of Trebjesa is mentioned, which has historical significance because the government of the former Yugoslavia, fleeing from the occupiers, hid paper money in it, which they took with them when leaving the then-occupied Belgrade.

In the period of publishing the work, activities were already started on the construction of the hydro system "Gornja Zeta", i.e., part of the digging of the tunnel from Slivlje to Perućica, where the construction of the hydroelectric power plant was planned, during the mentioned works, unexplored canals were found. Something similar happened during the bombing of the hill Voluica near Bar, during the construction of the "Port of Bar", where a cave was discovered whose canals were rich in jewelry and with a small lake. The paper mentions that during the blasting in the quarry above the Cetinje monastery, a cave was found which was difficult to enter due to the narrow entrance channel. The last object mentioned in the paper is the pit "Duboki do" on Njeguši, which was researched by Mr. Lahner and determined to drain water in the direction of Kotor. It was concluded that the water in that facility was in fact a former river that flowed through the Njeguši field when the evolution of the flow moved it to the underground. (Pejović 1955).

At the Second Congress of Speleologists of Yugoslavia, we have a great work by Egon Pretner, who presents in great detail the data on speleological objects in Montenegro that are mentioned in the list, researched and unexplored objects alike. Egon Pretner is one of the most famous Slovenian speleologists and speleobiologists, who was known for his results not only in his country and the former Yugoslavia, but also in international speleological and speleobiological circles. He researched speleological objects and specialized in insect research. He spent most of his working life at the Karst Research Institute of the Slovenian Academy of Sciences and Arts in Postojna, where he worked until his death in 1982. The introductory part of the paper comments on "Previous research" and mentions the most important names from the world of speleology, such as E.A Mrtel, G. Lahner, G.W. Gesman, B. Wolf, K. Absolon. Martel first commented on the works of P. Rovinski, who in his multidisciplinary research of Montenegro visited Lipska and Obodska and some other caves. At the end of the nineteenth century, he lived in Montenegro (1893) and explored caves along the road Kotor—Njeguši, the abyss under the Granica, Obodska cave, and mentions Sopot near Risan, Obodska, Strugarska, Lipska cave and abysses in Nikšić field. Gesman examined the Lipska Cave in 1905, and in 1907 published its description and draft in the article "Die Tropfsteinhöhle von Lipa in Montenegro" (Pretner 1961). During 1908 and 1909, the famous Austrian speleologist B. Wolf researched speleological objects in Montenegro at several locations: in the zone of the Cetinje fields, the Crnojević River, Podgorica, and in the zone of Slivlje in the Nikšić field. He published the results of the mentioned research as early as 1910 in the works "Höhlenforschungen in Montenegro" and in 1911 "Bericht über die 1910 in Montenegro vorgenommenen Höhlenforschungen". The works deal with the problems of water supply and drainage, but he notes that the cave on Pecko brdo in Zaćir and the Lipska cave have the potential for tourist valorization (Pretner 1961).

During the First World War, the Austrian speleologist Georg Lahner researched speleological objects for the purpose of finding drinking water for the Austrian army in the zone between the Bay of Kotor and Skadar Lake. During that period, he explored 23 objects in the Njeguš zone and explored the then-deepest pit in the world, Duboka do in Njeguško polje, with a depth of 340 m which was entered into world literature after his report. He published the results of his research in the works "Der westmontenegrinische Karst und sein hydrologischer Zusammenhang mit der Bucht von Cattaro" published in 1917 and "Im Karst der Schwarzen Berge" published in 1919 (Pretner 1961).

Research on the organization of the Austrian army during the First World War can be found in the records of the Czech speleologist Absolon, in works from 1943. He notes that the Austrian officers Hruboš, Novotny, Dobeš, Leifel and some others explored some 186 caves and pits in the period "from April 12, 1916 to July 14, 1917, in Krivošije". Unfortunately, no written traces of this research have been found (Pretner 1961).

Significant research was conducted in the period between 1955 and 1957, when researchers from Slovenia, together with the famous Yugoslav speleologist Jovo Petrović, explored the Nikšić field, in order to determine the number and characteristics of speleological objects for the construction of the Gornja Zeta hydro system. In the same period, organized by the Slovenian Academy of Sciences and Arts, bio-speleological research was organized in the zone of Skadar Lake, Rumija, Virpazar, Rijeka Crnojević and Njeguš (Pretner 1961).

The fact that bio-speleological research has an advantage over other studies of speleological objects in Montenegro is shown by the fact that back in 1864 the first insect (Neotruchus uturalis) from its underground was described thanks to a German naturalist named Ludwig Wilhelm Schaufuss. Apart from him, G. Paganetti-Hummler, J. Kratochvil and J. Mat'cha also dealt with the fauna of the Montenegrin underground. It is interesting to note that E. Pretner points out that most of the speleological objects in this area were investigated by the Austrian speleologist L. Weirather, but that his reports are inaccurate, because he stated wrong and fictitious names and descriptions of nearby or distant sites near which the object is located. Also, it should be noted that the objects in the north of Montenegro were examined by the French bio-speleologist P. A. Rémy, who in the period from 1930 to 1933 explored the areas around Berane and Pljevlja. It should be noted that E. Pretner himself researched speleological objects in Montenegro on several occasions in the Bay of Kotor, around Podgorica, Virpazar, Lovćen, Rijeka Crnojević, Nikšić, Šavnik and Žabljak.

The second part of the paper entitled "List of caves and pits" has 14 special units named as geographical units at the discretion of the author. The internal structure of the units is not the same for everyone, but is generally divided into three parts: A. Objects mentioned in the literature, with the name of the object, length, orientation toward the sides of the world (probably the entrance), and the name of the researcher from whom the data was taken; B. Unmentioned but researched objects in the literature, which also gives the name of the object, brief information about the location of the object and the name of the researcher; and C. Unexplored objects in which the names of objects with a short determinant of its locality are stated (Table 1).

As can be seen from the tabular overview, about two thirds of the listed objects have already been mentioned in the literature, while more than twenty objects are mentioned with the problem of listing objects from the works of L. Weirather due to poorly mentioned names and locations.

Mihail Vuckovic's work entitled "Study of the Montenegrin Underground" was published in the 1966 edition of "Our Mountains" in the journal of the Mountaineering Association of Croatia and the Mountaineering Association of Bosnia and Herzegovina No. 11–12 from 1966. It is important to single out the allegations from the paper about the stay in Montenegro of Henry Layard and Edvard Levich Mitword who came at Njegoš's request to solve the water supply problem by trying to find water underground, primarily from the Lipska Cave. It is also important to mention the Englishman Garden Wilkinson, who lived in Montenegro in 1844 and explored the Old Montenegrin area, Ostrog and Slivlje in the Nikšić field. Also, the research of P. Rovinski and his work with the Cetinje field, the Lipsko and Obod caves, also left a trace on the Gornjepolje vir, one of the most famous and largest estavels in this part of the Dinarides. The paper also highlights the contribution of the Italian researcher L. Baldacci, who stayed in Montenegro in 1886 and visited several localities. He stayed in Durmitor, Lukovo near Nikšić, where he explored the "Lepezica" cave, but also the Slivski abyss, which he describes as a large crack, irregular in shape with steep sides formed in limestone. In the continuation of the paper, M. Vučković superficially mentions several great researchers, such as the great German geomorphologist Kurt Haserat, who visited Montenegro on several occasions, but the paper mentions his movement around Durmitor and his interest in the Gornjepoljski whirlpool. Cvijić and his work on the glaciation of Durmitor is also mentioned, as well as the research of the Ice Cave on Durmitor and Obodska Cave. Also, in the form of enumeration, he mentions the French researcher A. Martel, whose movement and research coincide with those stated in the work of E. Pretner. There are two elements mentioned in this paper that are very important. The first is that on June 15, 1896, in "Glas crnogorca" it was written about the Lipska cave where its interior, decorations and water were described, which is probably the first mention of a speleological object in a local newspaper. The

Table 1 Tabular overview of caves and pits according to E. Pretner

No	Name of the locality	Objects mentioned in the literature	Unmentioned but researched objects	Unexplored objects	Total
1	Sandžak (Pljevlja)	9	–	1	**10**
2	Durmitor	6	10	14	**30**
3	Mt. Sinjajevina, Bjelasica, Mokra mountain	7	2	7	**16**
4	Maglić, Bioč, Ledenica i Golija	8	–	–	**8**
5	Vojnik, Moračka kapa, Komovi, Vizitor and Prokletije	2	–	9	**11**
6	Žijevo mountain and the Kučka plateau	7	–	9	**16**
7	Titograd	4	1	1	**6**
8	Ulcinj, Rumija i Virpazar	12	13	6	**31**
9	Lovćen	57	8	1	**66**
10	Bay of Kotor	17	2	1	**20**
11	Krivošije	64	–	–	**64**
12	Grhovo	16	2	1	**19**
13	Troglav i Njegoš	5	1	–	**6**
14	Nikšićko field	9	18	7	**34**
	Total	**223**	**57**	**57**	**337**

other important element is that on July 1, 1896, the first excursion to the Lipska Cave was organized, consisting of diplomatic representatives and some famous people of that time. It was recorded that they were "amazed by the beauty of the cave, so they compared it to the Postojna Cave". It is further stated that there are problems with the entrance to the cave, which was entered for a while by a manual lift, and the problems were caused by a strong wind blowing from the cave. Later, the entrance was rebuilt with the construction of 85 stone steps, after which it is stated that there is almost no day that someone does not visit the cave. Other statements in the paper largely coincide with the work of E. Pretner published at the Second Congress of Speleologists of Yugoslavia (Vučinić 1966).

At the Seventh Yugoslav Speleological Congress held in Herceg Novi in 1976, M. Lješević presented a paper which, represented the most accurate presentation of speleological objects examined in Montenegro to that day. The first part of the paper is an introduction to the basic topic of the paper, followed by a list of caves and pits. It is noted that the area of Montenegro was, according to its specifics, the subject of study of numerous categories of speleologists, from amateurs to the most famous world names related to speleology as a science and its related disciplines. There are many problems in finding and combining literature. The available literature is often scanty, inconspicuous, gives very little data and even such data is very difficult to come by. Plans and descriptions of facilities are very rare, as are precise explanations of the location. This data was necessary when registering in the cadastre and arranging it, so the data obtained did not give the possibility of forming a correct picture of speleological objects in Montenegro. A special problem, M. Lješević pointed out, is the records of researchers, mostly foreign ones, who gave objects arbitrary or fictitious names, inaccurate and incorrect data. As for the sub-areas that originated from speleology, based on the results known so far in the research of Montenegro, biospeleology has come the furthest, and speleoarchaeology is the least developed. The attached list of speleological objects includes 529 objects, most of which were taken over by E. Pretner presented at the Second Yugoslav Speleological Congress (Table 2). The author notes that he is aware of the numerous shortcomings that we can see in the paper, but that it represents a solid basis for a more solid file and a better cadastre that should be kept by a competent state institution (Lješević 1976). Although the author of the paper expressed modesty, it must be stated that this is the best presentation of speleological objects in Montenegro in one place. An overview of speleological objects in Montenegro is given by regional units, which are more precisely and logically formed, in relation to the presentation by E. Pretner. Also, a tabular overview of the presented objects is given incomparably more precisely, for twelve regional units in twelve columns. The first column provides information on the number of files; The second is the cadastral card number; The third is the name of the object; The fourth is the type of object (cave or pit); The fifth is the name of the settlement, (wider locality); The sixth is the locality (narrower, more precise positioning of the object in space); The seventh is the length of the object; The eighth is the depth of the object; The ninth is the absolute height of the object; The tenth is the hydrological property; The eleventh is the sea and river basin; and in The twelfth is a page number from the attached literature in which the object of study is mentioned (Lješević 1976) (Table 2).

Table 2 Tabular overview of caves and pits according to M. Lješević

No.	Regional unit	No. of objects
1	Bay of Kotor with the Montenegrin coast	16
2	Skadar valley and Bjelopavlićka plain	54
3	Rumija, Lovćen and the Katun plateau	89
4	Orjen, Krivošije, Grahovo, Rudine and Banjani	80
5	Nikšićko field	33
6	Kučka plateau and Žiovo	22
7	Prekornica, Moračke mountains and Sinjajevina	18
8	Vojnik, Golija and Ledenica	24
9	Piva, Volujak i Bioč	94
10	Durmitor, Jezera and Drobnjaci	71
11	Ljubišnja and the valley of Ćehotina	18
12	Polimlje, Prokletije and Ibar	10

At the next, Eighth Yugoslav Speleological Congress, M. Lješević and V. Barović presented the work "The largest speleological objects in Montenegro", (Table 3) which gave a brief overview of the objects that stand out from the others in terms of their dimensions, which have been established by previous research. It is important to point out that the authors state that "720 objects have been completely or partially explored in Montenegro". Also, they point out that they "recorded another 1264 speleological objects, in less speleologically examined terrains of northern Montenegro". Based on the known situation on the examined and unexamined terrains, they predict that there should be more than 5000 objects in Montenegro (Lješević and Barović 1981) (Tables 3 and 4). The statement presented in the paper, which can still be confirmed today, is that based on the available data, an uneven distribution of speleological objects in Montenegro is observed, which is not a real situation but a consequence of their unexploredness. The paper tabulates the data for the twenty longest caves and ten deepest pits that were explored until 1980 (Table 4). From both groups, a short textual presentation is given for four with the highest values from their group, and here we transfer data for five objects with the highest values.

The next, Ninth Yugoslav Speleological Congress, held in Karlovac, 1984, brings another work by M. Lješević with

Table 3 Tabular presentation of the longest caves explored until 1980 by Lješević and Barović (1981)

No.	Object name	Sliv	End	Location	Length in m
1	Grbočica	Skadarsko j	Crmnica	Trnovo	2650
2	Cetinjska cave	Skadarsko j	Cetinje	Cetinje	1420
3	Džupanska cave	Lim	Gornja sela	Lubnic	1187
4	Lipska cave	Skadarsko j	Dobrsko selo	Lipa	1140
5	Todorova pit	Piva	Pivska planina	Unač	900

Table 4 Tables of the deepest pits explored up to 1980 (Lješević and Barović 1981)

No.	Object name	Sliv	End	Location	Depth in m
1	Duboki do	Jadransko m	Njeguši	Njeguši	350
2	Todorova pit	Piva	Pivska planina	Unač	316
3	Gropa	Lim	Prokletije	Čardak	280
4	Djevojačka kapa	Zeta	Garač	Zagreda	365
5	Vjetrena pit	Jadransko m	Orjen	Malov do	226

Table 5 Tabular presentation of the longest caves explored until 1984 according to Lješević (1984)

No.	Object name	Area	Locality	Length in m
1	Lipska Pećina	Dobrsko selo	Lipa	3410
2	Grbočica	Crmnica	Trnovo	2650
3	Začirska pećina	Ceklin	Začir	2180
4	Cetinjska pećina	Cetinjsko polje	Cetinje	1420
5	Jama u sjever-noj Obručini	Durmitor		1224

Table 6 Tabular presentation of the deepest pits explored until 1984 according to Lješević (1984)

No.	Object name	Area	Locality	Depth in m
1	Jama u malom lomnom dolu	Durmitor	M.Lomni do	605
2	Jama u sjever-noj Obručini	Durmitor	Uvita greda	461
3	Jama u Vjetrena brda	Durmitor	Vjetrena brda	410
4	Duboki do	Lovćen	Njeguši	350
5	Todorova jama	Pivska planina	Unač	316

a similar title and content, but with new data and morphometric indicators for speleological objects that stood out from the others in terms of their dimensions. The paper in the first part treats two areas that have stood out so far in terms of the number of surveyed objects. It is about the area of the mountain Durmitor and Orjen with Krivošije. Durmitor is explained as a high mountain karst, which with exceptional tectonic activity and highly dissected terrain with dynamic hydrology of the terrain was a suitable soil for the formation of a large number of objects. Intensive tectonic activity is present near Orjen and Krivoši, with intense atmospheric precipitation, which conditioned the formation of a large number of speleological objects. The author tabulates the longest and deepest objects that have been measured so far (Lješević 1984) (Tables 5 and 6).

Compared to the previous congress, in just four years, there was a change in the order of the caves with the longest length, where Lipska was "extended" by more than two kilometers, Grbočica and Cetinjska cave kept the same length, and Džupanska cave and Todorova pit "dropped out" from the first five, and the Začirska cave and the pit in the Northern Obručina "jumped in" (Lješević 1984).

Comparing the data presented in this paper with the previous one, presented at the Eighth Yugoslav Speleological Congress, a significant difference can be noticed, where it can be seen that from the first five only the Duboka do pit retained its "placement" while the other four newly explored objects had significantly higher dimensions than the previous ones. This confirms that there were many intensive research procedures in this short period on the territory of Montenegro. The continuation of the work is dedicated to a brief description of several objects that have not been previously treated or there have been significant changes compared to the previous period. Additional explanations are given for the pit in Mali Lomni dol, pit in Northern Obručina, pit in Vjetreni brdi, Lipska cave and Začirska cave.

At various scientific, professional and other gatherings that dealt with speleological topics, but also at gatherings of some related disciplines, a large number of individual objects were covered, which we will not mention in this paper.

It took a long time for the national institutions to agree and decide who is responsible for everything that is important and related to speleological objects, as well as who will pass and formulate the legal regulations that should bring everything into the legal framework. After several attempts, finally this big responsibility went to the Environmental Protection Agency, which after its formation, but also after great organization and internal structuring, due to the large amount of work that it had, started collecting materials needed to form a cadastre of speleological facilities in Montenegro. A simple way of collecting data that was in the registered organizations that dealt with speleological research until then was attempted, which initially gave great results. Data was collected for close to 1600 objects that were entered into a database formed in Excel format. The database has seventeen columns that give quite detailed information about the objects that have been recorded, at least for those objects where all columns are filled. It starts from the ordinal number (1), cadastral number (2), followed by the name of the object (3), location (4), type of object—cave or pit (5), N—E—H, i.e., data on longitude, geographical latitude and altitude (6–7-8), then rectangular coordinates x, y and z (9–10-11), length of the object (12), depth of the object (13), degree of research, (14) draft, i.e., whether there is a topographic image of the object or not (15), the researcher, i.e., the name of the researcher of the object or club that conducted the research (16) and as the last column is a comment, in which some important facts related to the object in question are entered (17). Data are entered by municipalities, but the data is not complete for all entered objects. A total of 29 objects were registered for the municipality of Danilovgrad, Nikšić 183 objects, Žabljak 414 objects, Herceg Novi 44 objects, Kotor 24 objects, Cetinje 158 objects, Tivat 1 object, Plužine 121 objects, Šavnik 24 objects, Plav & Gusinje 71 objects, Pljevlja 43 objects, Bijelo Polje 114 objects, Mojkovac 36 objects, Petnjica 24 objects, Bar 26 objects, Budva 14 objects, Berane 90 objects, Podgorica 107 objects, Kolašin 64 objects (Table 7).

Table 7 Tabular overview of recorded speleological objects according to the cadastre of the Environmental Protection Agency

No.	Municipality	Number of objects
1	Bar	26
2	Berane	90
3	Bijelo Polje	114
4	Budva	14
5	Cetinje	158
6	Danilovgrad	29
7	Herceg Novi	44
8	Kolašin	64
9	Kotor	24
10	Mojkovac	36
11	Nikšić	183
12	Plav&Gusinje	71
13	Pljevlja	43
14	Plužine	121
15	Petnjica	24
16	Podgorica	107
17	Šavnik	24
18	Tivat	1
19	Žabljak	414
	Total	1587

Table 8 Tabular overview of the longest speleological objects according to the cadastre of the Environmental Protection Agency

No.	Object name	Area	Locality	Lenght in km
1	Cave above Vražiji firovi	Bijelo polje	Bistrica	20
2	Dvogrla pit	Lovćen	Majstori	7.5
3	Njegoševa cave	Lovćen	Njeguši	6.1
4	Željezna pit	Maganik	Trešteni vrh	4.1
5	Lipska cave	Cetinje	Lipa	3.4

Table 9 Tabular overview of the deepest speleological objects according to the cadastre of the Environmental Protection Agency

No.	Object name	Area	Locality	Depth in m
1	Željezna pit	Maganik	Trešteni vrh	1173
2	Pit on Vjetreni brdi	Durmitor	Vjetrena brda	775
3	Dvogrla pit	Lovćen	Majstori	715
4	Pala skala	Lovćen		667
5	Budna pit	Durmitor		657

Based on the available data, we can select the objects with the largest length and the largest depth according to length and depth (Tables 8 and 9).

Comparing this data with that presented in the work of M. Lješević at the 9th Yugoslav Speleological Congress, we notice that of the former first five objects, only one is now on the list, Lipska pećina, which moved from the first to the fifth place. The progress in research and discovery of objects is very obvious.

Something similar happens when it comes to objects with the greatest depth. And here only one object is repeated from the previously announced ones. This is the Pit on Vjetreni brdi, but its depth has also increased significantly, from 410 m in 1984 to 775 m according to the latest data. Other objects are "new" on the list of the deepest objects in Montenegro.

4 Conclusion

Based on the insight into the available documentation from the research conducted on the territory of Montenegro, it can be stated that a lot of work has been done and it can be concluded that the research of objects is at a high level, but of course there is room for further progress in all directions. Further research needs to be directed in several directions. One direction is to continue research in objects that have a further perspective of detecting channels that have not yet been explored, which was stated in the works that showed the objects explored so far. The second direction is to direct research in the direction of discovering new objects and research of some parts of the territory of Montenegro that are not or have been only partially explored. The third direction, but no less important than the previous ones, is the collection of material from research conducted by domestic and foreign individuals and clubs, and not submitted to the competent state authorities of Montenegro or partner clubs with which they conducted research. It is easy to notice that this text mentions data from many foreign researchers who dealt with Montenegro, which shows that the number of ascertained and researched objects is significantly different—higher than the one that is today in the database of the Environmental Protection Agency. Based on the above, we conclude that all those who speleologically researched Montenegro did not respond to the former invitation to submit data on the performed research to the Agency. The reasons for this move may be different, but it is still up to the competent state authorities to continue collecting existing material in order to at least approach the final list of investigated facilities.

References

Lješević M. (1976). Istraženost speleoloških objekata Crne Gore, Sedmi jugoslovenski speleološki kongres, Herceg Novi 1976.

Lješević M. (1984). Najduži i najdublji speleološki objekti u Crnoj Gori, Deveti jugoslovenski speleološki kongres, Karlovac/Zagreb,

Lješević M., Barović V. (1981). Najveći speleološki objekti Crne Gore, Osmi jugoslovenski speleološki kongres, Beograd,

Pejović B. (1955). Pećine u Crnoj Gori, Prvi jugoslovenski speleološki kongres, Slovenska akademija znanosti i umetnosti, Ljubljana (36)

Pretner E. (1961). Speleološka istraživanja u Crnoj Gori i spisak pećina i jama, Drugi jugoslovenski speleološki kongres, Zagreb.

Radojičić B. (1996). Geography of Montenegro, (1–228), Nikšić, UNIREKS

Vučinić M. (1966). Proučavanje crnogorskog podzemlja, Naše planine, PSH i PSBiH, Zagreb/ Sarajevo

Hydrogeological Characteristics of Montenegro

Milan M. Radulović

Abstract

Carbonate rocks (limestone and dolomite) cover over 60% of Montenegro. The high permeability of the karst aquifer is a consequence of a well-developed network of caves and conduits. Due to the extensive presence of karstified rocks, groundwater flow is often unpredictable, which has been shown many times through unexpected tracer test results. Generally, a carbonate aquifer is recharged by the infiltration of atmospheric water and discharged through the large springs distributed along the coast, valleys and canyons. These karstic springs represent the main drinking water sources, but wellfields in granular aquifers are also important for water supply and irrigation. Establishing adequate groundwater protection is currently one of the biggest environmental challenges in Montenegro.

Keywords

Groundwater · Aquifers · Hydrogeology · Karst · Montenegro

1 Introduction

Among other things, Montenegro is known as a karst country. The most important preconditions for the speleogenesis of Montenegrin karst systems probably were the great distribution of the soluble carbonate rocks, high degree of tectonic deformation and water abundance.

Over 60% of the territory consists of limestone and dolomite. Subsurface karst features are generally elongated along large faults. The density of caves and conduits is very high, so groundwater flows are often complex and unpredictable.

The hydrogeological characteristics of Montenegro have been studied by a large number of researchers (Cvijić 1899; Torbarov and Radulović 1966; Bešić 1969; Vlahović 1975; Radulović 1989, 2000, 2012; Radulović et al. 1989, 1998, 2013, 2015; Hrvačević 1999, 2004; Radulović and Radulović 1999, 2004; Lješević 2004; Burić 2010; Radojević et al. 2011; Dević 2011; Stevanović et al. 2016; Sekulić and Radulović 2019).

This chapter summarizes the geology, tectonics and hydrogeology of Montenegro. The hydrogeology, as the crucial part, is described separately for the main river basins.

The manuscript has been written (at the invitation of the editor) as one of the introductory chapters for the book entitled *Speleology of Montenegro* (the book series: *Cave and Karst Systems of the World*).

2 An Overview of the Geology and Tectonics of Montenegro

The geological exploration of Montenegrin territory began at the end of the nineteenth century. The first systematic investigations were carried out by foreign researchers (Tietze 1884; Baldacci 1886; Hassert 1895; Cvijić 1899; Nopcsa 1916; Bourcart 1926; Waisse 1948). From the '50 s of the last century, the number of geological studies increased sharply (Milovanović 1965; Bešić 1969; Grubić 1975; Vlahović 1975; Mirković et al. 1985).

Limestone and dolomite are the most widely distributed rocks in Montenegro. From the geological map (Mirković et al. 1985) it can be seen that the carbonate rocks are mostly marked as Triassic, Jurassic and Cretaceous

M. M. Radulović (✉)
Faculty of Civil Engineering, University of Montenegro, Cetinjski put bb, 81000 Podgorica, Montenegro
e-mail: radulovicmilan33@yahoo.com

G. Barovic (ed.), *Speleology of Montenegro*, Cave and Karst Systems of the World,
https://doi.org/10.1007/978-3-031-49375-1_2

formations. These sediments are characterized by great thickness, which can often exceed 5 km vertically.

Foliated metamorphic rocks and flysch sediments have significant distribution in north-eastern Montenegro (Mirković et al. 1985).

The large valleys, such as Zeta and Bjelopavlići, are filled with thick deposits of gravel, sand and clay.

The territory of Montenegro (Fig. 1) is divided into four separate tectonic zones (Mirković et al. 1985; Radulović et al. 2015; Sekulić and Radulović 2019):

1. The Paraautochton zone that is distributed along the coast (it is built by limestone and flysch);
2. The Budva–Cukali zone that occupies the narrow belt in the coastal area (it consists of marlstone, shale, sandstone, chert, limestone and igneous rock);
3. The Visoki Krš zone is distributed in the south and central Montenegro (it is built by thick deposits of carbonate rocks) and
4. The Durmitor zone occupies northern Montenegro (carbonate rock, flysch sediments and foliated metamorphic rock are most distributed).

The boundaries between tectonic zones are mostly represented by large reverse faults. The general direction of tectonic forms (folds and faults) is northwest-southeast, which is characteristic of the entire Dinaric region. Each tectonic zone is pushed over the southern neighbour zone (e.g., the Visoki Krš rides over the Budva–Cukali zone) (Mirković et al. 1985; Radulović et al. 2015).

3 The Hydrogeology of the Adriatic Sea Basin

The carbonate (karst) aquifer is widespread in Montenegro, especially within the Adriatic Sea Basin (Fig. 2). It is represented by limestone and dolomite. These rocks are very heterogeneous in terms of permeability, but generally, they belong to the group of highly permeable hydrogeological units. The granular aquifer has the largest distribution in the Zeta Valley, Niksić and Ulcinj. A fissured aquifer is characterized by low permeability. It is mostly distributed within the Budva–Cukali tectonic zone (Radulović and Radulović 2004).

Impermeable hydrogeological units are more present in the coastal area, upper parts of the Morača River basin, as well as along the Bjelopavlićka Valley and its northern edge (Sekulić and Radulović 2019).

3.1 The Carbonate Aquifer of the Adriatic Sea Basin

The main source of carbonate aquifer recharge is atmospheric water which infiltrates through the fissured and karstified aeration zone. The high karst plateaus, with altitudes of above 800 m asl, represent the main recharge zones. These areas are characterized by a large number of dolines and dry vertical caves. Settlements in the high karst plateaus often have problems with water supply, despite the huge amount of precipitation (e.g., the waterless karst area above Boka Bay has a precipitation rate of over 3000 mm/year). Most of the atmospheric water quickly percolates to the ground, recharging the carbonate aquifer at a certain depth and discharging only at much lower points (along the coast, canyons or large valleys). It is very demanding to find groundwater in these high karst plateaus. Sometimes even wells that are a few hundred metres deep are not sufficient to reach the water table. Periodical streams (that appear only after intense rainfall) sink very quickly through the swallow holes that are very frequent in these areas (Sekulić and Radulović 2019).

The infiltrated water flows relatively quickly from the recharge zone to a spring. Sometimes it takes a matter of days to reach a spring. The main reason for high groundwater velocity is a well-branched subsurface network of caves and conduits. One such karst network is developed in the catchment area of Crnojevića spring (Fig. 3), which consists of the Cetinjska cave, the Lipska cave and the Obodska cave (Petrović 1980; Radulović et al. 2015; Barović et al. 2018). According to previously performed tracer tests on the territory of Montenegro, the mean groundwater flow velocity is around 2.65 km/day (Djordjević et al. 2010; Sekulić and Radulović 2019).

Large karstic springs (Fig. 2; Table 1) are mostly located along the edges of valleys, canyons and the shores of lakes and the sea (Sekulić and Radulović 2019). The main such areas in Montenegro are Boka Bay, the Zeta Valley, the Bjelopavlića Valley, the Nikšić Valley; Lake Skadar, Lake Bilećko; the canyons of the Morača, the Mrvica (Fig. 4), the Mala Rijeka, the Cijevna, the Piva and the Tara.

Along the shores of Boka Bay and Lake Skadar, besides terrestrial springs, there are many vruljas (submarine or sublacustrine springs).

While sub-vertical corrosion caves are mostly distributed in the recharge zones (karst plateaus), the discharge zones are often characterized by large fluvial caves with karst springs at their mouths (e.g., Risanska Spilja, Sopot, Ljuta and Crnojevića spring).

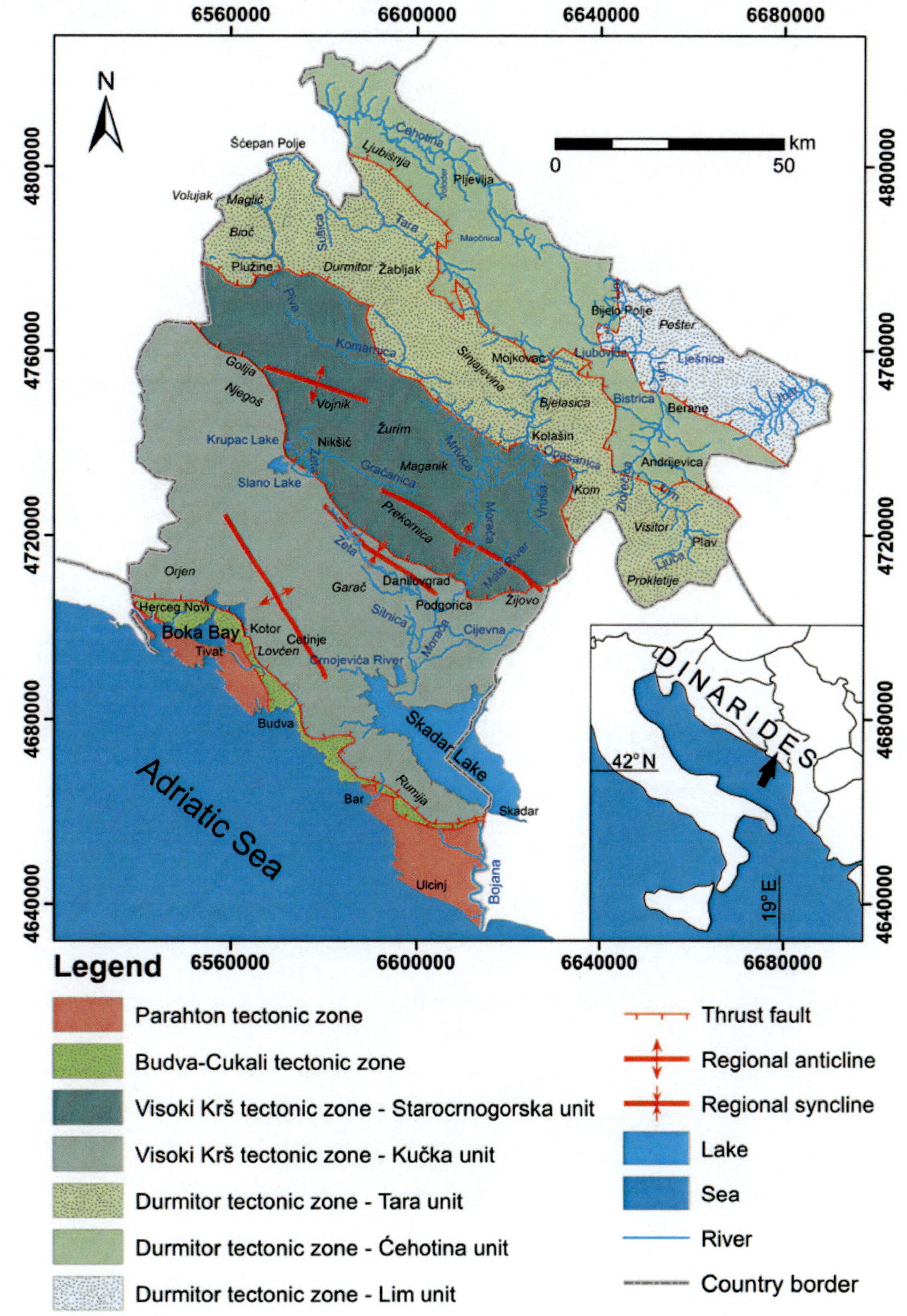

Fig. 1 Tectonic map of Montenegro (after Mirković et al. 1985; Radulović et al. 2015; Sekulić and Radulović 2019). Reprinted/adapted by permission from Springer, The handbook of environmental chemistry, The rivers of Montenegro by Pešić V, Paunović M, Kostianoy A (eds), vol 93, pp 13–42, Sekulić G, Radulović MM, The hydrology and hydrogeology of Montenegro, Fig. 2, Copyright 2019

The groundwater from the carbonate aquifer generally has low mineralization. The dominant ions are bicarbonate and calcium. Water quality problems appear mostly after extreme rainfall when there is increased turbidity in karst aquifer water. In coastal water sources, high salinity is a common problem during the dry period of the year, as a consequence of seawater intrusion.

3.2 The Granular Aquifer of the Adriatic Sea Basin

From the hydrogeological map (Fig. 2), it can be seen that the granular aquifers are present in the Ulcinj Valley, the Zeta Valley and Nikšić polje. Significant sources of granular aquifer recharge are infiltration of river water and

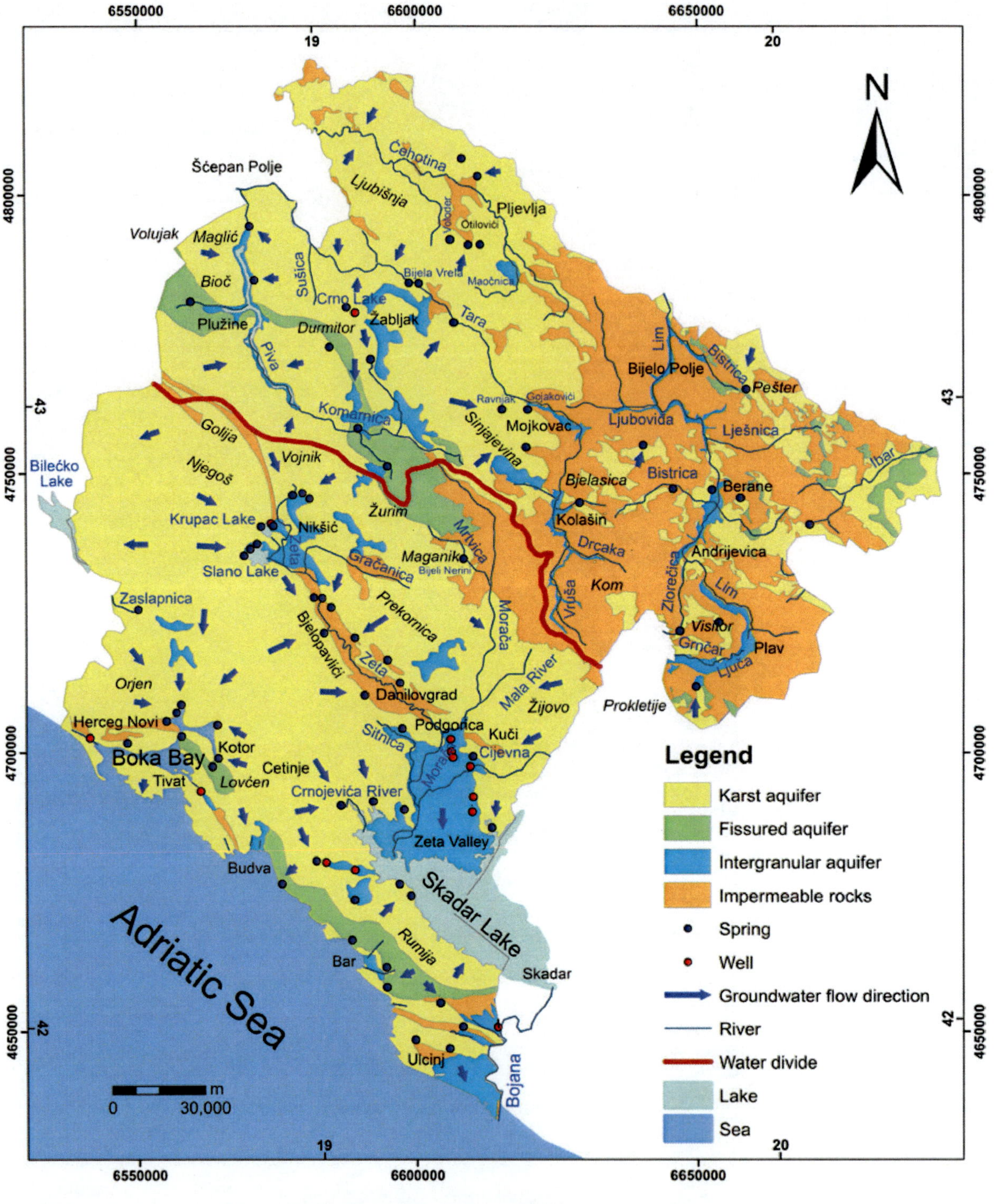

Fig. 2 Hydrogeological map of Montenegro (Sekulić and Radulović 2019). Reprinted/adapted by permission from Springer, The handbook of environmental chemistry, The rivers of Montenegro by Pešić V, Paunović M, Kostianoy A (eds), vol 93, pp 13–42, Sekulić G, Radulović MM, The hydrology and hydrogeology of Montenegro, Fig. 3, Copyright 2019

subsurface inflow from the neighbouring carbonate aquifer. The groundwater flow direction in the granular aquifers of the Adriatic Sea Basin is generally from north to south.

The Zeta Valley granular aquifer is one of the richest aquifers in the country. It is characterized by very high permeability. Groundwater from this aquifer naturally flows out via many small springs distributed along the northern shore of Skadar Lake. The granular aquifer of Zeta Valley is also known for its high yield. There are many production wells all around the valley which are used for different purposes such as water supply, irrigation and industry. In the industrial and agricultural parts of the valley (the Aluminium plant, the vineyards of "Plantaže", etc.), the groundwater quality is disturbed (Radulović 2018). In most

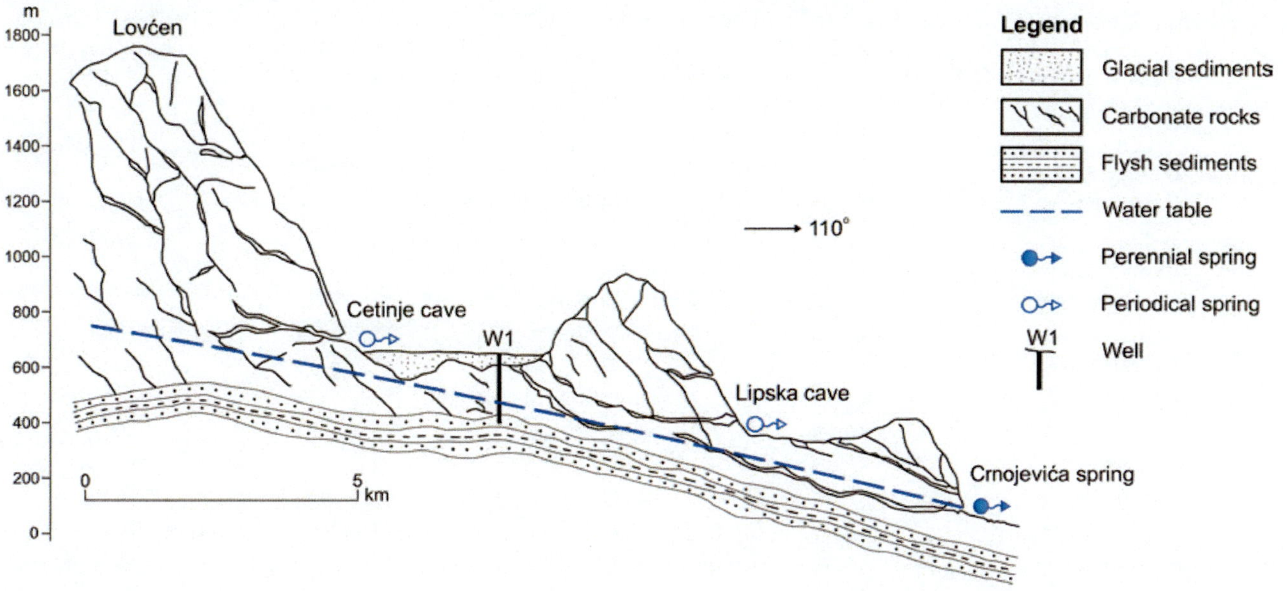

Fig. 3 Hydrogeological section Lovćen–Crnojevića spring (modified Radulović et al. 2015). Reprinted by permission from Springer, Environmental Earth Sciences, 74, 2015, 71–82, Radulović et al., Hydrogeology of the Skadar Lake basin (Southeast Dinarides) with an assessment of considerable subterranean inflow, Fig. 3, Copyright 2015

samples, increased concentrations of nitrates and phosphates are measured, but registered values are still below the threshold values (IHMS 2020).

4 The Hydrogeology of the Black Sea Basin

The carbonate aquifer is also widespread in the Montenegrin part of the Black Sea Basin (Danube River Basin) (Fig. 2). The granular aquifer is mostly distributed along alluviums of the Piva River, Tara River, Lim River, Ćehotina River and Ibar Rivers. A fissured aquifer is represented by volcanic rocks that are more common in the north of Montenegro (Mirković et al. 1985; Radulović and Radulović 2004).

Impermeable rocks have a much larger distribution in this basin, especially in the northeastern part of the country.

4.1 The Carbonate Aquifer of the Black Sea Basin

The type of carbonate aquifer recharge within the Black Sea Basin is similar to that in the south of Montenegro (it occurs mostly through the infiltration of atmospheric water), but the recharge rate is quite a lot lower due to the lower precipitation rate and rock permeability.

The karst plateaus represent the main aquifer recharge zones in the Black Sea Basin. Such typical areas are the Žabljak and Sinjajevina plateaus. From these zones, groundwater flows towards canyons, which cut deeply into the plateaus, and flows out through large karstic springs located along riverbeds and canyon slopes (Figs. 2 and 5). Karst conduits and caves are generally elongated along large faults, particularly along intersections of two tectonic planes.

The tracer test results in the highly karstified terrains of Montenegro are sometimes quite unexpected (Radulović et al. 2005). Good examples are the results of artificial tracer tests performed in Crno Lake (Durmitor Mountain). When the tracer was injected into the outflow stream of Crno Lake that sinks in the Žabljak area, the tracer flows out at the Bijela Spring located ten kilometres to the northeast, on the right side of the Tara River. It would be logical to appear on the left river bank which is closer to the injection point, but the obtained test results indicate that the marked water passes under the riverbed and flows out on the right side of the Tara River. Another test, when a tracer was injected into the swallow hole on the southwestern shore of Crno Lake, showed an opposite direction of groundwater flow. The tracer appeared at the Dubravska Springs, located around 19 km south of the lake (in Komarnica Canyon). According to the results of the two mentioned tracer tests, it could be concluded that a dividing line between river basins crosses precisely across Crno Lake (Radulović et al. 2005).

Karstic springs (Table 2), which generally demonstrate good water quality (Radulović 2000; MARD 2017), are the main source of drinking water. Frequent water quality problems occur only in the area of the thermal power plant (Pjlevlja), mainly as a consequence of polluted air.

Table 1 The main karstic springs in the Adriatic Sea Basin (after Djordjević et al. 2010; Sekulić and Radulović 2019). Reprinted/adapted by permission from Springer, The handbook of environmental chemistry, The rivers of Montenegro by Pešić V, Paunović M, Kostianoy A (eds), vol 93, pp 13–42, Sekulić G, Radulović MM, The hydrology and hydrogeology of Montenegro, Table 1, Copyright 2019

No.	Name of a spring	Sub-basin	Latitude (°N)	Longitude (°E)	Assessed minimal flow (m^3/s)
1	Opačica Spring	Coast	42.457691	18.578997	0.080
2	Morinj Springs	Coast	42.492694	18.663698	0.550
3	Sopot	Coast	42.506801	18.691179	0.250
4	Risanska Spilja	Coast	42.518874	18.696214	0.004
5	Ljuta Spring	Coast	42.486145	18.775222	0.100
6	Škurda Spring	Coast	42.432169	18.776423	0.200
7	Vrmac Spring	Coast	42.418913	18.763518	0.020
8	Plavda Spring	Coast	42.467958	18.696543	0.020
9	Reževića Rijeka	Coast	42.228544	18.911528	0.075
10	Brca Spring	Coast	42.136146	19.064733	0.040
11	Kajnak Spring	Coast	42.091800	19.137037	0.040
12	Zaljevo Spring	Coast	42.059624	19.137556	0.016
13	Kaliman Spring	Coast	42.033307	19.253623	0.005
14	Klezna Spring	Coast	41.994388	19.301018	0.015
15	Salč Spring	Coast	41.974083	19.198778	0.002
16	Gač Spring	Coast	41.959162	19.272157	0.030
17	Krnjice Spring	Skadar Lake	42.207904	19.196891	0.010
18	Raduš Spring	Skadar Lake	42.225931	19.164906	0.060
19	Velji Spring	Skadar Lake	42.201405	19.071393	0.050
20	Podgor Spring	Skadar Lake	42.264522	18.987501	0.230
21	Crnojevića Rijeka	Skadar Lake	42.354160	19.044434	0.240
22	Karuč	Skadar Lake	42.359544	19.114585	2.500
23	Bolje Sestre Spring	Skadar Lake	42.345822	19.181410	0.900
24	Vitoja	Skadar Lake	42.315335	19.370984	0.050
25	Ribnica Spring	Morača	42.431386	19.332059	0.010
26	Mareza Spring	Morača	42.476731	19.179737	2.000
27	Bijeli Nerini Springs	Morača	42.750539	19.318394	0.500
28	Oraška Cave	Zeta	42.530872	19.099285	0.150
29	Žarica Cave	Zeta	42.550508	19.175430	0.050
30	Springs in Slatina	Zeta	42.587652	19.148746	0.015
31	Viš	Zeta	42.624050	19.079322	0.010
32	Milojevića Spring	Zeta	42.632152	19.011314	0.050
33	Dobropoljska Vrela	Zeta	42.674750	19.028006	1.000
34	Glava Zete Spring	Zeta	42.689147	19.007361	2.000
35	Obošnica Spring	Zeta	42.690777	18.988346	0.200
36	Slansko Oko	Zeta	42.771999	18.847635	0.450
37	Krupački Spring	Zeta	42.806498	18.873984	0.130
38	Poklonci Spring	Zeta	42.810189	18.896455	0.200
39	Vir	Zeta	42.856480	18.944340	0.400
40	Vidrovan Spring	Zeta	42.855576	18.944680	0.350
41	Zaslapnica Spring	Trebišnjica	42.676009	18.603331	0.054

4.2 The Granular Aquifer of the Black Sea Basin

The granular aquifer is represented by the alluviums of the Lim River, Ibar River, Tara River and Ćehotina Rivers. The alluvial sediments consist of gravel, sand and clay, so permeability depends significantly on their composition.

The main source of granular aquifer recharge is the infiltration of surface water. That mostly occurs during the rainy period of the year, when the river level is higher than the groundwater level. The groundwater flow direction within the granular aquifer directly depends on this relationship between the water levels.

Fig. 4 Bijeli Nerini Springs located at the right canyon side of the Mrtvica River (photo by Milan Radulović)

Fig. 5 Gojakovića Spring (photo by Milan Radulović)

Table 2 The main karstic springs in the Danube River Basin (after Djordjević et al. 2010; Sekulić and Radulović 2019). Reprinted/adapted by permission from Springer, The handbook of environmental chemistry, The rivers of Montenegro by Pešić V, Paunović M, Kostianoy A (eds), vol 93, pp 13–42, Sekulić G, Radulović MM, The hydrology and hydrogeology of Montenegro, Table 2, Copyright 2019

No.	Name of a spring	Sub-basin	Latitude (°N)	Longitude (°E)	Assessed minimal flow (m^3/s)
1	Ibar Spring	Ibar	42.793380	20.071001	0.100
2	Alipašini Izvori	Lim	42.537531	19.816104	2.500
3	Krkor Spring	Lim	42.626476	19.784773	0.100
4	Merića Spring	Lim	42.856226	19.777519	0.200
5	Manastirsko Spring	Lim	42.853010	19.861362	0.080
6	Dapsića Spring	Lim	42.838862	19.922665	0.045
7	Bistrice Spring	Lim	42.927727	19.714015	0.500
8	Djalovića Spring	Lim	43.014576	19.941250	0.300
9	Bezarska Springs	Ćehotina	43.256061	19.368346	0.005
10	Zmajevac Spring	Ćehotina	43.255882	19.341511	0.045
11	Mandovac Spring	Ćehotina	43.264762	19.302337	0.010
12	Breznica Spring	Ćehotina	43.366361	19.365418	0.050
13	Jugoštica Spring	Ćehotina	43.395322	19.330314	0.005
14	Mušovića Rijeka	Tara	42.836678	19.573762	0.080
15	Plašnica	Tara	42.929164	19.459002	0.100
16	Gojakovića Spring	Tara	42.984200	19.499819	0.036
17	Ravnjak Spring	Tara	42.990465	19.409654	1.150
18	Ljutica Spring	Tara	43.131720	19.306882	2.000
19	Bijela Spring	Tara	43.196010	19.231727	0.100
20	Zminje Spring	Tara	43.157977	19.072226	0.040
21	Oko Bijele Spring	Piva	42.900763	19.155686	0.030
22	Šavnik Spring	Piva	42.962291	19.093184	0.100
23	Bukovica	Piva	43.074457	19.123005	0.200
24	Dubrovska Springs	Piva	43.094786	19.032942	0.500
25	Sutulija Spring	Piva	43.169747	18.727321	0.050
26	Medjedjak Spring	Piva	43.203892	18.864651	0.500
27	Kaludjerovo Spring	Piva	43.291125	18.856674	0.100

The granular aquifers of the Black Sea Basin have great potential to be used as sources for water supply. Results of hydrogeological investigations performed in Mateševo (10 km upstream of Kolašin) have shown that around 90 l/s can be pumped from alluvial sediments with negligible drawdowns in wells (Matović 2018).

The water quality of the granular aquifers is relatively good, but this depends significantly on the quality of the rivers that feed them. The Ćehotina River and its tributaries have the worst water quality in this basin, mainly because they flow through mining zones.

5 Conclusion

Karstic springs are the most common drinking water source in Montenegro. They flow out from carbonate aquifers which are characterized by large yields, but on the other hand, this type of aquifer is very sensitive to pollution from the surface. The main reason for both features is the well-branched network of caves and karst channels that allows for high permeability to these rocks.

One of the challenges of hydrogeological research in such karstic terrains is the reliable delineation of catchment areas of water sources and the further determination of their protection zones.

Future investigations should pay great attention to the protection of groundwater as one of the main natural resources of Montenegro.

Acknowledgements I am immensely grateful to my father, Prof. Dr Mićko Radulović (1948–2019), the author of the book *Karst Hydrogeology of Montenegro*, for the love and knowledge he transferred to me.

References

Baldacci L (1886) Ricognizionegeologico–minerariadel Montenegro. Boll R Comit Geol Ital Boll 17:9–10

Barović G, Spalević V, Pešić V, Vujačić D (2018) The physical and geographical characteristics of the Lake Skadar Basin. In: Pešić V, Karaman G., Kostianoy A (eds) Skadar/Shkodra Lake

Environment. The Handbook of Environmental Chemistry, vol 80. Springer I, Cham (Switzerland), pp 11–23, https://doi.org/10.1007/698_2018_276

Bešić Z (1969) Geology of Monetenegro–Karst of Monetenegro (in Serbian). Geological Survey of Montenegro, Podgorica

Bourcart J (1926) Nouvelles observations sur la structure des Dinarides adriatiques. In: Actasdel XIV Congreso Geológico Internacional, Madrid, pp 1899–1941

Burić M (2010) Atlas of waters of Montenegro (in Montenegrin). Montenegrin academy of sciences and arts, Podgorica

Cvijić J (1899) Glacial and morphological studies of the mountains of Bosnia, Herzegovina and Montenegro (in Serbian). Voice Serb Roy Acad Sci 57:1–196

Dević N (2011) Geochemical–environmental properties of the catchment area of the Morača River, upstream of its mouth to the Zeta River (in Serbian). MSc thesis, University of Montenegro

Djordjević B, Sekulić G, Radulović M, Šaranović M (2010) Water potentials of Montenegro (in Serbian). Montenegrin academy of sciences and arts, Podgorica

Grubić A (1975) Tectonics of Yugoslavia. Acta Geol 41:365–384

Hassert K (1895) Beiträgezurphysischen Geographie von Montenegro. Petermanns Mitteilungen Ergänzungsheft, Gotha

Hrvačević S (1999) Hydrogeological characteristics of Piva and Tara river basins from the aspect of utilization and protection of water potential (in Montenegrin). Master thesis, University of Belgrade

Hrvačević S (2004) Resources of surface water in Montenegro (in Montenegrin). EPCG AD, Podgorica

IHMS (2020) Report on water quality status in Montenegro for 2019 (in Montenegrin). Institute for meteorology and seismology of Montenegro, Department of Water Quality, Podgorica

Lješević M (2004) Karst of Piva (in Serbian). Montenegrin Academy of Sciences and Arts, Podgorica

MARD (2017) Strategy for water management in Montenegro (in Montenegrin). Government of Montenegro, Ministry of agriculture and rural development, Podgorica

Matović et al. (2018) Final report about results of detailed hydrogeological investigations of the aluvion of Tara River, downstream from Mateševo, for the purposes of water supply of facilities along the highway Bar-Boljare, section Smokovac-Uvač-Mateševo. Geoprojekt Ltd, Podgorica

Milovanović B (1965) Epeirogenic and orogenic dynamics in the area of External Dinarides and the problems of paleokarstification and geological evolution of holokarst (in Serbian). J Geol Surv Serbia 4(5):5–44

Mirković M, Žvaljević M, Đokić V, Perović Z, Kalezić M, Pajović M (1985) Geological map of Montenegro 1:200,000. The Republic Self Managing Community of Interest for Geological Exploration of Montenegro, Titograd

Nopcsa F (1916) Begleitwortzur geol. Karten von Nordalbanien, Rasien und Ostmontenegro. Földtani Közlöny 46:7–12

Petrović J (1980) Development of system of Cetinje caves (in Serbian). In: Proceeding of 7th Yugoslovenian speleological congress, Herceg Novi 9–14 October 1976, p 343–353

Radulović V (1989) Hydrogeology of Skadar Lake watershed (in Serbian). Geological Survey of Montenegro, Titograd

Radulović M (2000) Karst hydrogeology of Montenegro (in Serbian). Geological Survey of Montenegro, Podgorica

Radulović MM (2012) Multi-parameter analysis of groundwater recharge in karstic aquifers–case examples from Skadar Lake basin (in Serbian). Dissertation, University of Belgrade

Radulović MM (2018) Hydrogeology of the Skadar Lake Basin. In: Pešić V, Karaman G., Kostianoy A (eds) Skadar/Shkodra Lake Environment. Springer International Publishing AG, Cham (Switzerland), pp 25–45

Radulović V, Radulović M (1999) Groundwater bodies (in Serbian). In: Gomilanović M (ed) Mineral resources and mine production in Montenegro. Ministary of Industry, Energy and Mining of Montenegro, Podgorica, pp 595–760

Radulović M, Radulović, V (2004) Hydrogeological map of Montenegro 1:200,000 (in Serbian). Geological Survey of Montenegro, Podgorica

Radulović M, Danilović T, Radulovic MM (2005) Specifities of hydrogeologic watersheds and directins of movment of aquifer waters in karstic terreins of Montenegro. In: Stevanović Z, Milanović P (eds) Water Resources and Environmental Problems in Karst. Proceedings of the International conference and field seminars, Belgrade & Kotor, 13–19 September 2005, pp 437–442

Radulović MM, Novaković D, Sekulić G (2013) Geological and hydrogeological characteristics of the Montenegrin part of the Skadar Lake catchment area (in Serbian). In: Sekulić G, Bushati S (eds) Development of hydrogeological and hydraulic study of regulation of Skadar Lake and Bojana river water regime (IPA Project)–Volume 1. Montenegrin Academy of Science and Arts, Podgorica, pp 9–115

Radojević D, Nedić D, Blečić V, Zečević M (2011) Basic hydrogeological map of Montenegro 1:100,000, sheets: "Gusinje", "Ivangrad" and "Šavnik". Geological Survey of Montenegro, Titograd

Radulović M, Popović Z, Vujisić M, Novaković D (1989) Basic hydrogeological map of SRY 1:100,000, "Bar" sheet (in Serbian). Geological Survey of Montenegro, Titograd

Radulović M, Popović Z, Vujisić M, Novaković D (1998) Guide book for the Basic hydrogeological map of SRY 1:100,000, "Bar" sheet (in Serbian). Geological Survey of Montenegro, Titograd

Radulović MM, Radulović M, Stevanović Z, Sekulić G, Radulović V, Burić M, Novaković D, Vako E, Blagojević M, Dević N, Radojević D (2015) Hydrogeology of the Skadar Lake basin (Southeast Dinarides) with an assessment of considerable subterranean inflow. Environ Earth Sci 74:71–82. doi: https://doi.org/10.1007/s12665-015-4090-7

Sekulić G, Radulović MM (2019) The hydrology and hydrogeology of Montenegro. In: Pešić V, Paunović M, Kostianoy A (eds) The rivers of Montenegro. The handbook of environmental chemistry, vol 93. Springer, Cham, pp 13–42, https://doi.org/10.1007/698_2019_413

Stevanović Z, Kukurić N, Pekašš Ž, Jolović B, Pambuku A, Radojević D (2016) Dinaric Karst Aquifer – One of the world's largest transboundary systems and an ideal location for applying innovative and integrated water management. In: Stevanović Z, Krešić, Kukurić N (eds) Karst without boundaries. Taylor & Francis Group, London, pp 3–25

Tietze E (1884) Geologische Uebersicht von Montenegro. Hölder, Vienna

Torbarov K, Radulović V (1966) Regional hydrogeological research of Montenegro and Eastern Herzegovina (in Serbian). Geological Survey of Montenegro, Titograd

Vlahović V (1975) Karst of Niksic polje and its hydrogeology (in Serbian). The Society of Sciences and Arts of Montenegro, Titograd

Waisse JG (1948) Les bauxites de l'Europecentrale (Province dinariqueet Hongrie). Dissertation, University of Lausanne, Lausanne

Geology of Montenegro

Slobodan Radusinović and Marko Pajović

Abstract

The territory of Montenegro belongs to the Dinarides, a tectonically active area of complex tectonic structure and lithological composition, where four regional tectonic units are distinguished as follows: the Adriatic–Ionian zone, the Budva–Cukali zone, the High Karst zone and the Durmitor tectonic unit. The Durmitor tectonic unit is characterized by clastic sedimentary rocks of the Devonian–Carboniferous age, flysch sediments of the Carboniferous and clastic sediments of the Permian age. Mesozoic carbonate sediments predominate in the terrains of the High Karst and the Adriatic–Ionian zone, as well as in some areas of the Durmitor tectonic unit. Predominantly built of carbonate rocks—limestones and dolomites, the High Karst zone is an area of classic karst development in Montenegro having very emphasized karstification processes. The result of long-term chemical and mechanical erosion of carbonate rocks on the surface of the terrain and underground is the formation of speleological objects.

Keywords

Geology · Lithology · Tectonics · Speleology · Montenegro

1 Introduction

Speleological objects represent natural phenomena conditioned by the lithological composition of rocks, tectonics, erosion processes, relief, hydrological-hydrogeological conditions, as well as the climate of a certain area. They occur predominantly in carbonate rocks, limestones and dolomites, but also in rarely represented sulphate rocks such as anhydrite and gypsum, and in salt deposits.

About 70% of the territory of Montenegro is built of carbonate rocks, which are most represented in the central and less in the coastal and northeastern areas of Montenegro. During geological evolution, geotectonic processes in the Mediterranean area significantly conditioned the structural-tectonic structure of the Dinarides, including the terrains of Montenegro, where four following regional tectonic units are clearly recognizable (Bešić 1948; Mirković 1989): Adriatic–Ionian zone, Budva–Cukali, High Karst and tectonic unit of Durmitor. The Old Montenegrin overthrust and Overthrust of Kuči single out within the High Karst (Bešić 1948, 1969). However, Dimitrijević (1974) and Schmid et al. (2008) treated the Overthrust of Kuči as a special geotectonic unit, called "Sarajevo Sigmoid" that is to say "Pre-Karst".

The geology of Montenegro is mainly presented on the basis of the results of the Geological Map of the former Yugoslavia (Osnovna Geološka Karta SFRY, 1:100.000), for the area of Montenegro. The oldest rocks in Montenegro of Devonian–Carboniferous age have been determined in the area of the Durmitor tectonic unit that is to say in the wider area of Plav and Bijelo Polje. These are clastic sediments: sandstones, shales and conglomerates, with rare interbeds of metamorphosed limestones. Flysch sediments of Carboniferous and clastic sediments of Permian age have a significant distribution in this unit. Mesozoic carbonate sediments, starting from the end of the Early Triassic to the end of the Late Cretaceous, are predominantly represented in the High Karst terrains and Adriatic–Ionian

S. Radusinović (✉) · M. Pajović
Geological Survey of Montenegro, Kruševac bb, 81000 Podgorica, Montenegro
e-mail: radusinovic.s@geozavod.co.me

G. Barovic (ed.), *Speleology of Montenegro*, Cave and Karst Systems of the World,
https://doi.org/10.1007/978-3-031-49375-1_3

zone, as well as in some areas of the Durmitor tectonic unit. Deep-water carbonate rocks have a small distribution in the trench structures of the Budva–Cukali zone, as well as the Durmitor flysch.

Overall, the High Karst area, built of layered and very thick-bedded limestones, dolomitic limestones and dolomites, presents a space of classical karst development on the surface of very fragmented relief with various karst phenomena, as well as in its underground with numerous speleological objects—caves and cave systems.

According to the previous knowledge, the geological history of karst formation in Montenegro terrains began in the Middle Triassic, when geotectonic processes created land areas on the terrains of current Kuči tectonic unit (Pivska Župa and Nikšićka Župa). These terrestrial surfaces, in areas of tropical-subtropical climate, have been exposed to karstification for up to 10 million years, as well as the formation of karst bauxite of Middle Triassic age. The longest continental phase in the High Karst terrains, including the formation of deep karst marked by large deposits of karst bauxites, lasted from 5 to 40 million years, in the period from the Late Triassic to the Middle Kimmeridgian. The depth of karst depressions with red bauxite exceeds 50 m, in the area of Nikšićka Župa, and in Western Montenegro—up to 15 m deep (Pajović 2000; Pajović et al. 2017). The next continental phase, during the Early Cretaceous, was marked by white bauxites and shallow karstification on the terrains of the High Karst, that is to say the Old Montenegrin tectonic unit. Finally, in the Adriatic tectonic unit, from the end of the Late Cretaceous to the Middle Eocene, a period of emergence marked by paleocarstification and the appearance of bauxites of Paleogene age lasted in the terrains between Bar and Ulcinj, as well as the Bay of Boka Kotor.

Based on all the above, it can be concluded that the geological column of Mesozoic carbonate sediments in the High Karst unit, over 6,000 m thickness is characterized by paleocarstification and bauxites at three stratigraphic levels: Middle Triassic, Jurassic and Cretaceous, which further more complicates the state and image of overall processes that enabled the creation of karst.

2 Brief History of Geological Explorations

The terrains of Montenegro belong to the southeastern Dinarides. Geological studies began in the second half of the nineteenth century in Montenegro. The book "*Traganje za rudama u Crnoj Gori (Searching for Ores in Montenegro)*", (Kalezić and Gomilanović 2004) presents the work of 39 mostly foreign researchers, in the period from 1837 to 1920, and only the most significant will be listed. Amie Boue, a famous French geologist, is considered to be the founder of geological science on the Balkan Peninsula. His most significant work "La Turque Europe", containing a map *(A. Boue, 1840)* presents the basic geographical and geological characteristics of the Balkan Peninsula. *E. Tietze (1884)* made the first geological map at a scale of 1:400,000, where, among other things, he determined Cretaceous and flysch sediments in the syncline structure of the Zeta River, as well as Early Triassic sediments and so on. Italian mining engineer *L. Baldacci (1884)* visited Montenegro in order to determine the presence of "*mineral layers*". The famous German geographer *K. Hassert (1893)* published a paper about his exploration results, edited and supplemented the geological maps of E. Tietze and L. Baldacci according to five months long terrain explorations of Montenegro in 1891. Occasionally, in the following period 1897–1913, J. Cvijić explored the terrains of the Dinarides and for the first time on the Balkan Peninsula found glacial sediments on the Rila Mountains in Bulgaria (Cvijić 1897) and on the mountains in the Dinarides. His published books (Cvijić 1924, 1926) represent an indispensable literature in the study of karst and geomorphology in the Dinarides area. He founded the science of karst.

The first tectonic regionalization of the Dinarides and Albania was done by Nopsca (1921). He singles out four zones in the coastal parts of the Adriatic–Ionian area as follows: the Adriatic–Ionian, the Hellenid flysch zone, the Olinos–Cukali zone (the Budva zone in Montenegro), as well as the limestone zone of Albanian table. A little bit later, Kossmat (1924), within the tectonic regionalization of the Balkan Peninsula, from the Adriatic Sea toward the northeast singles out the following: the Adriatic–Ionian fold complex, the Pindos–Cukali zone and the Western Montenegrin–Croatian zone of the High Karst. The well-known tectonist L. Kober studied the tectonic structure of the Mediterranean for over 40 years, he also supplemented and changed perceptions in order to single out the "outer zone" (composed of the Adriatic–Ionian and Cukali–Meridita zone) within the Dinarides (Kober 1952) and the central zone of the Dinarides.

Z. Bešić has dedicated his entire fruitful creative life to the study of geological structure and tectonics, mainly in Montenegrin terrains. He singled out four overthrusts in the territory of Montenegro: Coastal, Old Montenegrin, Kuči and Durmitor overthrusts (Bešić 1948). It is interesting that Bešić did not accept the existence of nappes in the Dinarides, unlike many others (Petković 1961; Ćirić 1975; Anđelković 1981, 1982). Dimitrijević (1974, 1982), made a special contribution to the study of the tectonic structure of the Dinarides, who in the coastal part of Montenegro distinguishes the following zones: the Adriatic–Ionian belt, the Budva–Cukali zone and the Dalmatian–Herzegovinian zone. According to the same author, the mentioned tectonic units represent nappes, with vergence toward the southwest.

Perceptions and interpretations of the history of the formation of the tectonic structure of the Mediterranean and the Dinarides have changed significantly with the appearance of the plate tectonics theory. Schmid et al. (2008) gave an evolution overview of The Alpine–Carpathian–Dinaridic Orogenic System. They state that from the Triassic to the Cenozoic the carbonate platforms of the High Karst and Adriatic zone, in the proximal parts of the Adriatic margin interspersed by intervening narrow deep-water basins (the Budva zone in Montenegro). The Pre-Karst tectonic unit (Kuči unit) is treated as a transitional area between High Karst and Bosnian flysch (Durmitor flysch). Regarding the age of the Bosnian flysch, it is said to be variable, from the Earliest Jurassic up to the Maastrichtian and Paleogene. The Bosnian–Durmitor nappe (Durmitor tectonic unit) is a composite tectonic unit consisting of the basal part of Paleozoic and Mesozoic formations. Ophiolites, which were obducted during the Early Jurassic to the Early Cretaceous are in the upper part of this tectonic unit. Also, "second-class" Durmitor and Lim subunits are single out within this unit.

Geological studies of the Dinarides and the terrain of Montenegro after the Second World War were intensive and comprehensive. More detailed explorations of lithostratigraphic, hydrogeological, engineering-geological and geophysical characteristics, including explorations of mineral raw materials, enabled a high level of overall knowledge in all previously mentioned areas. However, it must be noted that the project of making the Geological Map of the former Yugoslavia, on the scale of 1:100,000, which was realized on the territory of Montenegro from 1962 to 1988 was crucial for getting to know the geological structure of all areas covered by this project. The project was realized by the Geological Surveys of all six republics of the former Yugoslavia. According to these results, a Geological Map of Montenegro was made on the scale of 1:200,000 (Mirković et al. 1985) with Interpreter (Živaljević 1989). During the period of making these maps, as well as later, specialist explorations were carried out and according to them geological knowledge from specific geological areas was supplemented.

It could be said that the geological explorations so far have generally solved the fundamental geology of Montenegro. Regardless of unresolved questions of a theoretical nature, especially those related to the complex geotectonic structure of the Inner Dinarides, four geotectonic units clearly stand out in Montenegro terrains: the Adriatic–Ionian zone, the Budva zone, the High Karst and the Durmitor tectonic unit. Within the High Karst, the Old Montenegrin and tectonic unit of Kuči were singled out, while within the Durmitor geotectonic unit, six tectonic units were singled out: Bjelasica; Komovi and Visitor; Sinjajevina, Durmitor and Ljubišnja; Lim; Ćehotina, Mihajlovica and Kovač mountains; Rožaje (Mirković 1989) (Fig. 1). Considering the fact that the composition and geological development of these units is significantly different, in the next chapter the geological presentation of the lithological and tectonic characteristics of the terrain of Montenegro is given according to the mentioned geotectonic units.

3 Geological Characteristics of Montenegro

It is stated, in the previous text, that geological formations were discovered on the terrains of Montenegro: Devonian–Carboniferous, Carboniferous, Permian, Triassic, Jurassic, Cretaceous, Paleogene, Neogene and Quaternary. In the relatively small land area of Montenegro, out of only 13,812 km^2, rocks formed in the last 400 million years have been discovered. Also, it was pointed out that the geotectonic structure of Montenegro is very complex, with different geological development, that is to say paleogeographic conditions of geological formations. Basically, all four geotectonic units represent the nappes formed by subduction in the direction of the north and northeast, that is to say by the vergence of the pulling onto the plane toward the southwest. Thus, observed from the southwest to the northeast, the Adriatic–Ionian table is subducted below Budva–Cukali, and this one under the High Karst, which is subducted under the Durmitor geotectonic unit belonging to the inner Dinarides. Figure 2 shows the lithostratigraphic composition and position of reverse dislocations of geotectonic units. The geological structure and structural-tectonic characteristics of these units are presented below in the text, from southwest to northeast.

3.1 Adriatic–Ionian Zone

The Adriatic–Ionian zone includes a protruding coastal belt in the area of Ulcinj and in the area of the Bay of Kotor. This zone is known as Paraautohton, Adriatic, South Adriatic and Dalmatian zone in the literature. Although the geological development of this unit is the same, the structure differs significantly according to areas in the mainland part of this unit. The front part of the Budva zone nappe is the northeastern border of this unit (Fig. 2).

The Area of Ulcinj. The Adriatic–Ionian zone in the area of Ulcinj is built of limestones, dolomites and dolomitic limestones of the Late Cretaceous age, over which are transgressive foraminiferal limestones of the Middle Eocene, and then Eocene and Oligocene clastic flysch sediments. The transgressive boundary between the Late

Fig. 1 General tectonic map of Montenegro. Based on Mirković (1989), modified and updated

Cretaceous and Middle Eocene sediments is marked by occurrences of red bauxites and intensive paleocarstification of the Late Cretaceous rocks on the surface and at the bedrock of the bauxite.

The structure of the Ulcinj area is characterized by the anticline structures: Volujica–Šasko jezero, Možura–Briska gora, Bijela gora and Mendra (Fig. 3), among which are syncline structures with flysch sediments. Anticline structures are manifested on the terrain surface as elongated hills built of carbonate rocks of predominantly limestone composition, and in some places of carbonate breccias. By deep exploration drill holes, it is determined that all mentioned anticline structures were pulled apart by reverse dislocations and overthrusted toward the southwest.

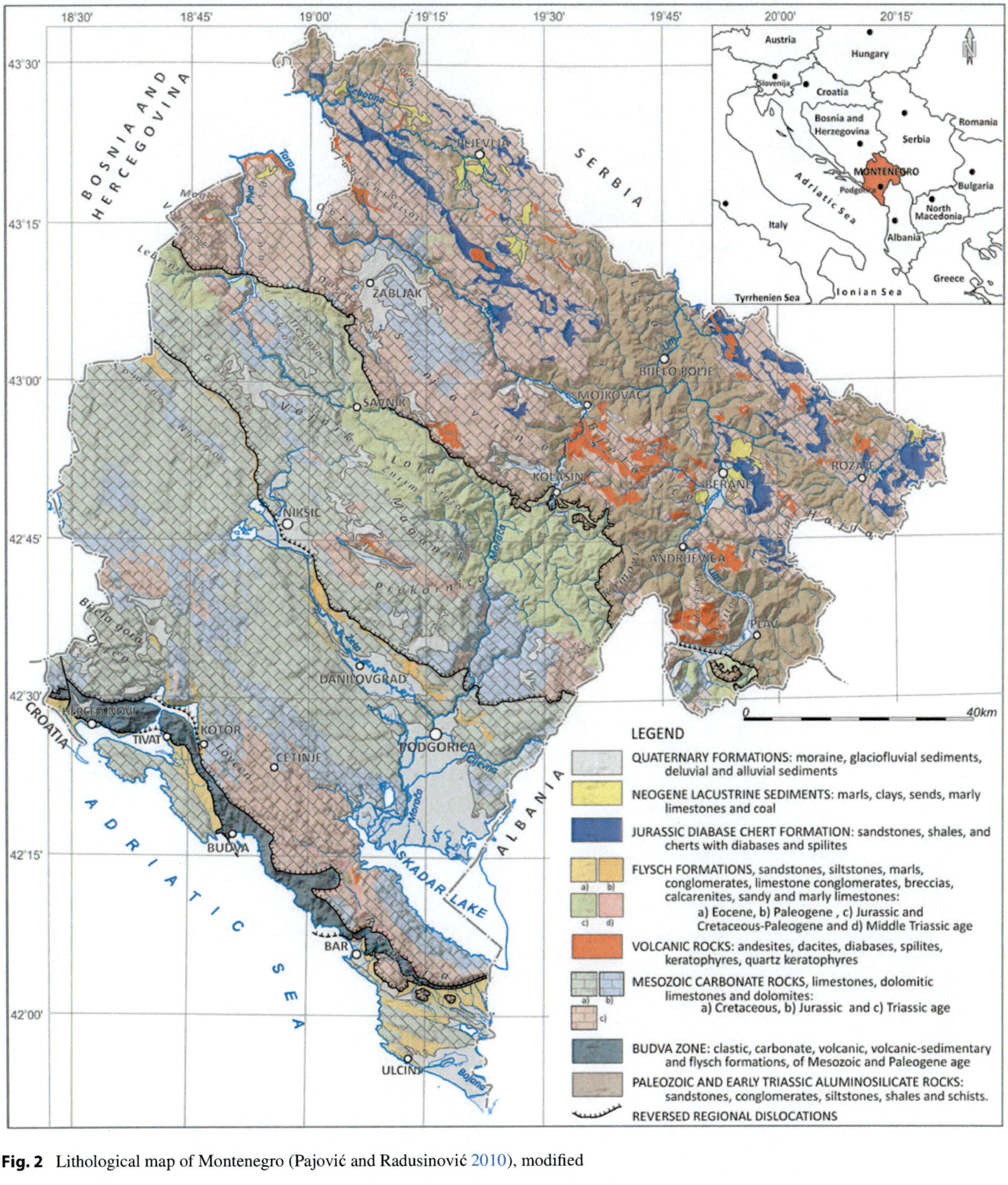

Fig. 2 Lithological map of Montenegro (Pajović and Radusinović 2010), modified

Thick-bedded and layered limestones, dolomites, dolomitic limestones and limestone breccias were discovered on the surface of these anticlines (Fig. 3). These rocks were formed by deposition on the Adriatic carbonate platform in the shallow tidal-subtidal environment. Due to intensive geotectonic processes in the Mediterranean, there was an emersion of certain parts of the Adriatic carbonate platform at the end of the Cretaceous. During the Paleocene and Early Eocene, Paleogene bauxites were formed on the mainland of this platform which appear in the area of

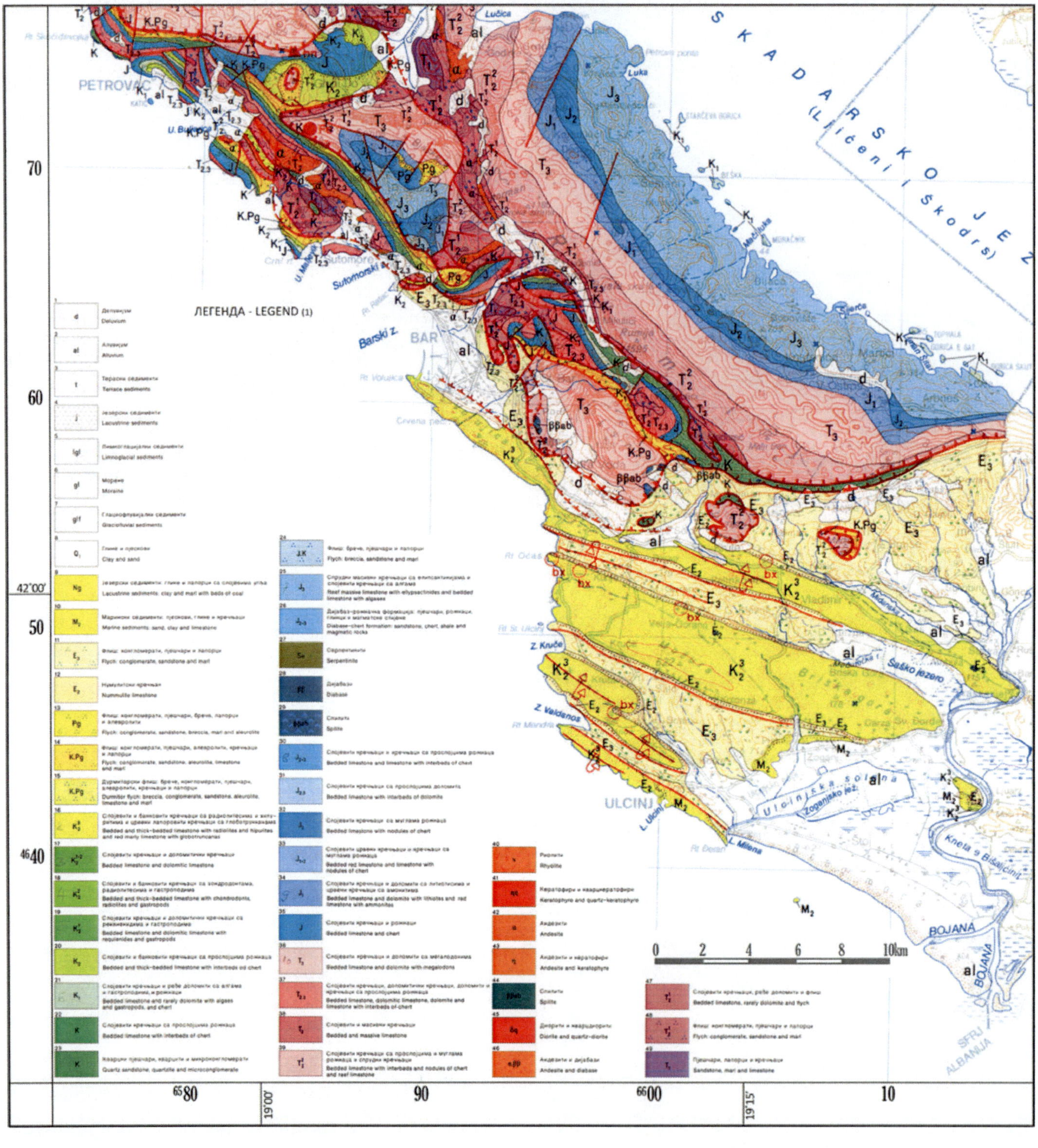

Fig. 3 Geological map of Montenegro, 1:200,000, (Mirković et al. 1985). Part of the map for the Bar and Ulcinj area, with the legend of the geological units for Figs. 3 and 4

Ulcinj and Boka Kotorska in the form of smaller lenses and traces along the entire erosion-discordant border, between the Late Cretaceous bedrock carbonates and the Middle Eocene hanging wall carbonates (Pajović and Radusinović 2015). Transgressive sediments over bauxite and paleorelief built of carbonate rocks, that is to say, layered limestones with numerous foraminifera, known in the literature as "foraminiferal limestones" or "numulitic limestones" (Antonijević et al. 1973; Mirković et al. 1978; Drobne et al. 2019). Their thickness is from 50 to 100 m. At the same time, these foraminiferal limestones mark the end of the existence of the Adriatic carbonate platform.

Geotectonic processes in the Late Eocene conditioned the deposition of marls, as well as thick *Eocene flysch* deposits over foraminiferal limestones. These flysch sediments form syncline structures between these anticlines in the area between Bar and Ulcinj, as well as most of the southern slopes of the Rumija and Lisinj mountains. According to Mirković et al. (1976a, b, 1978), flysch sediments in the Adriatic zone are represented by different types of sandstones, as well as conglomerates, calcarenites, clays, marls and clayey marls. This formation is characterized by rhythmic sedimentation, with the most common sequences sandstone-marl, and rarely conglomerate–sandstone–marl.

Over the flysch Eocene deposits and Cretaceous limestones in the area of Ulcinj field, in the erosion-discordant relation, *marine sediments of the Middle Miocene* were developed and they are represented by: sands and sandstones, over which clayey sands, sandy clays and, at the end limestones of the column were deposited, having overall thickness from 50 to 120 m (Mirković et al. 1978).

Quaternary sediments in the Ulcinj area are represented by *red soil (terra rossa)* on terrains built of karstified carbonate rocks in the hinterland of Ulcinj, where it covers the bottoms of karst sinkholes and coves, especially on the northern slopes of Možura and Volujica, as well as *alluvial sands, gravels and clays* in the field of Ulcinj, as well as in the surroundings of Šasko jezero, and *deluvial sediments* on the slopes of Rumija and Lisinj.

The carbonate rocks of the Late Cretaceous on the surface of the terrain are very karstified and difficult to pass, with typical karst relief, sinkholes of various dimensions and scrapes. It is interesting that the presence of caverns was proven by exploratory drilling at deep levels. Thanks to the watertight flysch sediments in the area of Ulcinj, a local hydrogeological network has been developed next to the Bojana river, as well as numerous springs of low yield.

The Area of Boka Kotorska. The Adriatic–Ionian zone was discovered in the area between the Gulf of Jaz in the southeast to Sutorina and Prijevor in the northwest, in the area of the Bay of Kotor. It includes the coastal hilly terrain of Grbalj, Luštica and Sutorina as well as the plains around the river Sutorina, Kumbor Strait, Tivat and Grbalj fields.

The tectonic structure of this zone in the Bay of Boka Kotor is simpler than in the area of Ulcinj (Fig. 4). In other words, the hilly part is formed by Late Cretaceous carbonate sediments and the flat part by Eocene flysch formation. The Cretaceous limestones of this area probably represent the NE wing of the anticline the core of which is mostly below the sea surface. Intensive karstification of carbonate terrains, as well as faults of different directions of expression, are expressed in hilly terrain, while in the flysch formation meter and decametre folds are observed, mostly torn or broken. Carbonate rocks in contact with the sea are susceptible to abrasion and erosion, the results of which are underwater caves and other karst formations. It is in these carbonate rocks of the Upper Cretaceous age that the famous Blue Cave was formed, the most attractive in a series of caves on the Luštica peninsula.

The geohistorical development of the Adriatic–Ionian zone in this area fully corresponds to that described in the area of Ulcinj. Namely, the oldest discovered geological formations are carbonate rocks of Late Cretaceous age (Fig. 4). Antonijević et al. (1973) describe the Late Cretaceous carbonates of this zone as Maastrichtian formations, built of dolomite, limestone dolomites, dolomitic limestones, limestones and carbonate breccias—which alternate laterally and vertically. Thick-bedded gray bituminous dolomites appear in the middle part of the geological column.

Traces of *Paleogene bauxites* (same as in the area of Ulcinj) are found on the Cretaceous limestones of Grbalj, Luštica and Sutorina, as well as transgressive foraminiferal limestones of *the Middle Eocene*, over which are *Eocene flysch deposits*.

The mentioned authors describe the *Eocene flysch* of the Adriatic zone as a change of sandstones and marls about 300 m thickness, over which conglomerates about 100 m thickness were developed. In addition to the above, the following are involved in the structure of flysch: micro-conglomerates, graywacke, sandy marls and clays.

Terrains built of flysch deposits around the river Sutorina, Tivat and Grbalj fields are covered in some places with *alluvial clays*, *sandy clays*, and rarely with *sandy-gravel material*. Alluvial sediment, mainly represented by sand, gravel and clay material, builds the lowest areas around rare streams and small rivers in Grbalj and Mrčevo fields, in Sutorina, near Zelenika and Meljine. *Red soil* occurs in karstified limestone terrains, especially in the area of Grbalj, southwest of Radanovići and Mrčevo field.

3.2 Budva–Cukali Zone

The Budva–Cukali geotectonic unit includes a narrow coastal belt on the southwestern slopes of Orjen, Lovćen, Sozina and Rumija mountains (Figs. 1, 2, 3 and 4). It is much more widespread in the terrains of Albania and Greece to the east, and that is the reason it is called by different names in the literature. According to its geological composition and structural-tectonic characteristics, this zone is one of the most complex geotectonic units in the Mediterranean.

Basically, Budva–Cukali zone is a complex trench structure that began to form at the end of the Middle Triassic and as such existed until the end of the Paleogene. According to Dimitrijević (1974), the width of this structure in today's

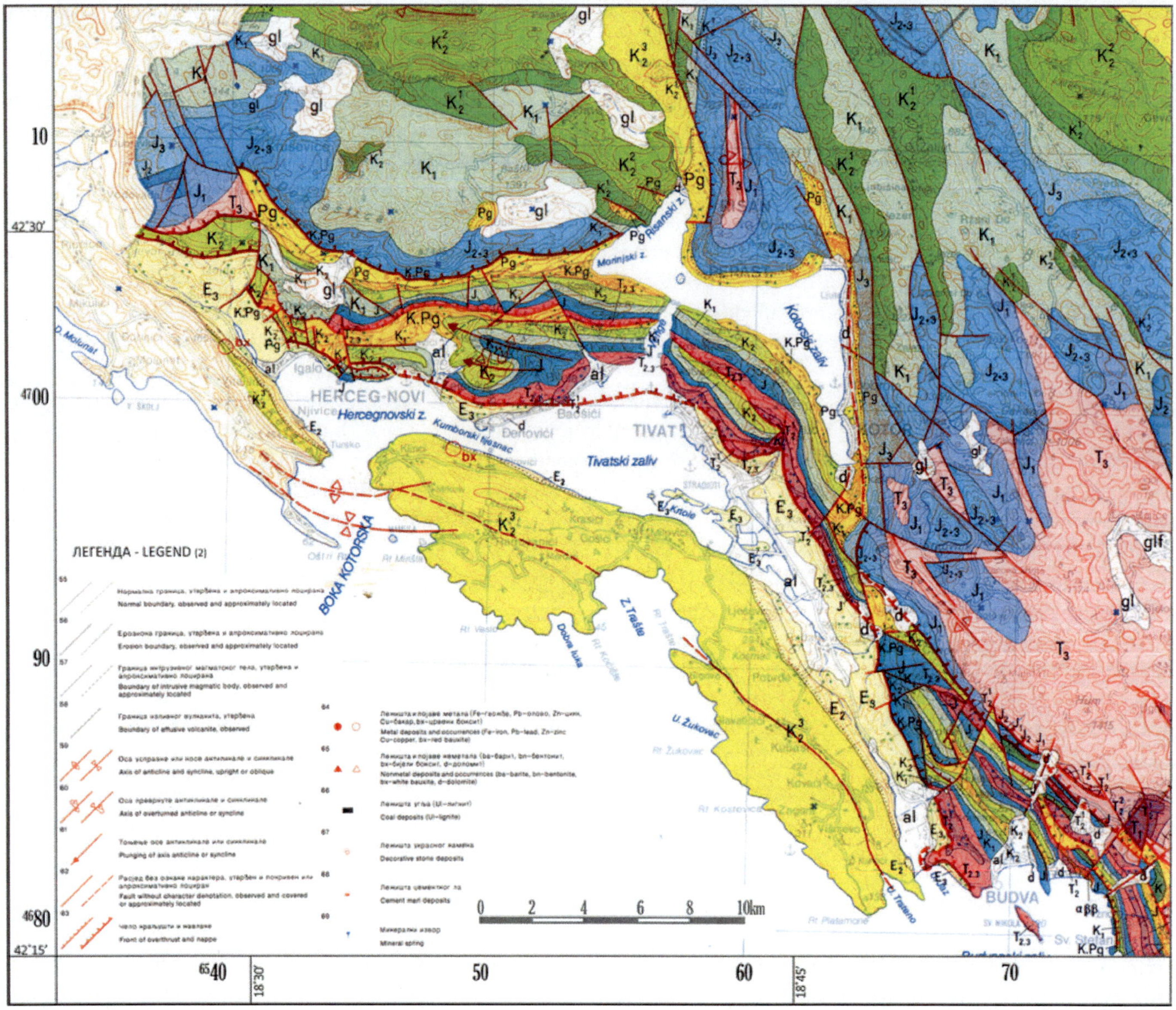

Fig. 4 Geological map of Montenegro, 1:200,000, (Mirković et al. 1985). Part of the map for the area of the Bay of Boka Kotorska, with a legend of standard simbolization for Figs. 3 and 4

Montenegro area was from 20 to 40 km. Geotectonic processes caused its narrowing, that is to say, folding and scraping at the end of the Cretaceous and during the Paleogene. The fold axes west of Verige mainly have a west–east orientation, from Verige to Petrovac first northwest–southeast, then north–south, and from Petrovac to above Bar northwest–southeast. The tectonic structure of Budva–Cukali zone consists of inverted isoclinal folds, mutually torn apart and separated by local overthrusts.

The general spreading of layers and folds is NW–SE, with occasional minor deviations. The intensity of tectonic disturbance in this zone also changes according to the spreading. Two monoclinic belts of Mesozoic and Paleogene sediments have been formed, separated by reverse dislocation in the area between Herceg Novi and Tivat.

There are several hectometric parallels overturned anticlines and synclines in the area between Tivat and Buljarica, mutually torn apart by reverse faults by SW vergence. The overturned anticlines of Miločer, Sveti Stefan and Petrovac stand out among them. The anticlines of Sustaše, Turčin and Rađen, which cores are built of sediments of the Early Triassic and Anisian floor are between Sutomore and Medjurijecje. They are separated by local overthrusts, that is to say, reverse dislocations.

The succession of geological formations of the Budva zone can only be traced within individual anticline or syncline structures (with vergence mainly toward the SW) which are mutually separated by tectonic dislocations, mostly of reverse character. Practically, along the entire length of Budva–Cukali zone of the Montenegrin coast, from Sutorina in the northwest to the river Bojana in the southeast, there is no profile without the repetition of separated folded structures. This is the reason why the presentation of the geological structure of this zone is only possible according to the geological structures that have a local distribution, or to describe the composition and development of geological formations participating in the structure of this geotectonic unit as done in this paper.

Triassic, Jurassic, Early and Late Cretaceous and Paleogene formations (Figs. 3 and 4) participate in the geological composition of the Budva–Cukali zone.

Triassic. The oldest rocks in the Budva–Cukali zone belong to the *Early Triassic*. They are represented by sandstones, sandy marls and marls, as well as by oolithic limestones, dolomites and marly limestones. They were discovered in the fields of Čanj, Spič and Sustaše. The thickness of these sediments is from 150 to 200 m. Carbonate and flysch formations of the *Anisian age* were formed over the Early Triassic.

Carbonate rocks belong to the Sparite type, gray and reddish in color, very thick-bedded in texture, which laterally and vertically turns into dolomitic limestones and dolomites. They were found in the vicinity of Bar, Sutomore, Budva and Herceg Novi. They represent smaller or larger hydrogeological collectors with springs of different yields, between the watertight rocks of the Early Triassic and the Middle Triassic flysch.

The Middle Triassic flysch was developed over the Anisian limestones, which contains conglomerates, sandstones, siltstones, sandy-clayey limestones, marls and clays. They have typical characteristics of flysch, in which Dimitrijević (1967) singled out 17 rhythmic sequences. Red ammonite limestones were developed over flysch formation and in some places they also were developed over Anisian limestones.

Over the flysch sediments, and in some places over the Anisian limestones there are volcanic rocks that are determined as porphyrites, diabases, spilites, andesites, dacites and keratophyres. They have the highest distribution on the southern slopes of Sozina and Rumija, as well as in the area of Bečići and Buljarica. They appear in the form of elongated lenses, as well as in the form of strips in some places.

A volcanic-sedimentary formation of Ladinian age was developed over volcanics and Anisian limestones. It is built of multicolored thin-bedded chert (radiolarites) in alternation with tuffs, tuffites, and bentonites in some places, as well as plate and layered limestones with lens of cherts.

At the end of the Ladinian and the beginning of the Late Triassic, there is a complete palaeographic and geotectonic differentiation between the Budva–Cukali zone and the High Karst. That is to say, during the Late Triassic, the sedimentation of deep-water sediments continues (as well as in Ladinian stage) in the Budva zone, while in the area of the High Karst, shallow-water carbonate sediments are deposited.

Jurassic. Jurassic sediments appear in the form of narrow and long strips along the entire geotectonic unit, in the Budva–Cukali zone. They are characterized by considerable facial diversity, which indicates different conditions for their formation. Reddish and red thin-bedded limestones alternate with cherts in the area from Sutorina to Bečići. In the upper parts of this geological unit, microbreccias, calcarenites and marly siltstones occur in places. Jurassic sediments are almost exclusively represented by cherts (radiolarites) in the area of Vrmac. Jurassic is represented by cherts limestones northward of Herceg Novi, recrystallized limestones, limestone breccias, thick-bedded detrital and oolithic limestones with lens of cherts. Very thick-bedded and thick-bedded dolomites of Jurassic age occur northeast of Herceg Novi. Paleontologically, it has been proven that in the Budva zone, a complete Jurassic has been developed.

Cretaceous. *The Early Cretaceous* is distinguished by its superposition, between the Late Jurassic and Late Cretaceous formations. The composition of this undivided formation includes characteristic thin-bedded multicolored cherts and silified marly carbonate rocks. It was discovered in the majority part of Budva–Cukali zone, in anticline and syncline structures, as well as in separated blocks.

Late Cretaceous. Late Cretaceous sediments in this zone are also developed as a whole but lithologically they differ in different parts of this structural-tectonic unit. Thus, in the northwestern part, this stratigraphic unit is built of layered and thin-bedded limestones, calcarenites and micrites with lens and interbeds of cherts, within which intercalations of rough breccias occur. In the central part (from Kotor to Petrovac) calcarenites and micrites with interbeds and lenses of cherts were developed. There are Late Cretaceous breccia limestones, breccias and limestones with globotuncana, which occur in almost all parts of Budva–Cukali zone, in the tectonic very complex terrain above Bar, that is to say on the southwestern and southern slopes of Rumija.

Paleogene. Paleogene sediments, represented by flysch of Paleocene–Early Eocene age, represent the youngest sediments in the geotectonic unit of Budva–Cukali zone. They occur in thin zones under the nappes of High Karst, as well as under frequent overthrusts within this unit. In some areas, however, under the Paleogene flysch, Danian stage sediments were discovered, and that is why a stratigraphic unit of Cretaceous–Paleogene was singled out on the Basic geological map sheet of Kotor and Budva. The sediments of this geological unit are represented by marls with pelagic fauna of globigerina, which gradually turn into flysch sediments. Paleogene flysch is represented by sandstones, siltstones, marls, clays, marly-sandy limestones, breccias and conglomerates.

3.3 High Karst Zone

The High Karst geotectonic unit was overthrusted onto Budva–Cukali zone. The front of this thrust stretches below the Orjen mountain massif and descends below the Morinj and Kotor bays, from where it climbs to the southeast below the Lovćen and Paštrovska gora up to the bend above Petrovac, from where it turns east and in the form of a tectonic half-window to the southeast it stretches below the peaks of Sozina and the mountain Rumija (Figs. 1, 2, 3, 4 and 5). In other words, the highest parts of the Montenegrin coastal area belong to the High Karst zone.

In the area of the Bay of Kotor, near the front of thrust, numerous springs, hot springs and speleological objects have been registered and studied, mostly for the needs of water supply. Thus, Radulović (2000) points out that frequent emersions, accompanied by karstification in several phases, which is intensified and often combined with tectonic and neotectonic activity, with significant sea level oscillations during the Quaternary, caused very complex hydrogeological relations in the coastal karst aquifers. The most famous springs are Gurdić (estavelle of sifonal character, length of examined canals 740 m), Škurda, Tabačina, Plavda, Orahovačka Ljuta (cave canal in the zone built of coarse-grained carbonate breccias of Paleogene age, which descends almost vertically to a depth of over 120 m), Hercegovina, Spila Risanska (occasional karst spring from the siphon cave), Sopot (cave formed in the Late Cretaceous limestones from which large amounts of water erupt in the hydrological maximum), then Morinj and Kostajnica springs. The two largest known submarine springs are in the bay near Sopot at depths of 28 and 36 m. All springs drain water from the hinterland, that is to say from the High Karst zone.

In the previous text, different understandings regarding the name and tectonics of this geotectonic unit are given. Bešić (1948) singles out the Old Montenegrin and Kuči overthrust within this unit, while Dimitrijević (1974) named these geotectonic units the Dalmatian–Herzegovinian and Sarajevo's sigmoid.

During the construction of Basic geological map, 1:100,000, it was determined that the geological composition and development of both tectonic units are very similar, and that two geotectonic units of lower order, the Old Montenegrin and Kuči tectonic units (according to the regionalization of Z. Bešić), are clearly visible within High Karst.

3.3.1 Old Montenegrin Tectonic Unit

This unit includes the areas of Rumija, Crmnica, Old Montenegro, Zeta-Skadar Plain and the area between the Zeta River and the Duga gorge in the northeast and Montenegrin Coast in the southwest (Figs. 1, 5 and 6). Structurally, it was built from the anticlinorium of Old Montenegro and the synclinorium of the Zeta valley and

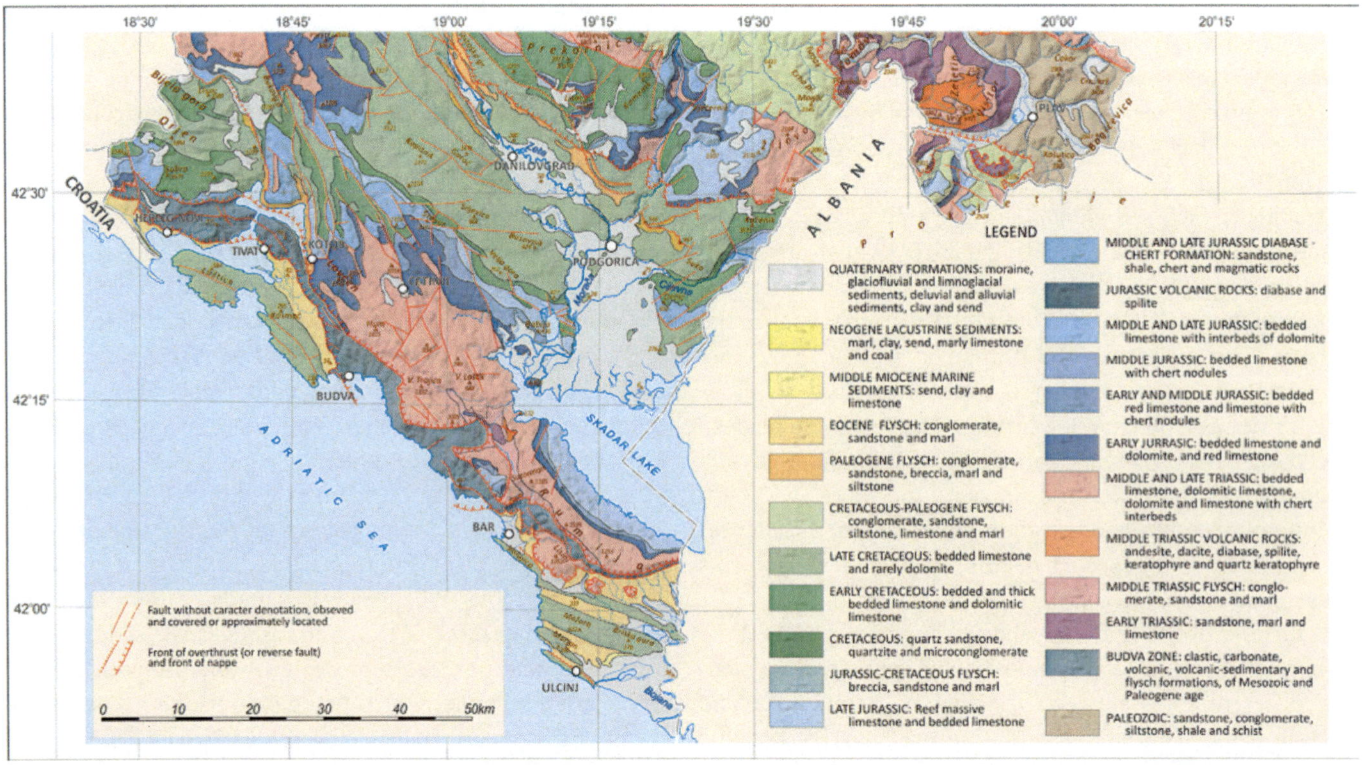

Fig. 5 Geological map of Montenegro, southern part. Based on Mirković et al. (1985) and Pajović (1999). Simplified, modified and updated

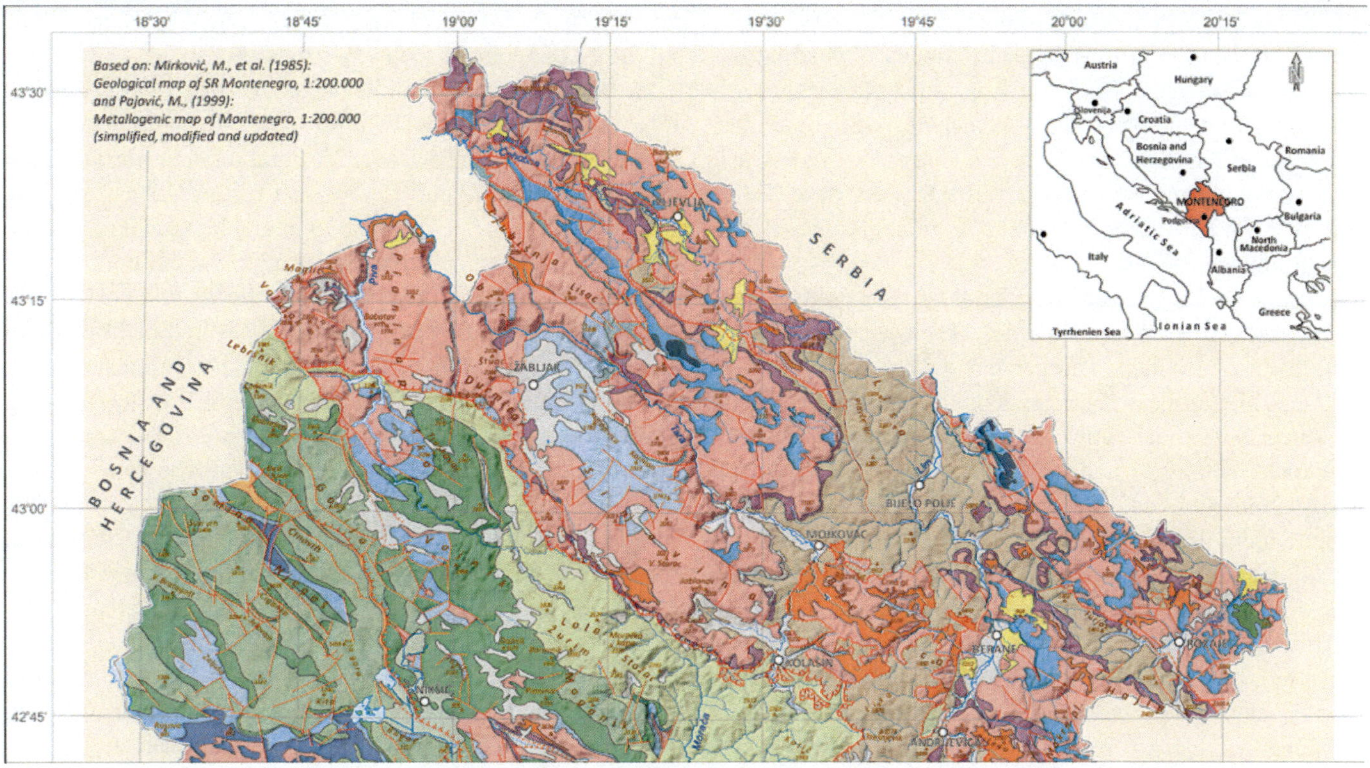

Fig. 6 Geological map of Montenegro, northern part. Based on Mirković et al. (1985) and Pajović (1999). Simplified, modified and updated

the Duga gorge. Basically it is made of the anticlinorium of Old Montenegro, which differentiates toward the northwest into a series of complex and most often reversely broken anticline-synclinal folds. In addition to the reverse ones, diagonal and transverse faults are present, which further complicate the tectonic situation of this area.

The synclinorium of the Zeta River valley to the northwest stretches across Nikšić and Duga gorge and further to the terrains of Bosnia and Herzegovina. It is characterized by narrow zones of Paleogene flysch in syncline structures, which are mostly reversely separated on the southwest side of the inverted anticlines. The core of the anticline folds is formed by younger Late Cretaceous limestones. In the area of the Bjelopavlići plain the structure of the synclinorium is largely masked by clayey-marly deposits of glacial origin, over 100 m thickness.

Within the Old Montenegrin anticlinorium, the following stands out: the complex synclinal structure of Orjen and Bijela gora, over which the anticlinorium of Old Montenegro is pulled onto. Both of these structures are intersected by numerous faults that have contributed to intensive karstification and difficult patency of carbonate terrains.

In the bauxite-bearing area of Bijele Poljane, a system of parallel isoclinic folds is noticeable, with SW vergence, on which wings of white bauxites appear, separated from each other by local reverse dislocations, transverse and tangential faults. Most folds have spread in NW–SE direction.

There are tectonic klippe of Vladimir, Vranje glijezdo, Rastiš, Lisinj and Ilino brdo in the extreme SE part of this geotectonic unit—overthrusted to the geological formations of Budva–Cukali zone.

The geological composition of this tectonic unit includes shallow-water carbonate sediments of the Mesozoic, flysch and volcanic rocks (Figs. 5 and 6).

Triassic. Geological formations of the Early, Middle and Late Triassic have been determined in the area of the Old Montenegrin tectonic unit. Sediments of the Early Triassic were discovered in Crmnica in a significant area and in a very small area in Brajići. The lithological composition of this stratigraphic unit is represented by reddish, gray-green and dark gray sandstones, clay sandstones, clays, marls and shales. In the upper part of the geological column of this unit, clay-sandy limestones and dolomites appear in places.

The *Middle Triassic* period is characterized by intensive geotectonic processes and the formation of various types of sedimentary, volcanic and volcanogenic-sedimentary formations. Thus, limestone, flysch sediments and volcanic rocks are formed during the Anisian stage. Anisian limestones were discovered in Rumija, in the area of Crmnica and Paštrovska gora. These are mostly very thick-bedded, and in the upper part of the column also layered limestones, in some places dolomitic limestones and dolomites. They represent the lateral facies of flysch in some localities. Their thickness varies from 50 to 300 m.

The Anisian flysch was discovered on the SW slopes of Rumija, the northern and eastern slopes of Sozina and in the area between Budva and Cetinje. The lithological composition of this formation includes conglomerates, sandstones, siltstones, sandy-clayey limestones, marls and clays. Hanbulog limestones of small thickness were developed over the Anisian flysch, and also in some places over Anisian limestones.

Volcanic rocks of the Middle Triassic have a significant distribution in the entire Dinarides. They were discovered in the form of smaller or larger masses on the southwestern slopes of Rumija and Crmnica on the terrains of the Old Montenegrin tectonic unit. According to the composition these rocks belong to andesites, dacites, keratophyres and quartz keratophyres, followed by tuffs and tuffites. Most volcanic rocks are pyritizated.

Ladinianian stage sediments are much more widespread than Anisian. Their greatest spreading is in Rumija and they occur in Crmnica and Obzovica. Bedded and thin-bedded limestones with cherts participate in the lithological composition. They were discovered in Crmnica through Hanbulog limestone or volcanic rocks, flysch sediments or Anisian dolomites. Layered stratified limestones with lens cherts, which gradually turn upward into very thick-bedded, on the surface into karstified rocks of younger Ladinian and older Carnian stage, over 400 m thickness, have been developed over this formation.

Late Triassic sediments are continuously developed over Ladinian carbonates. They have an incomparably higher presence than the other Triassic rocks mentioned above. They were discovered in the form of a wide belt on the terrains of Rumija, as well as on the terrains of Crmnica, Njeguši, Ćeklići and especially, in the area of Grahovo. Late Triassic carbonate sediments are from 500 to 600 m thickness. They are represented by the Carnian-Noric Lofer formation about 400 m thickness, characterized by rhythmic structure, over which light gray thick-bedded and very thick-bedded sparitic limestones of the Rhetian age were formed, with large remains of megalodons and involutinas, from 100 to 200 m thickness.

The described Triassic formations are often separated by a network of faults, dominated by NW–SE direction faults and NE–SW faults. The area of Old Montenegro is known for its exceptional karstification, with karst valleys, sinkholes and jagged ridges as well as numerous speleological objects. An exception is part of the Crmnica area, where, in addition to carbonate, there are also impermeable rocks represented by flysch sediments, volcanic rocks and Early Triassic clastites. Also, exceptions are karst fields—Cetinje, Grahovo, Dragalj and Nikšić fields, which are covered on the surface with fluvioglacial gravelly and sandy-clayey deposits of Quaternary age. Abyss, pits and caves regularly appear on the edges of karst fields. Such is the abyss of Slivlje on the edge of the Niksic field.

Jurassic. Jurassic sediments in the area of this geotectonic unit are exclusively represented by carbonate sediments. They were discovered continuously through the River of Crnojevići and Cetinje and further toward the NW in the wider vicinity of Grahovo and Herzegovina from the border with Albania and Lake Skadar. In the area of Old Montenegro and Banjani, within the Jurassic deposits there is a hiatus, marked by red bauxites, in which bedrock there are Early and Middle Jurassic carbonate sediments, while the Kimmeridgian-Tithonian limestones are in the hanging wall.

Early Jurassic. Early Jurassic sediments lie concordantly over the Late Triassic. They appear in the form of belts or isolated areas on the terrains of Krajina, Rijeka Crnojevića, Cetinje, Njeguši, Ceklin, Cuce and Grahovo. Early Jurassic sediments belong to different facies due to different paleogeographic conditions in which they were created: marly limestones with brachiopods, lithiothis limestones and ammonite limestones. Several types of facies have been developed, which alternate laterally and vertically on the mountain Rumija and in the area of Krajina. These are layered marly limestones with lens of cherts, very thick-bedded dolomites, lithiothis limestones, dolomitic limestones, cherts with limestone interbeds, oolithic and pizolithic limestones and red limestones with ammonites. Lithiotis limestones have been developed in the Krajina's region. The Early Jurassic is represented by limestones and dolomitic limestones with lithiothis on the terrains of Cuci, Trešnjevo, Meoce, Bijele poljane, Kosijeri and Čelinac. These layered limestones are bituminous in places. Brachiopod marl limestones also appear in some places in these areas. Marly limestones with brachiopods have been developed in the area of Lovćen (Fig. 7) and Njeguši.

Previous data indicate that Early Jurassic sediments were formed in the deep-water environment (marly limestones and limestones with cherts) and in the predominantly shallow-water environment (limestones with lithiothis, megalodus, etc.).

Middle Jurassic. Middle Jurassic sediments are represented by carbonates, mainly limestones. They appear in the form of separated zones or irregular surfaces. Thick-bedded and very thick-bedded reef limestones with corals and hydrozoa have been developed on the terrains of Rumija. Brachiopod marly limestones below the reef facies have also been developed in some places.

From Skadar Lake and Rijeka Crnojevića to the northwest, the Middle Jurassic sediments are located at the bedrock of the Jurassic bauxites (Lješanska and Katunska nahija, areas of Trešnjevo, Grahovo, Rudine and Banjani). These are thick-bedded and very thick-bedded limestones,

Fig. 7 Lovćen Mountain, view from Luštica peninsula

very karstified. These limestones contain rich gastropod fauna only in some places. Middle Jurassic limestones are very poor in fossils and it is difficult to distinguish them from the Late Jurassic and Liassic limestones, especially since Jurassic bauxites were not formed in the areas of Njeguši, Lovćen and Kosijeri.

The terrains in which Middle Jurassic limestones were discovered are intersected by numerous faults and very pronounced karstification with numerous sinkholes, lapis, less often caves. The terrestrial phase probably began during the Bajocian stage and lasted until the Middle Kimmeridgian (from 7 to 15 million years) in areas where bauxites occur and during that time intensive karstification was carried out. The same terrains were definitely the mainland during the entire Cenozoic, when the thick carbonate deposits of the youngest Late Jurassic, Early and Late Cretaceous have been eroded. In other words, the terrains where Jurassic bauxites occur in the Old Montenegrin tectonic unit suffered karstification during the Jurassic and during the Cenozoic.

Late Jurassic. Late Jurassic sediments have a wide development in the area of the Old Montenegrin tectonic unit. The Late Jurassic was developed into two facies: the facies of zoogenous reef coralligenous-hydrozoic limestones with ellipsactinia and the facies of stratified limestones with the remains of characteristic algae (*Clypeina jurassica*) on the terrains of Rumija and Krajina. Tintinnides are found in the youngest strata of the Late Jurassic. Reef ellipsactinia limestones appear in Crmnica. The Late Jurassic sediments are represented by stratified limestones, dolomitic limestones and dolomites, with clypeines and tintinides in the area of Dečic, Skorać and Cijevna, as well as in Lješanska nahija. There are Kimmeridgian-Tithonian limestones, dolomitic limestones and dolomites with remnants of algae (clypeines) and tintinides on all terrains of this geotectonic unit where Jurassic red bauxites are formed, that is to say in their hanging wall. Their thickness is from 150 to 300 m.

Cretaceous. Cretaceous sediments are widespread, from the Albanian border in the southeast to the Bosnia and Herzegovina border, in the northwest in the Old Montenegrin tectonic unit. Complete Early and Late Cretaceous have been developed in most of the terrains, and there is a hiatus in the area of Western Montenegro within the Early Cretaceous deposits marked by deposits and occurrences of white bauxites.

Early Cretaceous. Early Cretaceous sediments were discovered in the area of the Cijevna River, on the edges of

Fig. 8 Orjen Mountain, Siljevik peak, view toward Vratlo pass and Subra peak

Skadar Lake, as well as in the form of a belt from Grbavci over Orasi and Lješev Stup to Bijele Poljane, as well as in the area of Dobraštica, Orjen and Bijela gora. Early Cretaceous sediments are mainly represented by bedded, rarely thick-bedded grayish-yellow detrital limestones with rare microfacies of dasycladaceae, charophytes, algae and foraminifera. The thickness of the Early Cretaceous sediments is up to 350 m. Numerous speleological objects are known and partially examined, such as Kozja rupa with a depth of 654 m and PT4 cave 455 m deep in the area of Orjen (Fig. 8) and Bijela gora, as well as in other karst areas built of Cretaceous carbonate sediments.

White bauxites that have Liassic limestones in the bedrock (in the area of Bijele Poljane); Late Jurassic and Early Cretaceous sediments of Berissian–Aptian age were formed on the terrains of Western Montenegro, between Niksićko polje, Dragalj, Grahovo and the border with Bosnia and Herzegovina. There are transgressive limestones of the Late Cenomanian in the hanging wall of white bauxites. The entire terrain with white bauxites was folded, with the direction of spreading reversely broken folds NW–SE. Shallow karstification is present in the bedrock limestones—up to 10 m, except in the area of Bijele poljane.

Late Cretaceous. Late Cretaceous carbonates are the youngest stratigraphic unit of the carbonate platform of High Karst. They have a large distribution of complex anticlines on the NE wing, which many authors also call the anticlinorium of Old Montenegro. They were discovered in a wide belt from the Albanian border across Podgorica, Danilovgrad, Nikšić, Njegoš mountain and further across the border with Bosnia and Herzegovina. A complete Late Cretaceous was developed in this belt with a total thickness of up to 800 m and perhaps over 1,000 m.

Carbonate rocks of Cenomanian age in the vicinity of the Zeta plain, in the Morača canyon (Bioče) in Kuči and Cijevna's canyon are represented by gray-brown to dark gray layered bituminous limestones, and rarely by dolomitic limestones and dolomites. They are represented by bedded and thick-bedded oolithic limestones, as well as dolomitic limestones and dolomites in the area of Orjen and Lovćen on the SW wing of the anticline of Old Montenegro.

Late Cenomanian sediments are transgressive over white bauxite or their Early Cretaceous bedrock westward of Nikšić, in a wide area where white bauxite has formed. They include brown and dark brown, slightly bituminous limestones, dolomitic limestones and dolomites.

Concordantly over the Cenomanian, the Turonian carbonate sediments are developed, in the same areas. The Turonian sediments are represented by thick-bedded and very thick-bedded dolomites, dolomitic limestones and light gray limestones with fossil remains, chondrodonts, etc., in the areas of Kuči and Morača. The same Turonian sediments are less bituminous from Podgorica to Budoš. The

Fig. 9 Lovćen Mountain, view toward Dolovi, Štirovnik peak and Jezerski peak

Turonian limestones are thin-bedded and very thin-bedded on the terrains of Orjen and Western Montenegro, with interbeds of dolomites, with frequent chondrodonts and rudists, over which are very thick-bedded and thick-bedded limestones, with shell remains and layered gray-brown medium-bedded limestones at the end of the geological column.

The youngest Late Cretaceous sediments in the area of Old Montenegro have the greatest distribution within the syncline structure of the Zeta River, as well as in the extension in the area of the Duga gorge. They are developed in different lithological facies: Shallow-water stratified-thick-bedded limestones with numerous rudists; Very thick-bedded light gray microcrystalline limestones; Medium and thick-bedded deep-water light gray microcrystalline limestones; Marble thick-bedded limestones. Late Cretaceous carbonate sediments are located at the base of the Paleogene flysch in the Duga gorge.

Paleogene. Flysch Paleogene sediments of Paleocene–Early Eocene age were discovered in the narrow syncline zones in the area of Bjelopavlići, Nikšić and Duga and Krstac gorges. Flysch sediments were concordantly developed over pelagic marls and limestones of Danian stage. Flysch sediments consist of gray and red marls, clays and sandstones, with intercalations of breccias and conglomerates. The thickness of flysch deposits ranges from 20 to 120 m. Frequent springs appear at the contact of these sediments and limestones on the NE slopes of Bjelopavlići valley.

Speleological objects in the wider area of Njeguši and the Lovćen Mountain have been particularly well studied (Fig. 9) in the area of the Old Montenegrin tectonic unit, where systematic speleological explorations have been carried out in the National Park for years (Madžgalj 2013). The most famous are Njegoš Cave and Duboki do Cave, with over 5 km that is to say 2.5 km of explored canals. In addition to them, also significant caves according to their dimensions are Zečja rupa, 400 m deep, as well as Bogos, Žestoka and Ledena cave. They are formed in thick-bedded and very thick-bedded highly karstified Late Triassic and Jurassic carbonate sediments, mostly limestones. Also, in the previously described carbonate rocks of Triassic age in the wider area of Cetinje and Rijeka Crnojevića, famous caves were formed: Cetinjska, Lipska (3.4 km long), Obodska and Grbočica (2.65 km long).

3.3.2 Tectonic Unit of Kuči

The tectonic unit of Kuči includes the terrains of the high Montenegrin mountains: Golija, Vojnik, Lola, Maganik, Prekornica and Žijovo and essentially represents a complex anticline structure of the carbonate composition which is in an erosion-discordant relationship with the Durmitor flysch (Figs. 1, 2, 5 and 6). This unit is overthrusted onto the Old Montenegrin tectonic unit on the southwest side. The carbonate complex is marked by the anticlines of Nikšićka Župa, Komarnica and the complex structure of the Golija mountain. The anticline of the Nikšićka Župa was separated from the southwest in the direction of the Niksic field to the regional fault of Mrtvo Duboko. Several local overthrusts of the same vergence were found in the northeastern wing of this anticline. Several reverse dislocations also participate in the structure of mountain Golija.

In addition to the mentioned structures, smaller folded structures were also found: the Dobrelica anticline, the Mrtvo duboko anticline, the Šišman syncline, the Lebršnik syncline and others. Numerous transverse, diagonal and longitudinal faults have been found in the entire area of the carbonate Mesozoic complex of this tectonic unit. Carbonate terrains are characterized by intensive karstification, especially on plateaus in the form of coves, valleys, sinkholes, caves, lapies and microkarsts.

The syncline part of the tectonic unit of Kuči is built of Durmitor flysch sediments, which clastic facies are characterized by complex folds. The carbonate facies are characterized by hectometric and kilometer folds, of which the following stand out: the Lola syncline, the Ivanbegov Katun anticline, the Brštevac syncline (on the SW slope of Volujak mauntain), the Todorova do, Prutaš and Sedlo synclines (on Durmitor). The clastic flysch facies is overthrusted onto the carbonate in the area of Gornja Morača. The Durmitor tectonic unit is overthrusted onto the sediments of the Durmitor flysch.

The tectonic unit of Kuči is built of Permian, Triassic, Jurassic, Cretaceous and Paleogene geological formations. Apart from the sedimentary formations, volcanic rocks of the Middle Triassic and two red bauxite formations (Triassic and Jurassic) appear in this unit. Quaternary glacial sediments are particularly diverse.

Permian. Sediments of Middle and Late Permian age, from 200 to 300 m thickness, represent the oldest rocks discovered in the core of the Nikšićka Župa anticline. In their composition participate medium and thin-bedded gray and dark gray (bituminous) limestones, over which sandy-marly bituminous limestones with detritus of coal matter have been developed, as well as sandstones in alteration with marls and rare coal interbeds. The youngest Permian sediments include dolomites, dolomitic and oolithic limestones with interbeds of marls.

Triassic. Various geological formations of Triassic age, with clastic rocks, carbonates, flysch sediments, volcanites and red bauxites were discovered in the area of structural-tectonic unit Kuči. Paleontologically, sediments of the Early, Middle and Late Triassic age are proven.

Early Triassic. Early Triassic sediments were discovered in Piva and Nikšićka Župa. They are represented by thin-bedded gray-reddish and red mica sandstones having thickness from 60 to 100 m, as well as thin and medium-bedded sandy-marly limestones, alternating with siltstones, marls and clays, and finally sandy dolomites and oolithic limestones.

Middle Triassic. Anisian limestones, Anisian flysch, Hanbulog limestones, volcanic rocks, volcanogenic-sedimentary formation and the formation of reef limestones of the Ladinian age were singled out within the Middle Triassic (Pajović et al. 2017). They were discovered on the terrains of Piva and Nikšićka Župa.

Anisian thick and very thick-bedded limestones gradually develop in the Early Triassic. These are gray and gray-brown, detrital and marly limestones, as well as thick-bedded crystalline limestones and dolomitic limestones, reddish in some places.

Anisian flysch sediments were only discovered in Nikšićka Župa (Pajović et al. 2017). They have a small distribution. They formed on Anisian limestones. Their composition includes coarse-grained conglomerates, over which fine-grained conglomerates, sandy-marly limestones and marls have been developed.

Volcanic rocks of the Middle Triassic were discovered in Nikšićka Župa and in Pivska Župa. According to the composition, they belong to andesites, and less to dacites, diabases and spilites. They lie over Anisian limestones and Anisian flysch. It is believed that the volcanic rocks of Nikšićka Župa formed in three phases. Tuffs and tuffites occur almost regularly on eruptive rocks, and in some places, at the same stratigraphic level, there are tuffs and tuffites in alternation with cherts and limestones containing the lens of cherts. These sediments belong to the Ladinian stage. Sandstones of Wetterstein limestones, with corals, sponges and numerous microfossils of foraminifera, algae and harophytes belong to the same stage over the volcanics in Nikšićka Župa and in Pivska Župa. The Ladinian stage is also represented by layered gray limestones. On the southern and western slopes of Prekornica, bedded limestones, dolomites and dolomitic limestones were developed together with Late Triassic carbonate sediments of similar characteristics. A deposit of red Triassic bauxite of Gornjopoljski vir was formed on the reef limestones of Ladinian age. The erosion-discordant surface of Ladinian limestones is paleocarstified, with small sinkholes, up to 10 m deep where bauxites have formed.

Late Triassic. Transgressively, over bauxite and Ladinian limestones in Pivska Župa and Niksićka Župa, Raibel sediments, up to 10 m thickness were found in some places, represented by dark gray shale clays, siltstones and bituminous very thin and thin-bedded limestones. Carbonate sediments of the Norian stage were deposited over the sediments of this formation and Ladinian limestones, which are widespread in the tectonic unit of Kuči: in the area of Pivska Župa, in the northern part of Niksićko polje, in Nikšićka Župa, at the foothill of Maganik, Morača canyon,

Kuči area and Žijovo mountain. Late Triassic sediments are typical formations of the Triassic carbonate platform of the Dinarides represented by a Lofer formation. The thickness of this formation is about 500 m in the area of Nikšićka Župa. Thick-bedded and very thick-bedded light gray limestones with megalodons, up to 200 m thickness were developed over the Lofer formation, which also have the characteristics of cyclothems.

There are deposits of Jurassic bauxites located in karst depressions of the previously described megalodon limestones in the area of Nikšićka Župa. These deposits are among the most important in the Dinarides, in terms of quantities and quality of bauxites. According to the thickness and length of bauxite ore bodies, it can be seen that karst depressions had a depth of over 50 m and a length of up to 1.2 km (Biočki stan deposit). There are six large bauxite deposits: Liverovići, Zagrad, Kutsko brdo, Biočki stan, Štitovo 1 and Štitovo 2 on an area of about 15 km^2. This fact indicates that the paleocarstification of terrestrial carbonate terrains during the Jurassic was intensive. By the exploitation of bauxite since 1948, various speleological shapes and types, caves, numerous caverns of different dimensions, as well as the presence of numerous gravitational and reverse faults have been determined at the bedrock of bauxite deposits. The depth of Jurassic paleocarstification in the carbonate terrains of Nikšićka Župa based on indicators was estimated up to 100 m (Pajović et al. 2017).

Jurassic. The Jurassic period was marked by geotectonic and paleogeographic processes that significantly influenced the tectonic structure and formation of Jurassic formations in the entire area of the Dinarides, as well as Montenegro. At the end of the Late Triassic, the breaking of marginal parts of the carbonate platform in the area of tectonic unit Kuči separated and deep-water formations formed during Liasssic and Doggerian age in the area of Morača, and regression at the same time the anticline structure of Nikšićka Župa. Land areas during the Early, Middle and older Late Jurassic were constantly increasing in this unit (as well as in the area of the Old Montenegrin tectonic unit) until the middle Kimmeridgian, when regional transgression occurs on the terrains of High Karst and part of Durmitor tectonic unit. There are deposits or occurrences of red bauxites on all these surfaces. That is the reason why the formations of Liasssic, Doggerian and Oxfordian age do not have continuity on the terrains of the tectonic unit of Kuči.

Early Jurassic. Sediments of the Early Jurassic age were discovered in the area from Žijovo to Kamenik, in the canyon of Morača river and Piperi area. The second zone builds an area from Kamenik and Lutovske platije to Mala Rijeka. According to Živaljević (1989), the Early Jurassic sediments are represented by facies of lithiotic limestones where shales and shale marls appear. The Early Jurassic sediments are completely different in the areas of Vojnik, Pivska Župa and Treskavac. That is to say they are reddish, very thin and thin-layered limestones with ammonites. They lie concordantly over the megalodon limestones of the Late Triassic. The youngest layers of the Early Jurassic are built of layered gray-green, marly limestones with interbeds of green marls. The thickness of ammonite limestones is from 5 to 30 m. The Early Jurassic sediments are formed by laminated and fenestral limestones, which lie over the megalodon Late Triassic limestones in the area of Platije. Fossil-bearing limestones have developed, alternating with laminated dolomites, which contain reef microfossils in the younger part of the Early Jurassic column. The thickness of the Early Jurassic sediments is about 200 m.

Middle and Late Jurassic. The sediments of the Middle and Late (Oxfordian) Jurassic were singled out as one unit in the area of Kuči, the river Morača as well as on the western slopes of Vojnik mountain. They are represented by microbreccia and organogenic-detrital limestones on both sides of Morača river. This unit is built of limestones with interbeds of cherts and very thick-bedded zoogenic reef limestones with corals, hydrozoa and algae.

Late Jurassic. The sediments of the Late Jurassic are transgressive on the geological formations of the Late Triassic, Early and Middle Jurassic, as well as on the Oxford reef limestones in the Morača canyon. As mentioned above, the Jurassic land phase is marked by deposits and bauxite occurrences on the karstified paleorelief of the Late Triassic, Early and Middle Jurassic, and the oldest Late Jurassic (Oxford). The transgressive border, without the appearance of bauxites, is marked by breccias from 0.5 to 3 m thickness.

The composition of transgressive limestones of Kimmeridgian–Titonian age includes layers with bauxite grains, as well as layered limestones with harophytes, over which layered limestones (biomicrites) with *Clypeina jurassica* as well as numerous foraminifera. The Late Jurassic carbonate succession ends with thick-bedded light gray microcrystalline limestones with tintinides. The thickness of these sediments is from 100 to 200 m.

Cretaceous. Cretaceous sediments in this tectonic unit can be traced in an uninterrupted belt from the Albanian border to the border with Bosnia and Herzegovina. Early and Late Cretaceous were mapped.

Early Cretaceous. Marine shallow carbonate sediments of Early Cretaceous age were discovered on the southwestern and southern slopes of Maganik mountain, where six lithostratigraphic units were singled out (Pajović et al. 2017). The oldest part of the Early Cretaceous column, of the Berriasian–Hauterivian age is formed by layered late diagenetic dolomites of variable thickness or alternating change between limestones and dolomites, as well as of late diagenetic dolomites about 40 m thickness, over which thick-bedded ooid limestones are deposited. In addition to this, Hauterivian–Barremian sediments, represented by thick-bedded and very thick-bedded limestones with numerous fossil algae, foraminifera and hydrozoa are singled out in the column. Bituminous limestones of Barremian age up to 120 m thickness follow in the profile upward. The limestones are dark gray to black in color with shells, over which laminated dark brown micrite limestones have been developed. The youngest sediments of this stratigraphic unit are thick-bedded brownish-gray weakly bituminous limestones. Barremian–Aptian reef limestones have been singled out over bituminous Barremian limestones, which are easily recognizable in the field even by intensive karstification. The thickness of Barremian–Aptian limestones is from 250 to 400 m. Bituminous Aptian limestones are deposited over the reef limestones, which represent a benchmark in the determination and separation of Early Cretaceous succession on the Dinarides carbonate platform. The thickness of the Aptian limestones is from 200 to 300 m. The youngest part of the column of this unit is built of medium and thick-bedded bituminous limestones, with an abundance of microfossils and the characteristic alga *Salpingoporella Dinarica*. The youngest sediments of the Early Cretaceous, of Albian age, were developed over bituminous Aptian limestones. They are represented by gray-brown detrital and oolithic bituminous limestones in places, as well as limestones with numerous shell clasts and foraminifera. Micritic and oolithic limestones were developed in the middle and upper part of the column, with requienids, gastropods and ostracods. These limestones are often, in the upper part of the layers, conglomeratic.

The total thickness of the Early Cretaceous deposits on Mountain Maganik is from 1300 to 1500 m. They are of similar development on the terrains of Kuči, Prekornica, Kamenik, Vojnik and further on the SW to the territory of Bosnia and Herzegovina.

Late Cretaceous. Late Cretaceous carbonate sediments represent the youngest sediments on the Jurassic–Cretaceous carbonate platform of this geotectonic unit. The same limestones are partially or completely eroded in the base of the Durmitor flysch. That is to say, the largest distribution of Late Cretaceous carbonates is located along the NE boundary of this tectonic unit. The sediments of the Late Cenomanian are transgressive over the Early Cretaceous carbonates on mountain Golija. They are represented by dolomites and limestones, over which are chondrodont limestones, and then very thick-bedded intensively karstified limestones with shells. The youngest Late Cretaceous limestones are represented by dolomites and limestones and in the youngest part of the column light gray thick-bedded and intensively karstified limestones with rudists (hippurites) have been developed.

Cretaceous thick-bedded and very thick-bedded limestones are intensively broken and karstified. It is in these limestones on the ridge of the mountain Maganik, locality Trešteni vrh, where the deepest pit in Montenegro "Gvozdena jama", was discovered, with a depth of 1.027 m, where one of the deepest verticals in the world is located.

Durmitor Flysch. During the Cretaceous on the NE side of the tectonic unit of Kuči, a trench structure began to form in which flysch sediments were formed, known in Montenegro as the Durmitor flysch, and many scientists call it the Bosnian flysch. Durmitor flysch is composed of two types of sediments of the Cretaceous–Palogene age: terrigenous-clastic and carbonate composition. The boundary with the Mesozoic sediments of the tectonic unit of Kuči is of an erosion-discordant character. The sandy-terrigenous flysch facies were discovered in the following direction Gornja Moraca—the river Tara and from Pivsko jezero over Stabna and build terrains between Volujak and Lebršnik. These are typical flysch sediments that form flysch sequences, composed of conglomerates, sandstones and marls, or sandstones, siltstones and shales. These sediments are folded in meter and rarely decameter dimensions fold.

The younger, carbonate facies of the Durmitor flysch is in erosion-discordant contact with Cretaceous and Jurassic carbonate sediments. The oldest parts of this flysch are formed by limestone breccias and conglomerates, up to 20 m thickness, over which typical flysch sediments have been developed: limestone microbreccias, sandstones, siltstones and marls, as well as sandy and marly limestones and marls. The most common sequences are microbreccias, siltstones and marls. The carbonate facies of the Durmitor flysch are built the mountains: Žurim, Lola, Kapa Moračka and Prutaš and Šareni pasovi on Durmitor (Fig. 10).

The carbonate rocks of these mountain areas have faults, they are often picked and broken with intense karstification, caves, caverns and other karstic forms. Numerous caves and cave systems have been formed near the front of the Durmitor's nappe on Durmitor, where 775 m deep cave on the Vjetrena brda is among the most famous.

Also, more than 40 speleological objects formed in the carbonate sediments of the Durmitor flysch and in the Late Triassic and Jurassic carbonate sediments were explored at Prokletije (Fig. 11). The most famous are the Ledena pećina almost 2 km long, as well as the Giant 1.6 km long, the Nbyczarna cave system and others (Madžgalj 2013).

3.4 Durmitor Tectonic Unit

The northeastern part of Montenegro belongs to the Durmitor tectonic unit. It is overthrusted to the tectonic unit of Kuči on the southwest side. The front of thrust extends

Fig. 10 Durmitor Mountain, Sedlo, view toward Dobri do

Fig. 11 Prokletije Mountain, katun Roman and Karanfili peaks

from southwest to northeast from the mountain Volujak, over Piva Lake and SW slopes of Durmitor, where it turns east below Semolj and SE slopes of Sinjajevina mountain, over the river Tara below Trešnjevik, Komovi, Visitor mountain sand further over Prokletije mountain to the territory of Albania (Figs. 1, 2, 5 and 6). This unit, in terms of space, occupies almost half of the mainland territory of Montenegro. It is also the area with the highest mountains in Montenegro: Ljubišnja, Durmitor, Sinjajevina, Bjelasica, Komovi, Visitor, Prokletije and Hajla.

Numerous reverse dislocations, tectonic klippe and tectonic windows were found in the area of this unit, which testifies to a very complex tectonic structure, which is why opinions are still not consentient regarding the manner of separation and the number of lower-order tectonic units in this area.

Mirković (1989) singles out six lower-order structural-tectonic units within the Durmitor nappe (Fig. 1).

The tectonic unit of the Bjelasica mountain is a tectonic window where Triassic sediments and Triassic volcanic rocks are located under overthrusted Carboniferous and Permian shales, sandstones and conglomerates.

The tectonic unit of Komovi and Visitor includes the terrains of the mountains of the same name and the massifs of Bogićevica and Čakor. It is built of Devonian, Carboniferous and Permian clastic sediments and Triassic sedimentary and volcanic rocks. It represents a complex syncline in structural terms, deformed by gravitational and reverse faults.

The tectonic unit of Durmitor, Sinjajevina and Ljubišnja includes the terrains of the mountains of the same name, Piva and Bioč mountains and the wider area of Krupica. The synclines of Sinjajevina and the anticlines of Tara, Kruševo, Mratinje, Plašnica and Štitarica stand out within this unit. Reverse faults of Moticki Gaj were found in the area of this unit, where the terrains of Sinjajevina and Jezerska površ are overthrusted onto the Durmitor`s massif and where the numerous gravitational faults, the Sušica fault, the fault Koritnik, Mjedenik, Meštrovac and Obzir stand out.

The Lim tectonic unit includes the terrain of the mountains Sjekirica, Mokra, Turjak and the wider surroundings of Berane. It is overthrusted over the tectonic unit of Komovi and Visitor. The sediments of this unit are often folded and intersected by faults.

The tectonic unit of Ćehotina, Mihajlovica and Kovač mountains includes the terrains of the river Ćehotina basin area. It is overthrusted onto the tectonic unit of Durmitor and Sinjajevina along the reverse dislocation: Vitina–Glibaći–Lever Tara–Rudanci. The terrains of this unit have very complex structural-tectonic relations. Tectonic klippe and windows of Metaljka, Rajišici, Plakala and others were singled out in the area of Kovač mountain. It is interesting that the reverse dislocations of this area, only in Montenegro have SI vergence. Numerous reverse dislocations of local significance have been ascertained in the entire area of this tectonic unit, such as follows: Bušnje, Zenica, Brvenica–Mijakovići, Perotin–Bliškovo, Vrulja and Gradojevići–Šljivansko–Kruševo.

The tectonic unit of Rožaje includes the northern slopes of Žljeb, the wider surroundings of Rožaje and Savin bor, Korita, Bistrica and Hajla. It is overthrusted onto the tectonic unit of Lim along the reverse fault of Kalače. The entire area of this unit is intersected by numerous faults.

There are also Neogene sediments with coal in the wider area of Pljevlja and Berane, in the area of the Durmitor tectonic unit. Today's different hypsometric levels, as well as the spatial distribution of Neogene sediments with coal undoubtedly indicate significant gravitational movements of individual tectonic blocks during the Late Neogene, Pliocene and Quaternary.

The terrains of the Durmitor tectonic unit are built of sedimentary, volcanic and metamorphic rocks of the Paleozoic, Triassic, Jurassic, Cretaceous and Neogene. Also, glacial sediments of Quaternary age have a significant distribution.

Paleozoic. Paleozoic sediments in the area of the Durmitor unit have a large distribution on the terrains of the Lim river basin, as well as in the middle and upper course of the Tara river basin, as well as in the terrains of the Ćehotina river basin. Devonian–Carboniferous, Carboniferous and Permian sediments were determined by explorations.

Devonian–Carboniferous. Sedimentary rocks of this age are the oldest rocks in the territory of Montenegro. They were discovered in the wider vicinity of Plav and in the valley of Lim, downstream from Bijelo Polje. They are represented by metamorphosed sandstones, siltstones and shales of different composition, as well as limestones and conglomerates. Limestones are represented in small quantities, in the form of rare smaller lenses, within the series of sandstones and shales. Due to their composition, Paleozoic terrains are mostly without speleological objects. The thickness of Devonian–Carboniferous sediments is estimated at about 600 m.

Carboniferous. Carboniferous sediments were isolated in the Lim valley near Andrijevica, downstream from Ivangrad, as well as downstream from Bijelo Polje. They were also singled out in the Tara valley, near Kolašin, in Trebaljevo and Štitarička rijeka and in the vicinity of Mojkovac. Quartz-mica sandstones, shales and siltstones, conglomerates and dark gray limestones participate in their composition. Carboniferous sediments were folded and faulted. Limestones appear in the form of lenses at different levels. The thickness of the Carboniferous sediments was estimated at 300 m.

Permian. Permian sediments build up terrains in the area of Komovi, Trešnjevik, Bjelasica mountains and in the wider area of Rožaje. Minor occurrences of these sediments were discovered in the vicinity of Pljevlja and Bukovica on the SW slopes of Kovač mountain. They are represented by sand-shale and carbonate series. The first includes sandstones, shales, conglomerates, quartzites, siltstones and marls. The most common are quartz-mica sandstones, as well as also quartz-sericite and graphite shales. Limestones, dolomitic limestones and dolomites have a significant distribution on Bjelasica mountain, while in other terrains they appear in the form of interbeds or smaller and larger lenses. Recrystallized and karstified, thick-bedded and very thick-bedded limestones rich in microfossils, dolomites and dolomitic limestones form the western part of Bjelasica mountain.

Triassic. Triassic sedimentary and volcanic rocks have the largest distribution in the area of Durmitor tectonic unit. They mostly build mountainous areas, starting from NW to SE: Volujak, Maglić, Piva mountain, Durmitor, Sinjajevina, Ljubišnja, Kovač mountain, wider surroundings of Pljevlja, Lisa, Bjelasica, Visitor, Komovi (Fig. 12), Sjekirica, and wider surroundings of Rožaje. The presence of sedimentary rocks of the Early, Middle and Late Triassic and Middle Triassic volcanic rocks has been proven.

Early Triassic. Early Triassic sediments have a relatively small distribution compared to Middle Triassic. They were discovered in the valley of Piva, Tara, on the slopes of Sinjajevina, on Kovač mountain, the wider surroundings of Pljevlja, Bjelasica, Komovi, Visitor, around Rožaje, on Hajla, Sjekirica. They mainly appear in the form of narrow strips. The composition of the Early Triassic sediments includes gray, gray-green and red mica sandstones, quartz sandstones, bedded sandy and marly limestones and marls. The upper part of the column sandy-marly limestones contains traces of bioturbations. They contain characteristic macrofossils of shells and gastropods. In the source part of Vezišnica, in the Ćehotina valley, the Early Triassic sediments are overthrusted onto Ladinian limestones with cherts. The same is on the SW slopes of Kovač mountain. The youngest, Early Triassic sediments in the Kovač area are represented by gray-reddish layered quartzites and oolithic limestones. The Early Triassic rocks are located at the front of the reverse dislocation "Motički gaj" along the valley of the river Bukovica, then across Pošćensko and Crno jezero, all the way to Crvena greda on Durmitor. There are smaller deposits of barite in the Early Triassic sediments of the Kovač mountain that contain cinnabarite mineralization. Early Triassic rocks were discovered in Mratinje and Šćepan polje in the valley of the river Piva in some places. The thickness of the Early Triassic sediments is estimated from 100 to 200 m.

Middle Triassic. Geological formations of the Middle Triassic age are widespread in the terrains of Durmitor

Fig. 12 Komovi Mountain, katun Carine, view toward Mojan

tectonic unit. The Middle Triassic period is characterized by intense tectonic activity in the Dinarides and precisely that is the reason why the diversity of geological formations exists, represented by Anisian limestones, extrusive and intrusive igneous rocks, volcanogenic-sedimentary Ladinian formation, bedded and reef limestones of Ladinian age and reef limestones Late Triassic age.

Sediments of the Anisian stage are concordantly developed on the Early Triassic sediments. They were discovered in the areas of Bioč mountain, Piva canyon, Durmitor, Sinjajevina, Tara canyon, Ljubisnja, Pljevlja area and Kovač mountain. They are represented by gray and gray-brown, in some places bituminous limestones, which alternate with light gray and reddish very thick-bedded limestones, dolomitic limestones and dolomites. Red Hanbulog limestones with ammonites have been found in some places.

Anisian limestones occur as isolated masses built of very thick-bedded and thick-bedded limestones, dolomitic limestones and Hanbulog limestones in the Bjelasica terrains.

Anisian limestones are represented by gray sandy limestones and dolomitic limestones, with interbeds of gray-reddish marls in the area of Burenje, Sljemen, Sokolac, Stožer, Zuber, Kovren, Kozica, Crni vrh and Kamena gora. Anisian limestones in the area of Zeletin, Visitor, Sjekirica, the surroundings of Rožaje and Hajla are similar. The thickness of the Anisian limestones is about 300 m.

Magmatic Rocks. Extrusive igneous rocks of the Middle Triassic age were discovered in some places in the entire area of the Durmitor tectonic unit, and intrusive mostly dike rocks very rarely. Volcanic activity took place during the Anisian and the beginning of the Ladinian age. The largest distribution of volcanic rocks is in the area of Ljubišnja mountain, in Mratinje, on the northern slopes of Piva mountain, in the Tara riverbed, on the SW slopes of Sinjajevina (Semolj, Krnja jela, Bare, Bukovica valley), on Durmitor, Bjelasica (Fig. 13), Zeletin, Sjekirica, as well as in the vicinity of Rožaje, Bijelo Polje, Pljevlja and on Kovač mountain.

The volcanic rocks of the Middle Triassic in terms of petrographic composition correspond to the normal type of volcanic rocks (andesites, dacites and their transitions) and alkaline volcanics (spilites, kertofirs and quartz keratophyres). Almost regularly these effusive rocks are accompanied by tuffs. Andesites and dacites in which the well-known lead and zinc deposit "Šuplja stijena" is located are represented in Ljubišnja, then on the SW slopes of Sinjajevina, on Durmitor, Mratinje. Keratophyres and quartz keratophyres occupy much larger spaces than andesites. The origin of polymetallic (lead–zinc–copper) mineralization in the "Brskovo" deposit is connected with these volcanic rocks. These are rocks of fine-grade texture, shaled in places and often pyritized and hydrothermally altered. In addition

Fig. 13 Bijelasica Mountain, view from the Krstac pass

to these types, there are also acidic volcanic rocks—rhyolites, which are associated with lead–zinc mineralization in Kozica. Pyroclastic volcanic rocks (tuffs, breccias) almost regularly follow these types of extrusive rocks.

Intrusive igneous rocks were discovered on the northern slopes of Visitor, in Konjusi, Sjekirica, Šekularska River and in the vicinity of Bijelo Polje. According to the composition they belong to diorites and diorite porphyrites. They occur in the form of smaller intrusive bodies or in the form of dikes. These intrusive rocks made thermometamorphic alterations of the surrounding rocks into cornites, epidosites, and recrystallized carbonate rocks in the Konjusi area. Also, in the same rocks there are smaller occurrences of lead–zinc and copper mineralization. Diorite dikes have also been found in Permian and Early Triassic rocks in the vicinity of Bijelo Polje.

Ladinian Stage. Ladinian sediments have a significantly higher distribution compared to Anisian sediments. They were discovered over the described volcanic rocks and Anisian limestones in the same terrains. The volcanic-sedimentary formation and different types of carbonate rocks participate in the composition of Ladinian sediments. Carbonate facies, without volcanogenic-sedimentary formations, have developed over volcanite or Anisian limestones in some terrains.

Ladinian rocks, in the NW part of the Durmitor tectonic unit in the areas of Vučevo, Vlasulja, Bioč, Piva canyon, Piva mountain (Fig. 14) and Durmitor are represented by volcanic-sedimentary formation (very thin-bedded red and gray-green cherts, tuffs and tuffites in alternation with thin-bedded limestones with chert lenses), over which, gradually, stratified and then very thick-bedded zoogenic reef gray and light gray crystalline limestones are developed, in irregular alternation with dolomitic limestones and dolomites. The development of Ladinian sediments is the same on the terrains of the Ljubišnja and Kovač mountains, as well as Sinjajevina, Stožer and Burenje (in the wider vicinity of Pljevlja).

Volcanogenic-sedimentary formation and layered limestones with lenses of cherts are presented in the area of Bjelasica. The development of Ladinian is similar in the areas of Zeletin and Visitor. Ladinian sediments are present in a wide area in the area of Giljeva mountain and Korita. They are built of volcanogenic-sedimentary formation, over which bedded limestones with a characteristic fossil fauna of shells have been developed. The youngest member of this Ladinian succession are the reef very thick-bedded limestones with corals, bryozoa, brachiopods and crinoids. The thickness of Ladinian sediments ranges from 300 to 500 m. The cave above Vrazji firovi in Djalovića gorge, northeast of Bijelo Polje, the longest and one of the most famous and beautiful caves in Montenegro was formed in carbonate sediments of Ladinian and probably Anisian age, in the canyon of the river Bistrica. The length of the cave channels explored so far is about 17.5 km (Madžgalj 2013).

Late Triassic. Late Triassic sediments were discovered on the terrains of Piva Mountain, Durmitor, Sinjajevina and Ljubišnja in the vicinity of Pljevlja in Korita and in the vicinity of Rožaje.

Fig. 14 Piva Mountain

The Late Triassic is built of marly limestones up to 10 m thickness, over which mostly thick-bedded, gray and gray-brown detrital and oolithic limestones, thick-bedded dolomitic limestones and light gray dolomites are developed in an irregular alteration in the area of Piva mountain, Durmitor and Sinjajevina. A well-known Ledena pećina almost 2 km long was formed in the carbonate sediments of the Middle and Late Triassic on Durmitor.

The carbonate sediments of the Late Triassic age are concordant over Ladinian reef limestones in the area of Ljubišnja, Lisac mountain and the surroundings of Pljevlja. They are composed mainly of medium and thick-bedded detrital limestones, which contain smaller lenses of red marly limestones in the upper part of the column. Layered and thick-bedded banded limestones occur in some places. They contain the characteristic macrofossils of megalodons.

The Late Triassic sediments in the area of the Korita and the surroundings of Rožaje have a similar development of medium and thick-bedded limestones with megalodons, alternating with dolomitic limestones and dolomites. The overall thickness of the Late Triassic carbonates is from 350 to 500 m.

Jurassic. The Jurassic is represented with carbonates and a diabase-chert formation, in the area of the Durmitor tectonic unit.

Jurassic carbonate sediments were developed concordantly over the Late Triassic megalodon limestones, on the terrains of Piva mountain, Durmitor, Sinjajevina, the surroundings of Pljevlja and in the area of Rožaje. The oldest sediments of Liassic age are represented by medium-bedded gray and reddish microcrystalline limestones over which very thin and thin-bedded gray-reddish and red ammonite limestones have developed. Rough limestones appear in places, with interbeds of green marls in the final part of the Early Jurassic column. The Early Jurassic sediments in the vicinity of Pljevlja have a similar composition—from Ošlje to Krnjača, Slatina, Miljevig and Mihailovići. The thickness of the Early Jurassic deposits is up to 30 m. The Early Jurassic is represented by dark-red limestones with cherts, heterogeneous breccias and clayey limestones in the vicinity of Rožaje.

Middle Jurassic sediments on Piva mountain are represented by layered detrital and oolithic limestones, with chert lenses and interbeds. The limestone facies of the Late Jurassic on Piva Mountain, Durmitor (Bobotov Kuk, Prutaš), Sinjajevina, Tara Canyon, SE slopes of Ljubišnja, on Bunetina and Lisac Mountain—are represented by zoogenic reef, ellipsactinia and coral limestones. These reef limestones of the Late Jurassic have a significant development in the SW part of Sinjajevina (Korman, Kučajevica, Orujica, Mrčajevac), which thickness is estimated at about 400 m.

Diabase-Chert formation. This Jurassic formation in the terrain of the Durmitor tectonic unit occurs in the form of zones and surfaces of irregular shape, most often in tectonic, and less often in erosion-discordant relationship with neighboring formations. One of these zones stretches along the SW slopes of Kovac Mountain and passes through Hoćevina, Glibaći, Djurdjevića Tara, Njegovudja and SW along the edge of Šaranci all the way to the source of the river Bistrica on Sinjajevina. It has a significant distribution in the vicinity of Pljevlja (Kosanica, Barice, Varine in the valley of Bjelopoljska Bistrica, in the vicinity of Berane, Rožaje and Korita).

Its composition includes sandstones, siltstones, marls, clays, cherts, marly limestones, rarely breccias and conglomerates. Also, pieces, blocks or larger masses of basic effusive rocks of diabases and spilites, as well as pieces and blocks of peridotite and gabbro participate in the composition of this formation. Copper mineralization (chalcopyrite, pyrite, quartz) was discovered in diabases, locality of Varine.

Neogene. Neogene lacustrine sediments are located in the wider vicinity of Pljevlja, in the area of Berane and northeast of Rožaje, on the terrains of the Durmitor tectonic unit.

Neogene sediments with coal were formed in Pljevlja, Otilovići, Mataruge, Maoče and several smaller basins in the area of Pljevlja. The oldest sediments are clays, sandy clays and sands in almost all basins, followed by layers of coal of various thicknesses, which hanging wall is made of marls, marly limestones or clays.

There are two Neogene basins in the Berane area: Berane and Police. Two types of development have been identified by explorations in the Berane basin which depth is over 150 m: six layers of hard dark coal are formed on the left side of the river Lim in deep-water lake sediments and the main layer of hard dark coal below which there are two bedrock layers of small thickness is on the right side in shallow lacustrine sediments–marls. There is the Police Neogene basin with several thin layers of coal at a higher altitude, in the immediate vicinity of the Berane basin.

Neogene sediments NE of Rozaje (Dolovi and Vuča) on a smaller area developed Neogene sediments: conglomerates, gravel, sand, sandy clays. Neogene sediments were also developed and represented by marls and clays with traces of coal on the western slope of Sinjajevina, in Crkvicko field.

Quaternary Period. Quaternary sediments in the area of this tectonic unit, as well as the whole of Montenegro, have been developed as deluvial, alluvial and glacial sediments. Deluvial sediments have developed on the slopes of mountains and river valleys. Alluvial deposits are located in the flat part of all rivers and individual tributaries in

Montenegro area. Glacial sediments and glacial processes are particularly important from the aspect of the topic of this monograph and that is why special attention is paid to this issue with the presentation in the next chapter.

4 Pleistocene Glaciations and Karst Processes

Geological, geographical and other types of explorations of karst processes indicate that speleological objects are the result of chemical and mechanical erosion of carbonate rocks on the surface of the terrain as well as in their underground. Accoring to the geological studies of the terrains of Montenegro, it was also proved that the final regresssion of the terrain of the previously described geotectonic units was different. Thus, for example, most of the terrain of the Durmitor tectonic unit became land at the end of the Late Triassic, and a smaller part at the end of the Late Jurassic. The tectonic unit of Kuči definitely became the mainland at the end of the Cretaceous, except for the Durmitor flysch. It is similar with the Old Montenegrin tectonic unit and the Budva–Cukali zone, where Paleogene flyschs were deposited in narrow syncline structures. The Adriatic tectonic zone, however, became the final mainland only in the Oligocene. According to these data, it follows that the most complex and deepest karstification should be in the carbonate terrains of the Durmitor tectonic unit, then Kuči, etc. However, according to previous data, it is not so.

In addition to lithological, the tectonic factor is the most important. Namely, throughout the entire geological history, tectonic processes have taken place in continuity, separating or folding certain terrains. According to previous knowledge, the main tectonic processes in the Dinarides took place at the end of the Cretaceous and during the entire Paleogene, when the described geotectonic units "merged". And that means that the former land surfaces were also folded and, by reverse or gravitational faults separated, and faulted. Morphological and tectonic conditions for "new" karstification, previously karstified terrains and carbonate formations were created in this way. Somewise, reconstruction of these processes could only be done by dating cave sediments.

In the previous text, it is said that the glacial sediments in the area of the Balkan Peninsula were first recognized by Jovan Cvijić, on the Rila Mountain in Bulgaria (Cvijić 1897), and then on Prenj, Čvrsnica and Durmitor (Cvijić 1897). After that, glacial sediments were found and described on almost all Montenegrin mountains. Prior to this discovery, Pleistocene glaciations were only developed in the Alps. Cvijić (1913) also studied and described glaciations in the mountains Komovi and Prokletije.

Intensive geological explorations started in all geological areas, after the Second World War. The types of glacial sediments and glacial forms are briefly described and singled out in the interpreters on all sheets of the Geological Map of the SFRY, in the area of Montenegro. It is important to point out, on the basis of the mentioned geological maps, that moraines and glaciofluvial sediments are regularly present in the mountainous terrains of Montenegro, while in the valleys of all rivers, as well as on mountainous surfaces, in karst fields and in the depression of Zeta and Bjelopavlići, thick deposits of glacial conglomerates (over 100 m thickness in Ćemovsko field) and glacial lacustrine sediments, over 100 m thickness were deposited in the Bjelopavlićka dolina.

Milutin Milanković published the original scientific work "*Canon of Insolation and the Ice-Age Problem*" in German, in 1941. This book was published in Serbian (Milanković, 1997), as part of the book collection of this famous scientist. He proved by a mathematical procedure that the amount of heat received from the Sun on Earth depends on the inclination of the axis of rotation of the Earth, the longitude of the perihelion and the eccentricity of the Earth's orbit. In this work, he presented the solar curve for the last 600 thousand years (and later for a million years), on which the four Pleistocene glacial periods and their stages are clearly defined: Günc with two, Mindel with two, Riss with two and Würm with three stages (glacial phases).

Numerous explorers, including Milojević (1951, 1954), Marković-Marijanović (1961), Lješević (1996), Stanković (1975, 1996), Radoičić (2008), have dealt with the problem of the influence of Pleistocene glaciations on the geomorphological, hydrographic and speleological characteristics of the Montenegrin terrain.

Pajović et al. (2017) during the detailed geological exploration of the bauxite-bearing region Vojnik–Maganik, paid special attention to the extraordinarily discovered circuses, lateral and frontal moraines and other glacial forms in the area between Nikšićka Župa and the Vojnik and Maganik mountains. A vertical incision of fluvioglacial sediments, about 50 m thickness, with red bauxite was discovered on that occasion, at the Zagrad red bauxite deposit, during 1992–1995. Alterations of gravels of different granulation, that were formed in the glacial ages Mindel, Riss and Würm and their stages could be observed (Fig. 15) on this profile, which was formed by the exploitation of bauxite.

In addition to the above, the same explorers determined the snowline of the coldest ice age Riss 1 on the terrains of the Vojnik–Maganik region, which maximum was 230,000 years ago. The mean altitude level of the snowline in Montenegro was calculated for all four ice ages, which is shown in Table 1, according to Milankovic's data.

The average determined height of the snowline (lower limit of glaciation) was around 1550 m above sea level, based on the data on the height of the beginning of the

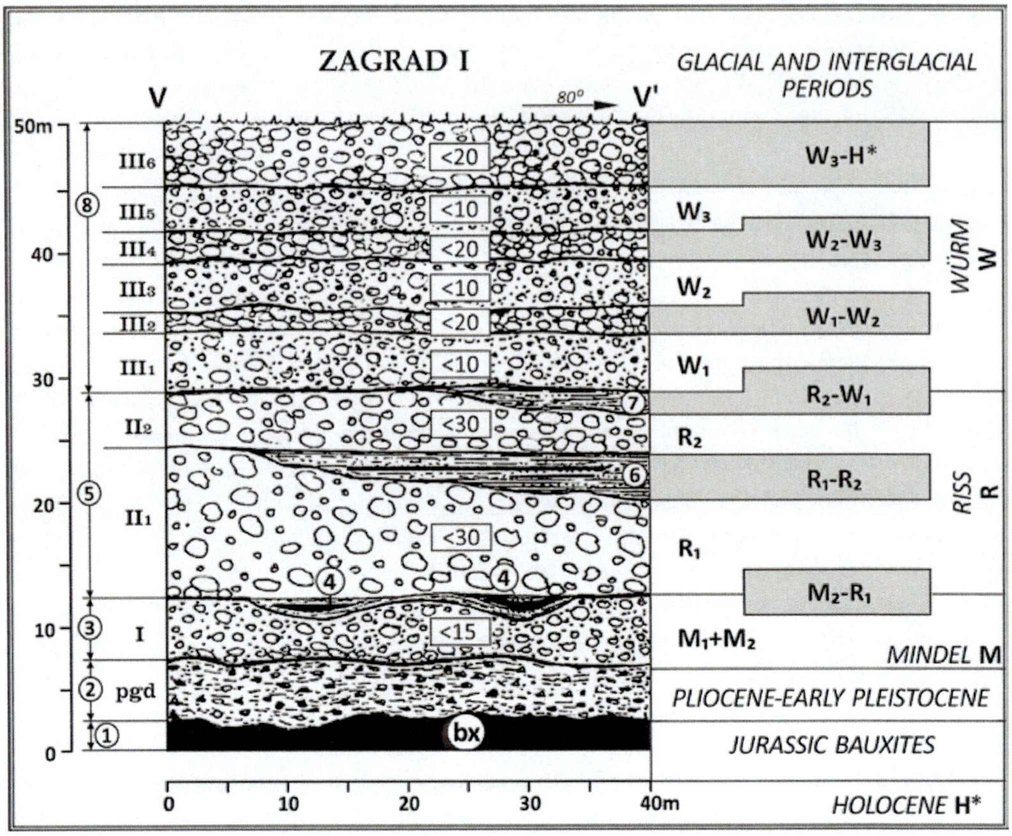

Fig. 15 Determination of Pleistocene Glacial and Interglacial periods in Central Montenegro. Glaciofluvial sediments in hanging wall of Zagrad I bauxite deposit (Pajović et al. 2017). **Legend**: 1. Jurassic red bauxites; 2. Coluvial preglacial deposit predominantly bauxitic composition (pgd); 3. Glacifluvial gravels/conglomerates (Mindel I phase); 4. Laminated carbonate silty sends with interbedds of glacial chalk; 5. Coarse-grained glaciofluvial gravels/conglomerates (Riss-two phases: II1 and II2); 6. Laminated carbonate send and fine-grained gravel with interbedds of glacial chalk; 7. Laminated carbonate send and fine-grained gravel; 8. Changing of medium and fine-grained glaciofluvial gravels/conglomerates (Würm-six phases: III1, III2, III3, III4, III5, III6)

Table 1 Altitude of snowline in four glacials of Central Montenegro area (Pajović et al., 2017)

Name of Glaciations									
Glacial	Würm			Riss		Mindel		Günc	
Stadial	3	2	1	2	1	2	1	2	1
Time in 1000 of years	25	72	115	188	230	435	476	550	590
Elevation of snowline a.s.l. (m)	2090	2000	1780	1910	1550	1960	2030	2400	2150

moraine on the slopes of Maganik mountain. It is probable that the height of this border increases toward the south, that is to say, decreases toward the north, but these differences, in any case, are not huge. Figure 16 shows the surface of the ice cover during the coldest ice age RISS 1, on terrains above the isohypse of 1500 m above sea level. Based on this picture and data on the distribution of moraine fluvioglacial and lacustrine glacial sediments, it can be concluded that almost the entire territory of Montenegro (and the Dinarides) was under the direct or indirect influence of glaciers and glacial processes during the Pleistocene.

Certainly, the question of the significance, that is to say the influence of certain ice ages and their stages on the terrains of Montenegro, is especially interesting. Explorers estimates of ice thickness during RISS 1 are different: from 10 to 50 m on mild slopes and in cirques from 100 to 200 m. Based on the morphological characteristics of cirques, they concluded that the thickness of the ice in some cirques reached a thickness of over 200 m in cirques on the NE side of Maganik (Pajović et al. 2017).

During the Günc ice age, the height of the snowline on the terrains of Montenegro was 2150 m, that is to say 2400 m. In other words, during this ice age, the ice cover was formed only on the highest mountains in Montenegro (Durmitor, Moračke planine, Sinjajevina, Komovi, Prokletije). The melting of these glaciers probably formed shorter glacial canyons in coves or valleys on the slopes of these mountains, the traces of which were probably erased by later glaciations.

The Mindel ice age lasted from 520 to 420 thousand years ago, with two stages and an altitude of 2030 and 1960 m above sea level, which means that the ice cover was above all these mountains in Montenegro. Glaciers from the mountains broke and along the bays and mountain depressions, slid in all directions and formed glacial canyons with lateral moraines, as well as thermal basins (location where the glacier stopped, surrounded by a frontal moraine) in

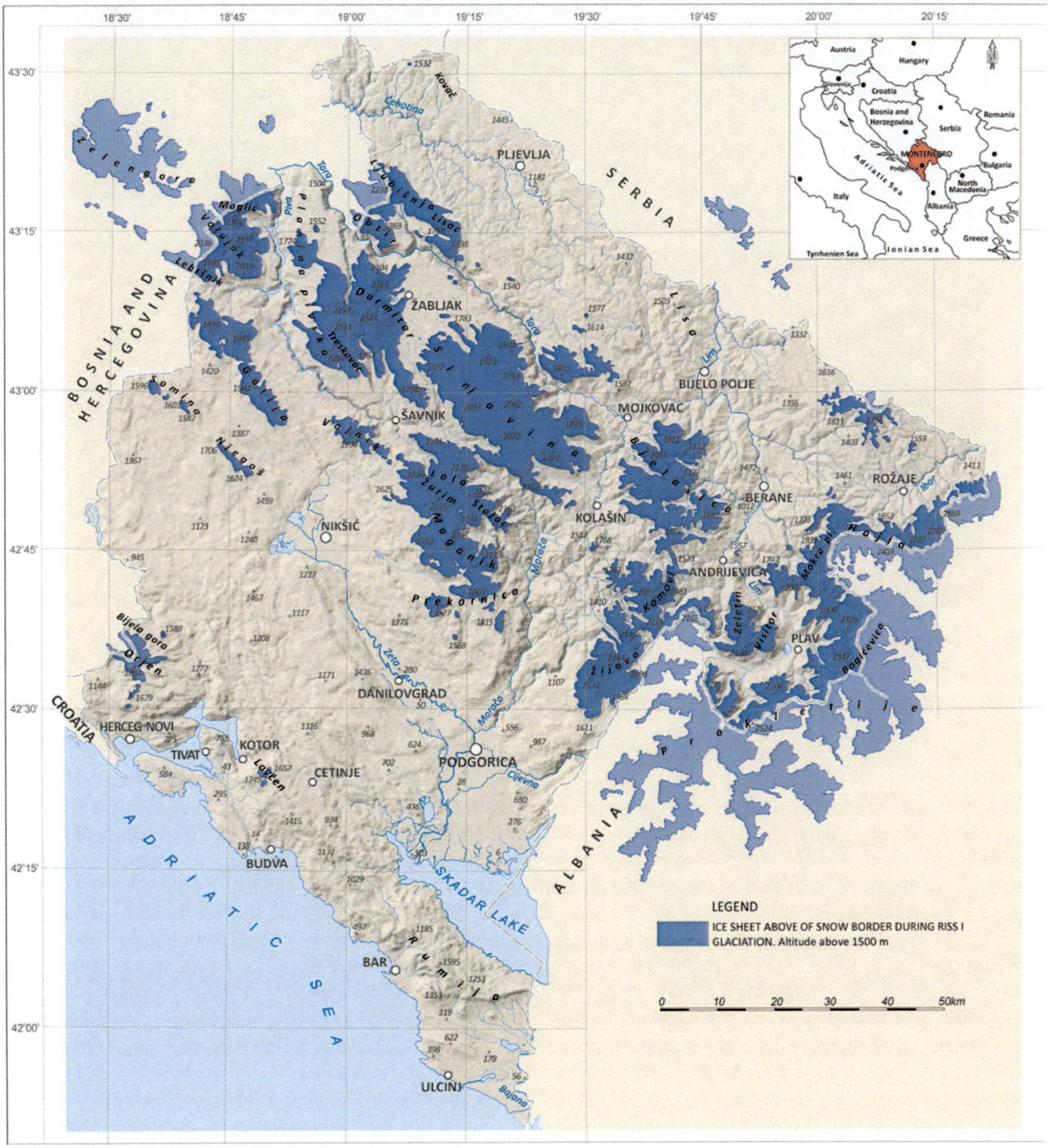

Fig. 16 Ice sheet in the Montenegrin mountains and surroundings, above 1500 m above sea level, during the RISS 1 Glaciation

the long interglacial period between Mindel 2 and Riss 1, which lasted about 170 thousand years. Some glaciers broke and descended into the rivers Morača, Tara, Lim, Ibar and Ćehotina, where the riverbeds were formed during the Miocene.

The Riss ice age lasted from 250 to 160 thousand years ago. The greatest cooling was 230 thousand years ago, in the first stage, when the snowline dropped to about 1550 m and in the second stage to 1910 m above sea level.

In the mountainous areas, in the interglaciation of Riss–Würm, the so-called "sub-bedrock" glaciers formed, which actually represent the areas where glaciers gathered. This ice age, without a doubt, caused the biggest changes in the morphology of the relief of Montenegro, the hydrographic

Fig. 17 Durmitor Mountain, Crno jezero (Black Lake)

network, the depth of karstification and the morphology of karst terrains.

The Würm ice age, with three stages, lasted from 130,000 to 11,500 years ago, with snowline heights, according to Milanković: 1780, 2000 and 2090 m. The formation of ice cover in this age, therefore, was on all the mountains of Montenegro, which domain of influence is most easily recognized, by glacial shapes along the waves, lateral and frontal moraines and glacial lakes.

Finally, it can be concluded that the Pleistocene ice ages influenced the change of relief, hydrographic network and objects, surface and underground karstification, the change of geological composition of certain terrains (by formation of glacial, fluvioglacial and limnoglacial sediments on the terrain) as well as the change of flora and fauna and finally—to changes in the human species.

Having in mind the topic of this monograph, the influence of the process of ice ages on the morphology of the relief, on the surface and underground karstification is especially emphasized. That is to say, glacial erosion has significantly changed all mountainous terrains in Montenegro, as evidenced by jagged mountain ridges, so-called "soe" (sohe) and needles, cirques with steep and vertical sections of carbonate rocks, canyons and extended valleys, flat areas of former karst fields and valleys in which different types of glacial material have accumulated (conglomerate, gravel, sand, clay), to deepen riverbeds and formation of terraces built of conglomerates.

In the canyon part of the rivers Morača, Tara, Lim, Ibar and Ćehotina there are numerous caves at different altitude levels. One of these speleological objects, at the lowest levels, probably is formed during the Pleistocene glaciations, when the level of subterranean spring deepened. Also, it is certainly proven that the sinkhole on the Jezerska surface below Durmitor, formed during the Pleistocene, through which groundwater flows into the Tara riverbed (the source of Ljutica and Bijele vode).

It is even more interesting that the famous glacial Crno jezero (Black lake) (Fig. 17), near Žabljak, is characterized by bifurcation and that its part (Malo Crno jezero) flows into the Komarnica River (Piva`s River basin), and that the water from the Veliko crno jezero over sinkhole of Žabljak springs above the Tara riverbed. The situation is very similar to the hydrogeological and speleological characteristics of other carbonate terrains in Montenegro.

Acknowledgements *Maps*: The authors of the maps printed in this chapter (Figs. 1, 5, 6 and 16) are Dr. Slobodan Radusinović and Dr. Marko Pajović. The maps were made on the basis of the mentioned references for the needs of the ongoing project Geological Survey of Montenegro: "*Geological Atlas of Montenegro*". The authors thank Prof. Dr. Gojko Nikolić for the relief "hill shade" map of Montenegro which was used as a basis for the maps presented in this chapter and to Prof. Dr. Milan Radulović for generating data from the DEM of Montenegro for Fig. 16.

Photographs: The author of the photographs printed in this chapter (Figs. 7, 8, 9, 10, 11, 12, 14 and 17) is Sandra Kapetanovic, an Architect conservator, who has been dealing with the protection of the cultural and natural heritage of Montenegro for many years and our special gratitude goes to her. The author of the photograph Fig. 13 is S. Radusinović.

References

Anđelković, M. (1981). Tektonika Spoljašnjjih Dinarida Jugoslavije. Geološki Anali Balkanskog poluostrva, knj. XLII, Beograd, 1–96.

Anđelković, M. (1982). Šarijaška građa crnogorskih Dinarida. Zbornik radova X jubilarnog Kongresa geologa Jugoslavije, knj. III, Budva, 515–525.

Antonijević, R., Pavić, A., Karović, J. (1973): Tumač za OGK SFRJ listova „Kotor“ i „Budva“, K 34–50, K 34–62, Savezni geološki zavod, Beograd, 1–64.

Bešić, Z. (1948). Geotektonska struktura severne Crne Gore (prethodno saopštenje). Glas. Prirod. Muz. Srp. zemlje, Ser. A, knj. 1, Beograd, 100–110.

Bešić, Z. (1969). Geologija Crne Gore, knj. II, Karst Crne Gore. Zavod za geološka istraživanja Crne Gore, Titograd, 1–304

Ćirić, B. (1975). Spoljašnji Dinaridi u Crnoj Gori – njihova građa i razvoj. Acta geologica, VIII/1 – 25, Zagreb, 317–345.

Cvijić, J. (1897). Tragovi starih glečera na Rili. “Glas” Srpske kraljevske akademije nauka, knj. LIV, 1–103.

Cvijić, J. (1899): Glacijalne i morfološke studije o planinama Bosne i Hercegovine i Crne Gore. “Glas” Srpske kraljevske akademije nauka, knj. LVII, Beograd, 1–196.

Cvijić, J. (1913). Ledeno doba u Prokletijama i okolnim planinama. “Glas” Srpske kraljevske akademije nauka, knj. XCI, Beograd, 188–267.

Cvijić, J. (1924). Geomorfologija - Morphologie terrestre, Knjiga prva. Državna štamparija Kraljevine Srba, Hrvata i Slovenaca, Beograd, 1–588.

Cvijić, J. (1926). Geomorfologija- Morphologie terrestre, Knjiga druga. Državna štamparija Kraljevine Srba, Hrvata i Slovenaca, Beograd, 1–506.

Dimitrijević, N.M. (1967). Sedimentološko–stratigrafski problemi srednjotrijaskog fliša uterenima izmedju Skadarskog jezera i Jadranskog mora. Geološki glasnik, Zavod za geološka istraživanja SR Crne Gore, UDK: 55/56, YU ISSN 0435–4249, knj. V, Titograd, 223–275.

Dimitrijević, M.D. (1974). Dinaridi: jedan model na osnovama “Nove globalne tektonike”, Savjetovanje “Metalogenija i koncepcija geotektonskog razvoja Jugoslavije”, Beograd, 119–151.

Dimitrijević M.D. (1982). Dinaridi – jedan pogled na tektoniku. Vesnik Zavoda za geološka i geofizička istraživanja, Knj. XL, ser. A, Beograd, 113–147.

Drobne, K., Pignatti, J., Ćosović, V., Trutin, M., Đaković, M. & Krašna, L. (2019). Giant *Nummulites maximus* D'Archiac, 1850 from the Upper Lutetian (Eocene) limestones of the Peri–Adriatic littoral of Montenegro. 6. Hrvatski geološki kongres s međunarodnim sudjelovanjem, Knjiga sažetaka, Zagreb, 59–60.

Kalezić, M., Gomilanović, M. (2004). Traganje za rudama u Crnoj Gori, CID, Podgorica, 1–449.

Kober, L. (1952). Leitlinien der Tektonik Jugoslaviens. Glavne tektonske crte Jugoslavije. Posebno izdanje SAN, knj. CLXXXIX, Geološki institut, knj.3, Beograd, 1–81.

Kossmat, F. (1924). Geologie der zentralen Balkanhalbinsel (Mit einer Übersicht des dinarischen Gebirqsbaues). Kriegsschauplatze 1914–1918. Geologisch dargestellt, Berlin, 1–198.

Lješević, M. (1996). Geomorfologija i morfogeneza područja Nacionalnog parka Durmitor. Geografski fakultet Beograd, Posebna izdanja br. 8, Beograd, 43–67.

Madžgalj, Ž. (2013). Speleologija: Kilometar pod zemljom. Moja planeta, br. 32, maj 2013, ISSN 2217-3307, Smart Art, Novi Sad, 11–16.

Marković-Marjanović, J. (1961). Kvartarni sedimenti Zetske ravnice u svetlosti Pleistocene klime. III Kongres geologa Jugoslavije (Budva, 1959), knj. I, Titograd, 252–283.

Milanković, M. (1997). Kanon osunčavanja Zemlje i njegova primena na problem Ledenih doba (2. Deo), Izabrana dela, Zavod za udžbenike i nastavna Sredstva, Beograd, 195–325.

Milojević, B.Ž. (1951). Durmitor – Regionalno-geografska ispitivanja, Zbornik radova Srpske Akademije Nauka, IX, Geografski institut, knj. 2, Beograd, 1–74.

Milojević, B.Ž. (1954). Boka Kotorska. Zbornik radova Srpske Akademije Nauka, XXVII, Geografski institut, knj. 5, Beograd, 1–52.

Mirković, M., Kalezić, M., Pajović, M (1976a). Osnovna geološka karta SFRJ, list “Bar” 1:100.000, K 34–63, Savezni geološki zavod, Beograd.

Mirković, M., Kalezić, M., Pajović, M., Živaljević, M., Đokić, V. (1976b). Osnovna geološka karta SFRJ, list “Ulcinj”, K 34–75, Savezni geološki zavod, Beograd, 1–61.

Mirković, M., Kalezić, M., Pajović, M., Živaljević, M., Škuletić, D. (1978). Tumač za OGK listova “Bar” i “Ulcinj”, K 34–63, K 34–75, Savezni geološki zavod, Beograd, 1–61.

Mirković, M., Živaljević, M., Đokić, V., Perović, Z., Kalezić, M., Pajović, M. (1985). Geološka karta Crne Gore, 1:200.000. RSIZ za geološka istraživanja, Titograd (2 lista u boji, format 92 x 60 cm).

Mirković, M. (1989). Litofacijalne i tektonske karakteristike terena Crne Gore. Geološki glasnik, Zavod za geološka istraživanja SR Crne Gore, UDK: 55/56, YU ISSN 0435–4249, knj. XIII, Titograd, 124–144.

Nopcsa, F. (1921). Geologische Grundzüge der Dinariden. Geologishe Rundscau, Bd. XII, Leipzig, 1–19.

Pajović, M., Radusinović, S. (2010). Mineralne sirovine Crne Gore; Crna Gora u XXI stoljeću u eri kompetitivnosti, Životna sredina i održivi razvoj, Posebna izdanja Crnogorske akademije nauka i umjetnosti, knj. 73, Sv. 2, Podgorica, 237–282.

Pajović, M., Radusinović, S. (2015). Stratigrafija boksita Crne Gore / Stratigraphy of bauxites in Montenegro; Geološki glasnik, Zavod za geološka istraživanja Crne Gore, UDK: 55/56, ISSN 0435–4249, COBISS.CG–ID 37922, knj. XVI, 27–57.

Pajović, M., Mirković, M., Svrkota, R., Ilić, D., Radusinović, S. (2017). Geologija boksitonosnog rejona Vojnik-Maganik (Crna Gora). Posebna izdanja Geološkog glasnika, knj. XXI, Zavod za geološka istraživanja, Podgorica, 1–431.

Pajović, M. (1999). Metallogenic map of Montenegro, 1:200,000. Separate Issues of Geological Bulletin, Vol. XVI, Geological Survey of Montenegro, Podgorica, Montenegro.

Pajović, M. (2000). Geologija i geneza crvenih boksita Crne Gore. Posebna izdanja Geološkog glasnika, knj. XVII, Zavod za geološka istraživanja, Podgorica, 1–242.

Petković, K. (1961). Navlake – kraljušti ili kraljušti u navlakama u tektonskom sklopu Crne Gore i Hercegovine. Geološki anali Balkanskog poluostrva, knj. XXVIII, Beograd, (157–176).

Radoičić, B. (2008) Geografija Crne Gore, Prirodna osnova, knjiga I, DANU, ISBN: 978-86-85779-08-4, Podgorica, 1–335.

Radulović, M. (2000). Hidrogeologija karsta Crne Gore, Posebna izdanja Geološkog glasnika, knj. XVIII, Zavod za geološka istraživanja, Podgorica, 1–271.

Schmid, S., Bernoulli, D., Fugenschuh, B., Matenco, L., Schefer, S., Schuster, R., Tischler, M.,Ustaszewski, K. (2008). The

Alpine–Carpathian–Dinaridic orogenic system: correlationand evolution of tectonic units. Swiss J. Geosci. 101, 139–183.

Stanković, S. (1975). Planinska jezera Crne Gore. Posebna izdanja Društva za nauku i umjetnost Crne Gore, knj. V, Odjeljenje prirodnih nauka, knj. 5, Titograd, 1–223.

Stanković, S. (1996). Vode Nacionalnog parka Durmitor. Geografski fakultet Univerziteta u beogradu, Posebna izdanja, knj. 8, ISBN: 86-82657-03-1, Beograd, 127–140.

Živaljević, M. (1989). Tumač Geološke karte SR Crne Gore, 1:200.000. Posebna izdanja Geološkog glasnika, knj. VIII, Zavod za geološka istraživanja, Titograd, 1–62.

Hydrology of Surface Water and Climate Characteristics of Montenegro

Golub Ćulafić and Jelena Krstajić

Abstract

Hydrological characteristics of surface watercourses will be presented in this paper as primarily determined by the geomorphological and hydrogeological properties of the terrain. Hereby only the general hydrographic properties of the main watercourses will be shown. The main hydrographic properties of Montenegro are presented by two, approximately equal catchment areas, where water flows through the surface and underground paths towards the Danube River basin (7260 km^2 or 52.5%) and the Adriatic Sea basin (6560 km^2 or 47.5%). According to the world standards, both catchment areas are rich in water (average water). However, a significant part of Montenegrin surface (about 60%) belongs to Dinaric karst area, which is without permanent streams and in which precipitation (liquid and solid) runs towards watercourses or the sea. Bearing in mind the fact that even 95.3% of watercourses in Montenegro are formed on its territory, i.e. with its source and catchment area on its own territory, it can rightly be said that water is our greatest natural resource, even though it is spatially and temporally very heterogeneous. Hydrological and hydrogeological factors have one of the most significant roles in the formation of underground morphological forms. The occurrence of underground morphological forms (caves, pits) in karst terrains, which can be found in most parts of the Montenegrin territory, is caused by the sinking of certain amounts of surface water and their movement through underground channels and caverns. Thus, the sparsest river network occurs in the area with the highest amount of precipitation (e.g. Orjen, Lovćen), while it is the densest in the area with the lowest amount of precipitation (northeastern part of Montenegro). This spatial distribution caused the development of a large number of speleological objects in terrains without surface runoff and direction of movement of sinking waters towards the main erosion bases (Lake Skadar and the Adriatic Sea). Another hydrological characteristic, related to the significant spatial and temporal unevenness of water resources throughout the year, is that yields of some rivers (e.g. Zeta, Morača) and karst springs (e.g. Crnojevića River) are often several hundred times higher during the rainy season than in summer (dry) months. Montenegro is a very complex climatic area characterized by pronounced variations in time and space, primarily thanks to its geographical position, proximity to the sea (Adriatic and Mediterranean) and orographic features—mountain ranges, which prevent maritime influences from penetrating deeper into the interior of the land, canyon valleys, etc.

Keywords

Hydrology · Surface water · Karst · Montenegro · Discharge · River basin · Climate · Precipitation

1 Introduction

Water is a unique and irreplaceable natural resource, with limited quantities and uneven spatial and temporal distribution. From the fact that all forms of life and all human activities are more or less connected to water, the relationship towards it, is of huge importance. Water resources are valorized through three basic components: quantity, quality

G. Ćulafić (✉)
Institute of Hydrometeorology and Seismology of Montenegro, IV Proleterske 19, 81 000 Podgorica, Montenegro
e-mail: golub.culafic@meteo.co.me; culle2461@gmail.com

J. Krstajić
Department of Hydrogeology, Faculty of Mining and Geology, Centre for Karst Hydrogeology, University of Belgrade, Đušina 7, Belgrade, Serbia

G. Barovic (ed.), *Speleology of Montenegro*, Cave and Karst Systems of the World,
https://doi.org/10.1007/978-3-031-49375-1_4

and spatial location. These three components represent the basic characteristics of the water resources of an area.

Identification of the river flow regime fluctuation, caused by natural or anthropogenic factors, is today one of the main issues of hydrology as a science. Modern physical-geographical research is based on the analysis and comparison of numerous indicators of hydrological and climatic elements, on the one hand, and orography, geology, hydrogeology (speleology), pedology, vegetation, morphometry etc., on the other hand.

On the rather small area occupied by Montenegro (13,812 km^2), one can encounter extremely diverse and specific forms of relief, as well as phenomena and related processes, which is a consequence of the long geological evolution of the terrain and pronounced endogenous and exogenous forces. The total area of the territory from which waters flow towards the Danube basin is about 7260 km^2 (52.5%) and the main watercourses are Lim, Tara, Piva, Ćehotina and Ibar. The catchment area from which waters gravitate to the Adriatic Sea is about 6560 km^2 (47.5%). Morača, Zeta, Cijevna, Rijeka Crnojevića and Orahovštica, as well as a number of smaller watercourses flow towards the basin of Lake Skadar, and from there through the Bojana River into the Adriatic Sea. In addition to the mentioned watercourses, this catchment area also includes the terrains of the Montenegrin coastline (without constant runoff, mostly torrential in nature), the western and southwestern parts of the Orjen mountain (which provide water to the Croatian coast), the western and northwestern karst terrains of the municipality of Nikšić (which provide water to the Trebišnjica basin, Bosnia and Herzegovina), the eastern slopes of Mount Čakora and Bogićevica (which provide water to Pećka Bistrica and further to the Drim River).

From regional perspective, the Danube basin occupies the area of the Central High Plains (1200–1800 m a.s.l.) and the area of northeastern Montenegro (600–1500 m a.s.l.). This area is characterized by high mountains (Komovi, Prokletije, Durmitor, Bjelasica, Sinjajevina, etc.) which are dissected and intersected by the basins of the rivers Lim, Tara, Piva, Komarnica, Ćehotina and Ibar and their tributaries. Considerable parts of the terrain are made of clastic and flysch-like clayey-sandy-marly sediments, in which occur phenomena of accelerated washing, dredging, tearing and sliding. In these areas, the process of river erosion progresses faster than the process of karstification. That is the reason why very strong karst springs appear along the river canyons (Ravnjak, Ljutica, Ali-pašini izvori, Oko Skakavica, Dubrovska vrela etc.), which ensure the permanence and continuity of the flow also during the dry period of the year.

The Adriatic Sea basin occupies the area of the Montenegrin coastline, the Deep Karst Plain (both without major surface runoff, mostly torrential in nature) and the Valley of Central Montenegro (Dinaric direction). This is an area of pronounced karst, in which there is a constant struggle between the process of karstification and river erosion. Numerous rivers in this area have been already disintegrated and gone underground (Cetinjska river, Karučka, Sinjačka river, rivers in Boka Kotorska), while some of them almost dry up in the summer period of the year (Gornja Zeta, Rijeka Crnojevića).

The specificity of Montenegro's position is that it is located in a zone of very pronounced thermal asymmetry between cold northern Europe and very warm northern Africa. Just above Montenegro, there is an intensive exchange of warm air masses that go to the north and cold air masses that go from the north to the south. Very often, these air masses with extremely different physical-meteorological properties collide and mix over Montenegro. The position of Montenegro is such that these systems strongly influence the weather and define the most diverse types of climate, from extremely harsh to very favorable, in all seasons.

The Montenegrin coast, together with the Zeta-Bjelopavlić valley, represents areas with a Mediterranean climate, characterized by long, hot and dry summers and relatively mild and rainy winters. This area is particularly known for its high summer temperatures, and the absolute maximum air temperature in Montenegro as well as the highest average number of tropical days (Podgorica, Danilovgrad).

Also, karst fields (*polja*), where a slightly harsher climate occurs, are also specific, bearing in mind that they are usually located at higher altitudes, and are on average 20–60 km away from the Adriatic Sea. In winter, during anticyclonic weather conditions, cold air settles in these fields, which descends on the sides of the surrounding mountains, while in the summer months, the ground layer of air here gets quite warm, as a result of which the annual fluctuation of air temperature is increased and uneven.

The central and northern part of Montenegro has the characteristics of a mountain climate, which is somewhat modified by the influence of the Mediterranean Sea, primarily reflected in the rainfall regime and the higher average temperature of the coldest months. The far north of Montenegro belongs to the continental type of climate, which is characterized by large daily and annual temperature amplitudes and a small annual amount of precipitation (with a fairly even distribution by month). In the mountainous areas in the north, summers are characterized by a relatively cold and humid climate, while winters are long and harsh, with frequent frosts and low temperatures, which drop sharply with the change in altitude.

According to the rainfall regime on the territory of Montenegro, we distinguish the Mediterranean and mild-continental regimes. The Mediterranean regime is characterized by the maximum amount of precipitation in November and December, and the minimum in July and August. The temperate-continental regime is characterized by more frequent precipitation in the second half of summer, a secondary maximum in October and a minimum in February. In the largest part of the geographical area of Montenegro, the maritime pluviometric regime is somewhat modified by the mountain and continental rainfall regime.

2 Overview of the Earlier Investigation

Characteristics of surface waters in Montenegro was a subject of an earlier research for different purposes and uses. Water regimes were mostly studied from the aspects of utilization for energy purposes, which were dealt with by Hrvačević (2004), while Radojičić (2005) provides only basic balance and morphometric data. Đorđević et al. (2010) provide an overview of the water potential of Montenegro, with the analysis of characteristic flow values. The research and data mentioned mainly refer to the observation period 1961–1990. All relevant data were processed through the preparation of the Water Management Master Plan of Montenegro (2001), but the observation period is the same as for the aforementioned research. In recent years, a more detailed study of water regime oscillations and the possible impact of climate change on them has begun in Montenegro, as Burić et al. (2012) analyze floods in Montenegro, with special reference to fluctuations in the flow of the Morača River. A more detailed analysis of hydrological characteristics is given by Sekulić and Radulović (2019), where, in addition to mean monthly yields, they also analyze some of the characteristic values and the probability of large and small water occurrences. Then Ćulafić (2020) in his research determines negative trends in the flows of Lim (HS Plav and HS Bijelo Polje), especially in terms of seasonality (summer) and deals with the ranking of years by water level (SDI index). Stevanovic and Blagojević (2021) analyze the hydrological characteristics of rivers in the Šavnik basin (Bukovica and Bijela) where they also provide projections of future flow fluctuations in the light of climate change.

Greater attention must be paid to future research, especially from the aspect of investigating the movement of water in the karst and determining hydrogeological divides and adequate water balancing based on the data obtained.

3 Data and Methods

During the analysis for this paperwork, the data of the Institute for Hydrometeorology and Seismology of Montenegro (www.meteo.co.me) were used as given in Tables 1 and 2. The analysis only refers to flow values (monthly average and annual) because they represent the reaction of the entire river basin, and even though together with the water level they constitute the main characteristics of the water regime of a watercourse, flow values are still much more reliable for displaying certain characteristics. The annual flow is an indicator of the amount of water in a given river basin, because if the annual flow increases, the risk of flooding also increases. Low annual flow could lead to a series of causal events that adversely affect human activities and the aquatic ecosystem. The thirty-year series (Crnojevića River) was taken as the shortest acceptable measurement period, while for the other hydrological stations, the reference period 1961–2015 was taken. Choosing the profile for analysis was done with special care and certain criteria which had to be met were set (that the location was not changed), so that the obtained data would not be doubted and the display would be as representative as possible.

Climate, in a narrower sense, represents the average state of the atmosphere. Here is a presentation of spatio-temporal variables (air temperature, amount of precipitation, relative humidity, wind), i.e. of those elements that have the most influence on corrosion processes in karst. For the description and analysis of the climatic characteristics of Montenegro, data were used on measurements and observations of meteorological elements and phenomena from meteorological stations that are part of the observation system of the Institute for Hydrometeorology and Seismology of Montenegro (www.meteo.co.me), for the climatic period 1991–2020 year, and a comparative analysis is also given for the period 1961–1990 year.

Table 1 Hydrological stations in the Adriatic basin

Hydrological station	River	Sub-basin	F km^2	Geographical		Elevation "0" m a.s.l	Period
				Latitude	Longitude		
Podgorica	Morača	Skadar lake	2627.7	42° 27′ 11.89″	19° 15′ 42.65″	24.7	1960–2015
Zlatica	Morača	Morača	772.4	42° 29′ 01.64″	19° 18′ 25.04″	53.0	1983–2015
Međuriječje	Mrtvica	Morača	207.7	42° 43′ 13.05″	19° 22′ 01.31″	187.5	1960–2015
Duklov most	Zeta	Morača	342.2	42° 47′ 35.41″	18° 56′ 03.10″	615.2	1960–2015
Brodska Njiva	Rijeka Crnojevića	Skadar lake	79.3	42° 21′ 24.46″	19° 00′ 44.73″	8.3	1991–2020

Table 2 Hydrological stations in the Danube River basin

Hydrological station	River	Sub-basin	F km^2	Geographical		Elevation "0" m a.s.l	Period
				Latitude	Longitude		
Trebaljevo	Tara	Drina	247.0	42° 46′ 29.55″	19° 32′ 37.70″	965.8	1960–2015
Plav	Lim	Drina	363.6	42° 36′ 29.90″	19° 55′ 37.40″	906.6	1960–2015
Ravna Rijeka	Ljubovida	Lim	351.0	42° 59′ 21.73″	19° 44′ 17.27″	582.0	1960–2015
Bijelo Polje	Lim	Drina	2182.7	43° 01′ 54.84″	19° 44′ 49.07″	559.7	1960–2015
Gubavač	Bjelopoljska Bistrica	Lim	492.0	43° 06′ 05.41″	19° 47′ 36.83″	545.0	1960–2015
Gradac	Ćehotina	Drina	809.0	43° 23′ 30.89″	19° 08′ 56.09″	665.0	1960–2015
Rožaje	Ibar	Velika Morava	94.7	42° 49′ 51.75″	20° 08′ 43.01″	1035.1	1968–2003

4 Hydrographic and Hydrological Characteristics of the Rivers in the Adriatic Basin

The geological composition of the terrain, the soil and the tectonic relations, caused the drainage of waters of the topographic basin of the ***Morača River*** towards Lake Skadar, either by underground or surface paths. Morača River source is above the village of Ljevišta (970 m a.s.l.), formed by joining waters coming from the Grlo spring (located in the Pleistocene cirque Vragodo, at cc 1370 m a.s.l.) and the Koritski and Rupochajski creeks, while after 2.5 km, it receives the water of the Javorski creek, on its left side. In addition to the numerous smaller streams that feed Morača, especially in the rainy season, its left tributaries are significant: Koštanica, Sjevernica, Kruševački potok, Mala Rijeka, Ribnica and Cijevna, as well as the right tributaries: Ratnja Rijeka, Požnja, Mrtvica, Ibrištica, Bogotovski creek, Zeta and Sitnica. Nowhere is the relief as dynamic and detailed as in this relatively small area. The lowest point of the researched area is the confluence of Morača into the Skadar lake (8 m a.s.l.) while the highest point is Torna (2277 m a.s.l.).

Only in the upper part of the Morača Valley River has the Dinaric direction of flow, northwest-southeast, while in the middle part, the river flows in a north–south direction, and similarly direction of flow through the Zeta Plain to its mouth in the Skadar Lake. This area is decorated with canyons: Morača–Platije (38 km long and an average depth of about 1000 m), Mrtvica, Ibrištica and Mala Rijeka, as well as smaller ones that are not far behind in their beauty and attractiveness: Bogotovski potok, Kruševački potok, Trebješica, etc. The length of the Morača stream from its source to the inflow of Zeta (Vranićka Njiva) is 64 km, and 98 km to its confluence into Lake Skadar. The area of the basin, excluding the Zeta River, is 1031 km^2, while the total area of the basin up to the confluence with Skadar Lake is 3260 km^2 (Fig. 1) (Hrvačević 2004).

The hydrological characteristics of Morača are shown on two profiles, the Vezirov Most (Podgorica) profile which collects the waters of the Morača and Zeta basins, and the Zlatica profile, which drains only the Morača basin. The multi-year average flow of the Morača at HS Vezirov Most is 157.3 m^3/s and at HS Zlatica is 58.9 m^3/s. Both profiles are characterized by the appearance of maximum water in November (243.8 and 97.9 m^3/s), December (269.3 and 102.0 m^3/s) when absolute maxima occur, and in April (242.5 and 102.0 m^3/s), while on both profiles the minimum water occurs during the summer months, with the main minimum in August (26.4 and 4.9 m^3/s). The high degree of karstification that occurs in the stretch around Duga monastery causes that this part of Morača River almost dries up (coloring of the Manastirski Mlin sinkhole showed the connection with the Piletić spring, where a fictitious speed of 1.64 cm/s was achieved (Radulović 2000)). One part of the river continues to flow directly, and the other part goes indirectly through underground paths and through highly porous alluvial sediments.

Mrtvica is the right tributary of Morača (flows into Morača River at 182 m a.s.l., in the town of Međuriječje, Kolašin municipality), which is formed in Velje Duboko place (Pleistocene cirque) by joining of several streams that flow from the slopes of Stožac (2141 m a.s.l.), Lubanje glave (1930 m a.s.l.), Rogođeda (2017 m a.s.l.) etc. The terrain substrate in which Mrtvica cuts its imposing canyon is made up of dolomitic megalodon limestones and Upper Triassic dolomites. Observing the profile of this area, one can see the first development of the complete lias in the form of limestone and dolomite up to 200 m thick, while the upper part of the profile consists of banked and massive dogger and Oxford-Kimmeridgian limestones (up to 200 m thick). At the very top of the profile are massive and banked dolomitic limestones and Tithonian and lower cretaceous dolomites (more than 50 m thick). Cretaceous sediments exposed west of the canyon reach a thickness of about 1000 m. Međeđi vrh (2130 m a.s.l.), Kokotov vrh (2000 m a.s.l.), Petrov vrh (2124 m a.s.l.) and Trešteni vrh (1980 m a.s.l.) rise along the edge of the canyon. This dynamic appearance of the relief was shaped by fluvial, karst and glacial erosion. The geological composition (limestone) resulted in a great fragmentation of the relief with numerous speleological objects (mostly of vertical development),

Fig. 1 Adriatic River basin

such as the "Iron deep" with a depth of 1027 m and a length of 2820 m, whose research has not yet been completed (www.speleologija.me).

The catchment area of Mrtvica River (topographically) covers an area of 207.7 km^2 and has a stream length of 16.5 km. Primarily thanks to the karst spring Bijela Nerini, which has a minimum of 800 l/s, (Ćulafić et al. 2020), Mrtvica River has a constant flow. Mrtvica River is also receiving water from the "Jama" spring and the Dubočnjak stream ($F = 24.4$ km^2), which begins above the village of Mrtvo Duboko, at about 1100 m a.s.l. Based on the analyzed data, Mrtvica River has an average annual flow of 14 m^3/s, an average maximum of 23.7 m^3/s (November) and an average minimum of 2.7 m^3/s (August).

Zeta is the most important tributary of Morača and is divided into "Upper" and "Lower". Upper Zeta includes the source and part of the flow through the Nikšić field up to the Slivlje sinkhole. As the largest watercourse, it is formed in Gornje polje from the rivers Sušica and Rastovac. Sušica, which is an occasional watercourse, originates from the Vidrovanska river (which one branch during the rainy season flows underground through the Vidrovanska cave) and Boljašnica river. Zeta River generally flows south to the village of Zavrh, where it loses part of its water to the existing sinkholes (estavelas), and from there water appears again at watersources Poklonci and Blace (Radulović 2000). After Zavrh, Zeta River turns to the east, towards Glibavac, where it meanders along almost the entire course. From the Vukov bridge, the Zeta flows through a concrete channel on south towards Budoš, where it turns southeast and east to the compensation pool from where the water is brought to HPP "Perućica" through a tunnel and pipelines. Before the construction of the hydropower system and the regulation of the Zeta bed, this watercourse flowed along the southern edge of Nikšić polje (Budoška bara and Slivlje) and after a few kilometers the water appeared at the spring of Glava Zete (72 m a.s.l.), with a drop of 523 m. From this spring, and the springs of Perućica and Oboštica, starts the Lower Zeta watercourse. Today, on its right side, in the Vrtac retention area, the regulated flow of Zeta also receives the regulated flows of Opačica and Moštanica. The Opačica canal is a continuation of the Slana River, which is now flooded by the Slano reservoir, while Moštanica represents a continuation of the river of the same name from the Krupac reservoir. During larger rainy periods ($Q > 60$ m^3/s), Zeta, Opačica and Moštanica are flooded by the Vrtac reservoir (retention). Lower Zeta starts from Glava Zeta spring and HPP "Perućica" system, flows (meanders) through the Bjelopavlićka plain until its confluence with the Morača River (Vranićka Njive). Catchment area of the Zeta River is 1216 km^2 or 18.35% of the Adriatic Sea catchment area in Montenegro. The entire length of the Zeta stream, from its origin in Nikšić polje to its mouth in Morača, is 72.50 km.

Zeta River (the upper course in the Nikšić polje) is observed on the HS Duklov bridge and shows an extraordinary fluctuation during the year. The average annual flow is 18.9 m^3/s. The highest average monthly flows occur in spring (April 35.8 m^3/s), at the end of autumn (November 31.3 m^3/s) and at the beginning of winter (December 34.0 m^3/s), while they are the lowest in the summer months, with a minimum in August of 1.4 m^3/s. Due to the geological composition of the watershed, there are very pronounced relationships between small (probability of occurrence 95%) and large waters (probability of occurrence 1%), where the ratio exceeds the difference of 1:2000. Unfortunately, the profiles on Lower Zeta, Rošca and Danilovgrad do not have continuity in the observations, so they were not taken into account during this consideration and analysis.

Crnojevića River is a karst spring, which territorially belongs to the municipality of Cetinje, while geologically and tectonically it belongs to the High Karst zone and according to hydrogeological reionization it belongs to the area of the Outer Dinarides (Radulović 2000). The Crnojevića River is formed from the waters from the Obod Cave (81 m above sea level) given during the maximum rainfall in the basin, as well as its downstream zone (during the recession period of rainfall). It flows into Lake Skadar after 13 km. Bearing in mind the seasonal fluctuation of the water level of Lake Skadar (1–588 cm), the middle and lower parts of the stream are flooded during the winter and spring months, while the upper stream (1.4 km downstream from HS Brodska Njiva) is always above the maximum water level of the lake and without the influence of water stagnation. According to the hydraulic conditions of the discharge, a large fluctuation is noticeable, and the coefficient of hydraulic unevenness is approximately 210. The value of the mean annual flow is 7.2 m^3/s. Large yield values start from November (12.3 m^3/s), reach their maximum in December (15.8 m3/s) and slowly decrease in January (13.6 m^3/s). The minimums coincide with minimum precipitation in the basin, and increased air temperature and evaporation (July 0.8 m^3/s). Analyzing the intra-annual distribution and based on the pluviometric regime of the catchment area, large fluctuations are observed, especially during the spring–summer period.

5 Hydrographic and Hydrological Characteristics of the Danube River Basin

Lim is the largest tributary of the Drina river and the most hydrographically significant river in Montenegro. The total length of the stream is 219 km, of which 123 km is on the territory of Montenegro and covers a basin area of 2805 km^2 (state border with Serbia). It starts its flow from Lake Plav at an altitude of 908.9 m a.s.l. (at the middle level of the lake). A large part of the Lim basin consists of the Plav-Gusinj, Andrijevica, Berane and Bjelo Polje valleys. Some of the more important left tributaries are Murinska rijeka, Zloredčica, Kraštica, Trepačka, Bistrica, Brzava, Ljubovida and Lješnica, while the most important right side tributaries are: Komaračka, Velička, Piševska, Dapsička, Crnča and Bjelopoljka Bistrica. This area belongs to Northeastern Montenegro, and is a tectonic-erosive extension, bordered by the slopes of Prokletije, Visitor, Bogićevica, Komovo, Mokra, Bjelasica and Cmiljevica, which sides descend quite steeply into these valley bottoms and widenings. Average annual flows at the analyzed stations ranged from 19.3 m^3/s in Plav to 62.7 m^3/s in Bijelo Polje. The highest mean monthly flows occur during the spring months (Plav: April 32.5 m^3/s and May 42.4 m^3/s; Bijelo Polje: April 121.7 m^3/s and May 126.3 m^3/s). The lowest flow values coincide with the lowest amount of precipitation and the highest air temperature, i.e. during the summer months. The minimum occurs at both stations in August (Plav 6.6 m^3/s and Bijelo Polje 17.9 m^3/s). According to the classification, this flow regime is characteristic of the nival-pluvial regime, the Dinaric-Macedonian type (Dukić and Gavrilović 2008), which is characteristic of almost all rivers in the Danube basin (on the territory of Montenegro).

Ljubovida is one of the longest tributaries of Lim and its largest tributary on the territory of the municipality of Bijelo Polje. It has a very large watershed, which is separated from the Tara watershed by the northeastern branches of Bjelasica, then Burena and the Baričko-Stožerska area. From Ćehotina river watershed it is separated by the Bliškovo and Kovren areas, while from the Lješnica river watershed, it is separated by Lisa and many smaller hills. River Ljubovida is formed by joining the springs located at the foot of Stuga (1240 m) and the Kovren pass (1062 m) at 980 m a.s.l. Length of its course, from the source to the confluence with Lim, is 35.8 km, and the catchment area is 351 km^2. It flows in a north–south direction to the place Sljepač Most, from where it continues to flow parallel to the course of Lim. Hydrological station Ravna Rijeka is located about 0.8 km upstream from the confluence of Ljubovida and Lim. The average annual flow is 9.0 m^3/s. Minimum flows occur in the summer months (August 2.5 m^3/s), while maximums occur twice a year, in December (13.0 $m^{3/}s$) and April (17.5 m^3/s). The maximum coincides with the secondary maximum of precipitation in the basin, as well as with the increase in temperature, which conditions the melting of the snow cover (Fig. 2).

Bjelopoljska Bistrica is a right tributary of the Lim, 8 km downstream from Bijelo Polje (520 m a.s.l.). Part of the basin belongs to the territory of Serbia, around the Mačkovac mountain (Jelenak 1617 m a.s.l.). On its way, the river cuts through the Đalovića gorge, which is protected as a natural monument, with an area of 1,600 ha, and III management category ("Official Gazette of FRCG", no. 30/68). The most attractive part of the gorge is the cave above Vražiji firovi or Đalovića cave. The average annual flow is 5.9 m^3/s. Minimum flows occur in the summer months (August 2.1 m^3/s), while maximums occur at the beginning of spring (March 9.7 m^3/s, April 12.0 m^3/s and May 8.0 m^3/s). The maximum occurs because of the snow cover melting in combination with precipitation because the maximum precipitation in Bijelo Polje occurs in November (106 mm) and December (90 mm).

Tara is formed near Han Garančić (1100 m a.s.l.) by the confluence of the rivers Veruša and Opasanica. The highest springs of Veruša are located below Maglič, on the Bušat mountain (1622 m a.s.l.), while Opasanica is formed by several streams and springs, of which the most generous is the "Bijela voda" spring on the slopes of Komovo, which represent: "one of the most important hydrological hubs of the Balkans" (Radojičić 2005). The length of the Tara flow is different in the geographical literature, but we have adopted the value of 146 km, as stated in the Water Management Master Plan of Montenegro (Hrvačević 2004).

Tara generally flows in a southeast-northwest direction, i.e. along the Dinaric direction, then it breaks through the mountains of Bjelasica, Sinjajevina, Ljubišnja, Durmitor, Piva mountain and near Šćepan Polje at an altitude of 434.8 m a.s.l., meets the Piva river and thus forming the Drina river. The area of the Tara basin is 2040.2 km^2. Unfortunately, there is no continuity in the measurements and observations of hydrological parameters on such an important watercourse, except on the upper part of the stream (HS Trebaljevo). Therefore, all analyses of the entire basin are based on stochastic methods. The average annual flow at the mentioned station is 24.6 m^3/s. The highest mean monthly flows occur in spring (April 51.0 m^3/s and May 41.5 m^3/s) as well as at the beginning of winter (December 38.8 m^3/s), while the lowest are in the summer months (August 4.2 m^3/s).

The basin of the ***Piva*** river was changed and modified by the construction of HPP "Piva" downstream from the village of Mratinje in the municipality of Plužine. Before the construction of the reservoir, it was created by joining the Sinjac spring (574 m a.s.l.) and the Komarnica river and flowed cutting a deep canyon (in carbonate rocks) to Šćepan Polje, where together with Tara it formed the Drina river. The Piva basin covers an area of 1784 km^2 (Hrvačević 2004). ***Komarnica*** is the main tributary of Piva river, which (with its tributaries) drains the area of Sinjajevina and Durmitor. Its most important tributary is Pridvorica (left), which originates near Šavnik (833 m a.s.l.) by joining the Bukovica, Bijela, Šavnička glava (intermittent spring, which is used for the water supply of Šavnik) and Pošćenski potok. Of the left tributaries, the most important is Grabovica (during the summer months in the lower part of the stream it flows indirectly through the alluvium), which just before flowing into Komarnica, creates the magnificent Skakavica waterfall, about 51 m high. Komarnica, as a breakthrough river, cut the "Nevido" canyon, about 4 km long, which attracts a lot of attention from tourists and canyoning enthusiasts. Downstream of the village of Duži, Komarnica receives water from the Dubrovska vrela (karst spring) on the left side, which has an estimated minimum yield around 500 l/s (Hrvačević 2004). It is important to note that the hydrogeological connection of this spring with the water of Little Black Lake on Durmitor has been established. That is a very characteristic phenomenon of natural underground bifurcation, which is characteristic of terrains built out of carbonate rocks, as during geological evolution they were tectonically

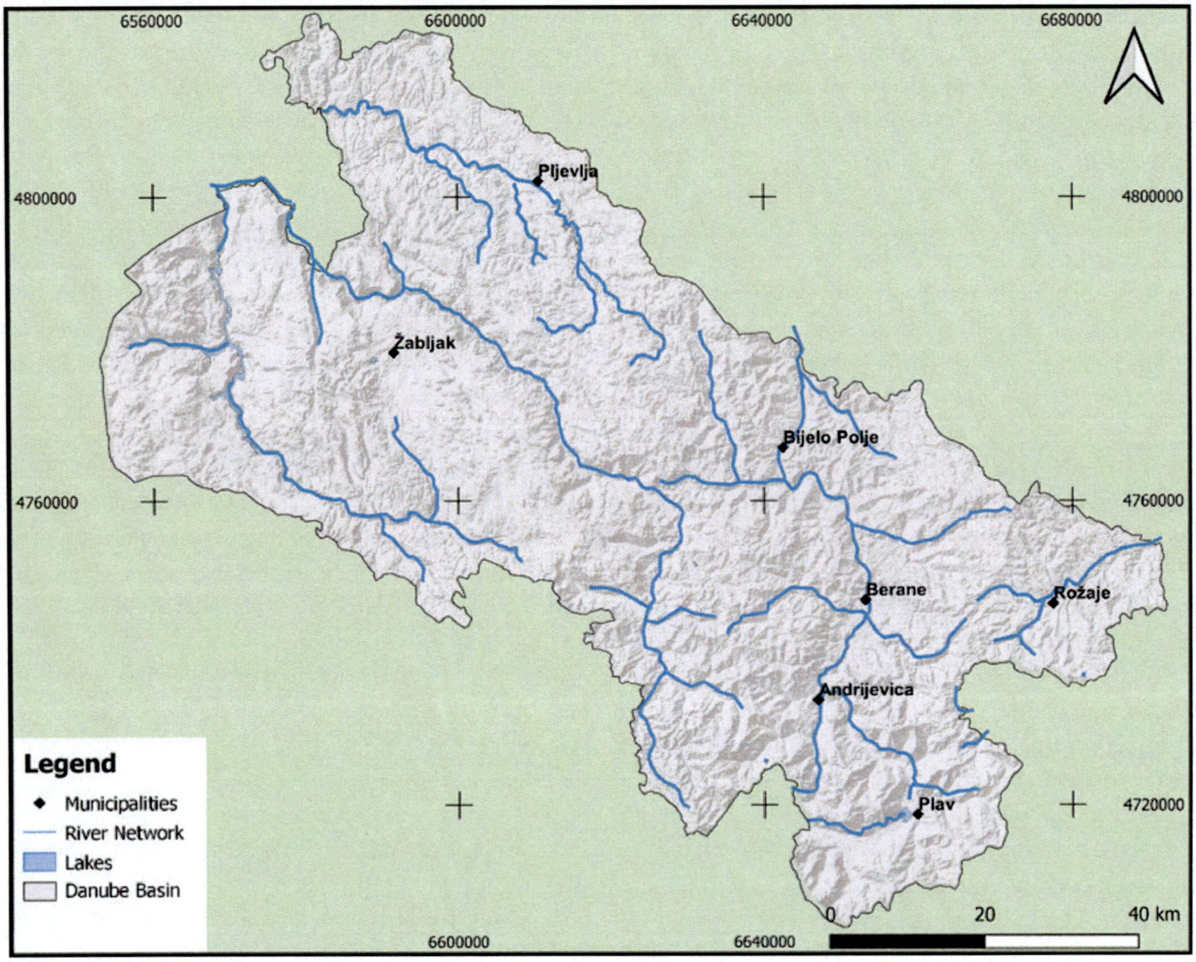

Fig. 2 Danube River basin

fractured, and intensively karstified to great depths. This was confirmed by the results of the dye test, which showed that the Little Black Lake is a watershed between the basins of the Tara and Komarnica rivers (the connections with the Bijele Vode spring in the Tara Canyon and the Dubrovska vrela spring in the Komarnica Canyon were established). This means that these waters circulate below the deep canyon of the Tara river and appear at springs on the other side of the watercourse (Radojičić 2005).

Ćehotina is a watercourse that drains the area with the lowest rainfall in Montenegro (Pljevlja 797 mm). It begins on the slopes of Stožer, at the source of Glava Ćehotine (945 m a.s.l.), where it forms a narrow topographic watershed between its basin and the Lim basin (the source of the Ljuboviđe River). It flows in a southeast-northwest direction (like almost all tributaries of the Drina) to the mouth of the Drina river near Foča (Bosnia and Herzegovina). The upper part of the stream is a characteristic of its meanders and has the characteristics of a gorge, where surface erosion is dominant on the sides, while the lower part of the stream has a plain character, with a more uniform flow. In the village of Otilovići, a dam of the same name was built, which reservoirs are used for the needs of the "Pljevlja" TPP. Hydrological characteristics were analyzed based on the most downstream (outlet) profile Gradac, which includes almost the entire Ćehotina basin with an area of 809.9 km^2 (Hrvačević 2004). The average annual flow is 12.9 m^3/s. The highest mean monthly flows occur in spring (March 20.3 m^3/s and April 21.7 m^3/s) and early winter (December 18.2 m^3/s), while they are the lowest in summer (August 5.9 m^3/s). This flow regime occurs because of the dry summer period, which is characterized by the highest air temperatures and the minimum amount of precipitation. The highest flows occur in the spring season, which is conditioned by the melting of the snow cover and precipitation (in liquid form). It is important to note that the natural flow regime has been modified by interventions in the area, primarily with construction of the Otilovići reservoir as well as the diversion of the Ćehotina stream through the Velika Plijeĺ hill (99 m a.s.l.) at the entrance to Pljevlja.

Ibar belongs to the territory of Montenegro only with the upper part of its course (37 km to the border with Serbia), and covers a basin area (up to HS Bać) of 405.2 km^2, while the hydrogeological basin covers an area of 413.6 km^2

(Hrvačević 2004). The highest springs in the basin are on the northern slopes of Hajla (1365 m a.s.l.) and the north-western slopes of Smiljevica (1963 m a.s.l.). Suvovar stream (1436 m a.s.l.) receives on its right side waters from the spring of Ibar (karst spring, broken type) and the Bjeluha stream, after which the mainstream of Ibar river is formed. The average annual flow (HS Rožaje) for the analyzed period is 2.5 m^3/s. The highest mean monthly flows occur in spring (April 5.2 m^3/s and May 5.8 m^3/s), while the lowest are in the summer months (August 1.3 m^3/s and September 1.0 m^3/s). According to the classification, this flow regime is characteristic of the nival-pluvial regime, Dinaric-Macedonian type (Dukić and Gavrilović 2008).

6 Hydrological Properties of Certain Speleological Objects

From the speleology perspective, it is important to classify objects according to their hydrological properties and function. Thus, Milanović (2012) divides them into: (a) speleological objects with constant underground flow; (b) speleological objects with occasional underground flow; (c) speleological objects with ponded water and (d) dry speleological objects. In the following paragraphs, examples of the most interesting objects according to their function are given.

One of the speleological objects with a constant underground (outflow) flow is the Obod cave, which represents the source of the Rijeka Crnojevića. The water that appears at this spring originates from atmospheric waters, which are formed in the area of Cetinje field and Dobrsko village. Water on the edge of the field sinks and moves mainly through underground caverns and channels, formed in the upper triassic limestones (Cetinjska-Lipska-Obodska cave system) where they emerge again on the surface of the terrain, as a gravity spring, which drains the aquifer formed in the rocks of karst-fissure porosity, presented by highly karstified limestones and dolomites.

Another such object is also the Grbočica cave in Trnovo, which, according to Radulović (2000), has a minimum volume of about 0.5 m^3/s, and whose waters are lost in an active sinkhole and appear again at the oko spring in Lake Skadar. In the limestone terrains of the coastal part of Montenegro, especially in the Bay of Kotor, there are a certain number of caves with water, which have entrances below the sea surface or just above the sea level. Their openings are large due to contact with flysch sediments. Typical examples of such forms are Risanska Spila and Sopot, whose waters become salty during the summer months, i.e. in the dry period of the year. Also, the Opačica cave in Herceg Novi is another such example, whose waters are used for water supply.

The periodic appearance of water in speleological objects (caves) is primarily influenced by hydro-meteorological conditions. The flow regime is not uniform, but varies throughout the year, i.e. very often has a seasonal occurrence, depending on the pluviometric regime of the climate in which it is located. One such object is the Vidrovan Cave, which is located in the northwestern part of the Nikšić field, on the right side of the river Boljašnica, and in the immediate vicinity of the Vidrovan springs (caprified for the needs of Nikšić's water supply). The cave belongs to the series of branched—storied speleological objects, and so far 136 m have been explored.

A typical representative of an object with lake water is the Slivski ponor, the largest sinkhole in Nišićko polje. Slivski ponor (sinkhole) is located in the extreme south-eastern part of the Nikšić field, at the extreme border of the former course of the Upper Zeta River, which was sinking there (during heavy rainfall), before hydrotechnical operations were carried out and a cylindrical dam was built around the sinkhole. It is estimated that its capacity was around 150 m^3/s (Radojičić 2005). The opening of the sinkhole generally has an elliptical shape. Its perimeter edge is located at an altitude of 594 m a.s.l., so it represents the lowest altitude in the Nikšić field and all the water gravitates towards it. The top of the dam is located at the level of 612 m a.s.l., while the highest recorded level of deceleration (which also represents the highest recorded level of flooding in Nikšićko polje) was 613.01 m a.s.l., during the floods of 2010 (Vlahović 2019). The sinkhole has the form of a vertical pit, which continues towards the north-east with channels in the form of "pots", which have been polished by erosion, and in which there is a certain amount of ponded water, which is being renewed only at maximum water levels. Investigations of this sinkhole date back to 1939, but unfortunately there are no written traces of them today. Only in 1955, as part of the Nikšić field research for the needs of construction of hydro-systems and reservoirs, speleologists from France, Slovenia, Serbia and Montenegro descended into the ponor (up to a depth of around 150 m). More recent research was done in 2005 and 2007, organized by the Association of Speleological Societies of Montenegro and Slovenia, during which 252 m of channel was explored and recorded (Mijušković and Ćulafić 2007). Slivski ponor is connected to the spring of Perućica, which was proven by tracing experiments. However, tests have shown that when the sinkhole is closed, or its natural congestion occurs due to a large inflow of water (e.g. 2010), the Perućica spring still works normally, almost as if the sinkhole had not been closed. This is because the water sinks into the neighboring groups of sinkholes and peripheral sinkholes.

Dry speleological objects represent the most widespread group of objects. Due to the lowering of the erosion base, caused by various factors, they remained completely out of their function, i.e. they became "fossil". During the long evolution of karst aquifers, the aquifer zones changed spatially and temporally, adjusting their levels to erosion bases. Such processes are active until they come into contact with an impermeable substrate. One of the distinct representatives of "dry" speleological objects is the Cetinje Cave, which once was hydrologically active, but today is a fossil channel, which is out of function. Due to the pronounced evolution of the karst aquifer, its level was lowered from 640 m a.s.l. (Cetinjska) to Lipska Pečina (525 m a.s.l.), then to the level of Dobrsko village (250 m a.s.l.) and then to Obodska cave (81 m a.s.l.).

7 Climatic Characteristics of Montenegro

The territory of Montenegro has extremely heterogeneous climatic characteristics and there is almost no area where different climates do not intertwine. In accordance with the spatial–temporal distribution of climatic parameters, several climatic regions with their own distinctive characteristics can be formed. Dominant climate types are maritime, continental and mountain. On the territory of Montenegro, it is very difficult to make a clear border between different types of climate. Thus, the maritime type of climate penetrates deep into the continental part, modifying the classic continental and mountain climate that prevails in those areas. So, as a consequence of the modification, we can have a continental type of climate modified by a mountain type or a continental type modified by a maritime type and even a mountain type of climate modified by a maritime type.

Air temperature is one of the most important climatic elements because it has great importance and influence on the climatic characteristics of a place or area. One of the basic goals of climate studies, whether it is the study of macroclimate, mesoclimate or microclimate, based on the analysis of climatic elements, their extreme values and specific climatic indicators, is to investigate the climate of a place or territory, analyze and present as faithfully as possible. The temperature regime of an area changes depending on the distance from the sea, the absolute height of certain parts (adiabatic cooling) as well as the breakdown of the relief.

On the territory of Montenegro, the warmest months are July and August, while the coldest month is January. The average mean annual air temperature gradually decreases moving from south to north. The Montenegrin coast and the area of the Zeta-Bjelopavlić plain are areas with a Mediterranean climate (16.4 °C), with characteristically long dry and hot summers and relatively mild and rainy winters. Places in the basins, such as Podgorica, have a lower temperature in January (by about 1.9 °C) than coastal places such as Ulcinj, which is located at approximately the same latitude, while during the summer they have a slightly higher temperature (by about 1.7 °C). In hot summers, the area of the Podgorica-Skadar basin and the Bjelopavlić plain stand out, and the absolute maximum air temperature in Montenegro was registered here (Podgorica 44.8 °C/08/24/2007) as well as the largest number of tropical days (90) and nights (64), while the absolute minimum was registered in Rožaje (−30.5 °C/01/26/2006), which gives an amplitude of temperature fluctuations of 75.3 °C. The absolute minimum temperature in Montenegro, for

the instrumental period of measurement and observation, was recorded in Rožaje of −32 °C (January 13, 1985). On the coast and in the Zeta-Bjelopavlić plain, days with an air temperature above 30 °C occur from April to October. Days when the minimum air temperature does not fall below 20 °C were not recorded in Kolašin, Plav and Rožaje. The average annual number of frosty days (days when the minimum daily air temperature is below 0 °C) ranges from 3.5 days in Bar and Budva to 151 days in Žabljak, where frosty days, as in other higher areas, were not recorded only in July and August.

The central and northern part of Montenegro has some characteristics of a mountain climate, but the influence of the Mediterranean Sea is also evident, which is reflected in the rainfall regime and the higher average temperature of the coldest month. The far north of Montenegro has a continental type of climate, which, in addition to large daily and annual temperature amplitudes, is characterized by a small annual amount of precipitation with a fairly even distribution by month. In the mountainous areas in the north of Montenegro, summers are relatively cold and wet and winters are long and harsh, with frequent frosts and low temperatures that drop sharply with altitude (Fig. 3).

By analyzing and comparing the obtained data for the reference climate periods 1961–1990 and 1991–2020 years, we see that the mean values of the annual air temperature have been increasing for the last 30 years by some 1.1 °C. The stated deviations are positive and range from 0.8 °C in Herceg Novi and Cetinje to 1.6 °C in Bijelo Polje. It should be noted that local microclimate conditions can have an impact on the differentiation of the average temperature as well as other climatic elements of a place, especially when orographic factors are very pronounced, as is the case with the area of Montenegro. The temperature regime and, therefore, other climatic elements change depending on the distance from the sea, the altitude and the variety of landforms. On an annual average, the lowest air temperatures are in areas at higher altitudes and in valley sides of rivers and basins.

7.1 Precipitation Regime

Precipitation is a variable category in space and time and, together with air temperature and represents the most important climatic elements. The annual amount of precipitation and its distribution by seasons and months (pluviometric regime) is one of the main climatic characteristics of an area. Almost all social and economic activities, to a greater or lesser extent, depend on the amount and regime of precipitation. The amount of precipitation during the year depends on and is related to annual changes in temperature, relative humidity and cloud cover. The mentioned parameters determine the convective processes in the atmosphere, the condensation of water vapor and its discharge to the ground in a liquid or solid state. In addition to the mentioned parameters, in the researched area, the orography has a significant influence on the discharge of precipitation. During the summer months, convective precipitation also occurs. The most significant forms of precipitation are rain, snow, sleet and hail.

The geographical area of Montenegro lies between 41° 30′ and 44° latitude, exactly in the belt of moderate

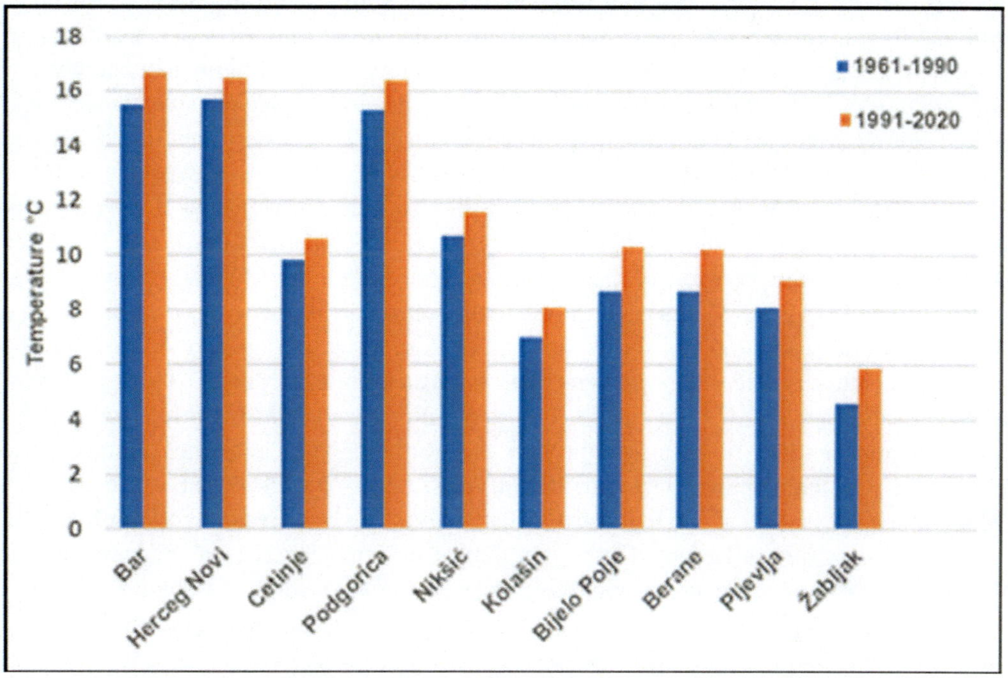

Fig. 3 Comparative overview of the average annual air temperature

latitudes, and this geographical position conditions the occurrence of seasonal precipitation and its specific regime. Precisely because of its location, there are very frequent penetrations of air masses from the Atlantic Ocean, which represent an extremely important factor in the precipitation regime in the central and northern areas of the territory of Montenegro. In addition, the area of the Western Mediterranean is a unique cyclogenetic area, which also directly affects the rainfall regime in Montenegro. During autumn and winter, there is a significant influence of southwesterly currents, which bring enough moisture from the Mediterranean Sea to the area in the southern parts. The direct influence on the precipitation regime is reflected in the orographic rise of moist and unstable air in the northwest-southeast direction (the Dinaric creeping direction), which contributes to an increase in the amount and intensity of precipitation. This direction of movement has the consequence that the mountain massifs in the hinterland of the coast, with prevailing southwesterly currents, condition the occurrence of non-windy and windy orographic precipitation. In addition, this direction of the mountain crowns represents a natural barrier to the influence of the Adriatic Sea towards the north as well as the continental influence on the southern areas.

The average annual amount of precipitation in the territory of Montenegro is very heterogeneous, with an extremely pronounced rainy and less rainy region. The distribution of the average annual amount of precipitation, thanks to the above, is very uneven and increases moving from the southwest (Crkvice–Orjen, cc 5000 mm) to the east (Pljevlja cc 800 mm. Analyzing the spatial and temporal distribution of precipitation in Montenegro (period 1960–2015), Ćulafić et al. (2020), applying PCA analysis and varimax rotation, identified four different clusters with different spatial patterns of precipitation in Montenegro, namely: C1 cluster with 4 stations (Bar, Budva, Podgorica and Ulcinj); C2 cluster with 5 stations (Berane, Bijelo Polje, Pljevlja, Rožaje and Žabljak), C3 cluster with 7 stations (Cetinje, Danilovgrad, Grahovo, Herceg Novi, Kolašin, Nikšić and Velimlje) and C4 cluster with 1 station (Crkvice).

According to the above and thanks primarily to the cyclonic activity in the Mediterranean, i.e. moist currents from the southern quadrant in the winter half of the year and orographic obstacles, the southernmost, southwestern and southeastern parts of Montenegro have a significantly higher annual amount of precipitation than the northernmost parts. On the slopes of Orjen in record years, 8000–9000 mm can fall (9079 mm fell in 2010), which makes this area the rainiest area in Europe. Other very rainy areas are the areas of Lovcen and Rumija, with more than 3500 mm, and Prekornica and Žijovo with more than 2500 mm.

The coastal zone, as well as the central part of Montenegro, is characterized by a Mediterranean and modified Mediterranean precipitation regime (maximum amounts of precipitation occur in November and December, and minimum in July and August). In the narrow coastal strip, the average annual rainfall ranges from 1360 mm in Bar to 1799 mm in Herceg Novi, with about 115–127 rainy days. Cetinje is the rainiest city in Montenegro with a long-term average of 3,348 mm of water precipitation, while in Podgorica this value is 1,696 mm, which falls on an average of 121 rainy days.

The temperate-continental regime is characterized by more frequent precipitation in the second half of summer, a secondary maximum in October and a minimum in February. This feature is characteristic of areas and places in the valleys of Lim (Berane 900 mm and Bijelo Polje 917 mm), Tara (Kolašin 2067 mm), Ibar (Rožaje 945 mm) and Ćehotina (Pljevlja 797 mm). Žabljak, as a city at the highest altitude (1450 m a.s.l.) has an average of about 1505 mm of precipitation and it falls on average over 182 days. Most of the territory of the Piva and Tara basins has a modified Mediterranean type of rainfall. Average monthly rainfall for the area has a certain regularity in terms of the amount of atmospheric precipitation released during the year. Pronounced maxima occur during October and March, and minima during July and August. The border between the modified Mediterranean rainfall regime and the continental rainfall regime stretches from the Ljubišnja mountain in the northwest, through Sinjajevina and Bjelasica to the Prokletija mountain range in the south. The rainiest months have an average of 13–17, and the driest 4–10 rainy days. The number of days with somewhat heavier daily precipitation (over 10 mm) ranges from 25 (Pljevlja) to 59 (Kolašin). Nevertheless, the highest number of days with heavy precipitation occurs in Cetinje, 74 days.

A permanent snow cover forms at altitudes above 400 m above sea level, while at altitudes above 600 m a snow cover greater than 30 cm can be expected, and at altitudes above 800 m a.s.l. and over 50 cm. The absolute maximum height of the snow cover was registered in Žabljak, namely 209 cm (1961–1990) and 230 cm (1991–2020). The characteristic of this climate is that almost every year the height of the snow cover exceeds 1 m in height, while, e.g. the average number of days with a snow cover greater than 50 cm is 76. Kolašin has about 10 such days and the other analyzed locations less than 4. By analyzing and comparing the data obtained for the reference climate periods 1961–1990 and 1991–2020, we see that in the last 30 years, there has been the largest increase in the average amount of precipitation in Cetinje by about 3.5% compared to the climate period 1961–1990, while the biggest decrease in the average annual amount of precipitation is in Herceg Novi by 6.3% (Fig. 4).

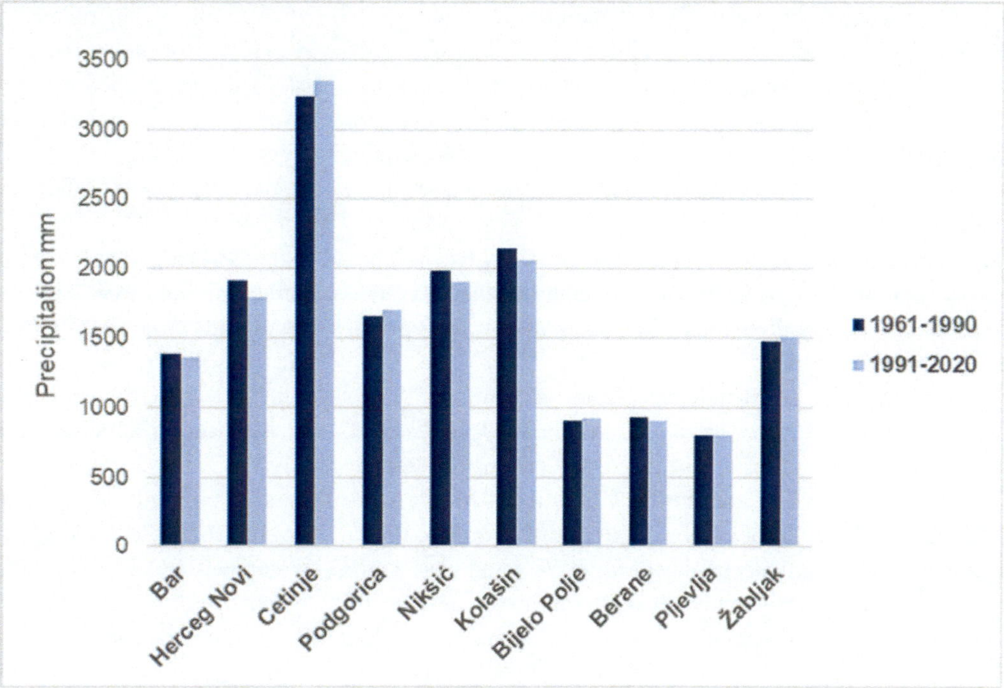

Fig. 4 Comparative overview of the average annual precipitation

7.2 Cloudiness and Duration of Sunshine

The entire human life depends to a large extent on the length of time the Sun shines. The length of time the Sun shines is a climatic element, which directly depends on latitude, season, cloudiness, terrain configuration, urbanization, air pollution and other factors. The duration of sunshine is also of great importance for vegetation (growth, flowering, germination, etc.). The intensity as well as the duration of the sunshine in a certain locality depends on the angle at which the rays fall and the degree of cloudiness.

The highest values of monthly, and therefore daily sums of the duration of sunshine were recorded in the summer period, and are primarily conditioned by the length of the day itself (daylight) and reduced cloudiness. The annual sum of the number of hours of sunshine in the Montenegrin coast ranges from 2418 h in Herceg Novi to 2537 h in Ulcinj. Moving further inland, the value slowly decreases and in Podgorica it is 2462 h, Nikšić 2233, while in mountainous regions, which are far from the sea, this value ranges from 1921 in Žabljak to 1629 h in Pljevlja. The geographical area around Ulcinjsko polje stands out as the sunniest area in Montenegro. In all parts of Montenegro, July and August are the months that have about 4–5 times longer duration of sunshine than the winter months. The longest duration of sunshine is in the summer months 250–350 h, while in the winter months, it is on average below 100 h, and in some localities it drops below 50 h. These values are affected by increased cloudiness and the appearance of fog (which is quite common in the area of the valleys), as well as shorter days.

Cloudiness represents the degree of sky coverage by clouds and is expressed in tens in climatology. It is an important climatic element and affects the solar radiation and radiation of the Earth's surface and atmosphere, i.e. affects the heat balance. This is also reflected in the temperature regime, so that cloudy days are characterized by small daily fluctuations in air temperature, and on clear days, temperatures reach extreme values, minimum in winter and maximum in summer. The average annual cloud cover in Montenegro ranges from 4 tenths in Tivat to 6.5 tenths in Bijelo Polje, while in Podgorica it is 4.6 tenths. The average cloud cover is highest in the winter months, which are also the months with the highest amount of precipitation, and the lowest in July and August, which are also the driest months. A clearer idea of the regime and character of cloudiness is obtained by displaying the number of clear days, i.e. days when the mean daily cloud cover is less than 2 tenths of the sky covered by clouds and cloudy days, days when the mean daily cloud cover is greater than 8 tenths. In Montenegro, there are on average 38 clear days per year in Pljevlja to 163 days in Tivat, and in Podgorica there are an average of 132 clear days. The average annual number of cloudy days is from 89 in Tivat to 228 in Bijelo Polje, 226 in Pljevlja and 104 days in Podgorica.

7.3 Wind Characteristics

The direction and speed of the wind depend on the distribution of air pressure and the intensity of the barometric gradient, but also on the shape of the relief (orography), so that local winds represent one of the basic climatic features of many of our regions (Bura, Jugo, North, etc.). Milosavljević (1990) states that "wind, as a climatic element, is so important that it is often considered as a factor that determines the climate in certain cases". In other words, intrusions of air masses (wind) condition, sometimes sudden, weather changes—changes in temperature and air humidity, evaporation, cloud cover, occurrence of precipitation, etc. Wind is also a very important bioclimatic element because it affects human activity and the subjective feeling of comfort (physiological temperature).

In the territory of Montenegro, each narrow area shows certain specificities in the direction of wind movement, mainly due to the influence of the orography of the terrain. Over the northwestern and northern regions of Montenegro (Pljevlja, Žabljak, Kolašin, Nikšić, Podgorica), north, northeast and south winds occur most often. West and east winds occur most frequently above Rožaj, southwest and northwest winds above Cetinje, and northeast winds above Bar and Ulcinj. In the central and Primorje areas, the gusts of north and northeast winds very often reach storm force, and the higher areas in the north of Montenegro as well as on the coast have storm gusts of south wind that reach a speed of around and more than 100 km/h (Buric et al. 2012). In the area of Bar, the most dominant winds are from the northeast 35%, while in the area of Podgorica, the most dominant winds are from the north 35% and from the south 25%. In the higher areas in the north of Montenegro, the dominant winds are from the southern quadrant, in the area of Žabljak with a frequency of about 36% and in the area of Pljevlja about 25%. Representation of cases when there is no wind, the so-called silence, on the coast is small, in Bar it is 1.4%, in Podgorica about 28%, in Žabljak about 19%, while in Pljevlja due to the configuration of the terrain and the location of the town where measurements of wind speed and direction are made, the percentage the occurrence of silence is high, 43%.

8 Conclusion

The hydrological characteristics of the terrain of Montenegro are presented with the desire to bring the spatial arrangement of the most important surface watercourses closer to the reader. The paper presents the basic morphometric and hydrological characteristics, with special reference to some specificities (primarily hydrogeological) and their connection with speleological objects. Since Montenegro is a distinct country of karst, the directions of surface streams are determined by the geological structure and tectonic structure of the terrain. The aforementioned factors and agents have led to the fact that the unevenness of available water resources is very pronounced in the territory of Montenegro. A significant part of the surface of Montenegro belongs to the karst area, with very pronounced karstification and a wealth of speleological objects (pits, sinkholes, caves), while some parts of the country (especially the southwestern part) are without constant surface runoff, but with pronounced pluviometric aggressiveness. Here, one encounters a "hydrological resource paradox" that rarely exists in the world where the amount of precipitation deviates from 2500 to 4500 mm, without surface runoff but immediate infiltration of precipitation (rain and snow) to the underground, and then as groundwater flows through karstified forms towards the Adriatic sea or towards some other watersheds outside the country's borders (such as the case of the underground drainage of part of the Stara Crna Gore overthrust towards the Trebišnjica watershed (B&H) or part of the waters of Orjen mountain which are drained towards Cavtat (CRO).

A special problem lies in the fact that, due to increasingly pronounced climatic, spatial, geotechnical, ecological, urban, sociological and other limitations, only a small part of the water present in the surface catchments, can be used as an adequate water resource. It can be concluded that despite the data on average values which showed a significant wealth of water resources, due to unfavorable hydrogeological and geomorphological phenomena, as well as large spatial and temporal unevenness of precipitation, Montenegro has evident problems caused by insufficient amounts of water.

Montenegro is a very complex climatic area characterized by pronounced variations in time and space, primarily due to its geographical position, proximity to the sea (Adriatic and Mediterranean) and orographic characteristics—mountain ranges, which prevent maritime influences from penetrating deeper into the interior of the land (canyon valleys). The Montenegrin coast, together with the Zeta-Bjelopavlić plain, represents areas with a Mediterranean climate, characterized by long, hot and dry summers and relatively mild and rainy winters. The central and northern part of Montenegro has the characteristics of a mountain climate, which is somewhat modified by the influence of the Mediterranean Sea, which is primarily reflected in the rainfall regime and the higher average temperature of the coldest months. The far north of Montenegro belongs to the continental type of climate, which is characterized by large daily and annual temperature amplitudes and a small annual amount of precipitation (with a fairly even distribution by month). In the mountainous areas in the north, summers are characterized by a relatively cold and humid climate, while winters are long and harsh, with frequent frosts and low temperatures, which drop sharply with the change in altitude.

According to the rainfall regime on the territory of Montenegro, we distinguish the Mediterranean and temperate-continental regimes. The Mediterranean regime is characterized by the maximum amount of precipitation in November and December and the minimum in July and August. The temperate-continental regime is characterized by more frequent precipitation in the second half of summer, a secondary maximum in October and a minimum in February. In the largest part of the geographical area of Montenegro, the maritime pluviometric regime is somewhat modified by the mountain and continental rainfall regime.

Bearing in mind the geological predisposition of Montenegro (more than 60% of the territory consists of limestone and dolomite), i.e. that it is known as the land of karst, its great richness in speleological objects and other karst landforms is not surprising. One of the most important prerequisites in terms of speleogenesis is the extremely large distribution of carbonate rocks (which are subject to dissolution) and the tectonic predisposition of the relief and an extremely high amount of precipitation. All of the above indicates that the geographical area of Montenegro, especially its southern, central and northwestern parts, is very suitable for the development of speleological facilities (caves, pits). This is supported by the fact that a very large number of speleological objects were explored and recorded precisely in areas characterized by large amounts of precipitation (Orjen, Lovćen, Maganik, Durmitor, Prokletije, etc.).

References

Buric, M., Micev, B., and Mitrovic, L. (2012). Climate Atlas of Montenegro (in Montenegrin). Montenegrin Academy of Sciences and Arts, Podgorica.

Culafic G., Popov T., Gnjato S., Bajic D., Trbic G. and Mitrovic L. (2020). Spatial and temporal patterns of precipitation in Montenegro, IDŐJÁRÁS – Quarterly Journal of the Hungarian Meteorological Service, Budapest. pp. 499–519 DOI: https://doi.org/10.28974/idojaras.2020.4.5.

Đorđević B., Sekulić G., Radulović M. i Šaranović M. (2010). Vodni potencijali Crne Gore, CANU, Odjeljenje prirodnih nauka, Knjiga 32, Podgorica.

Dukić D. i Gavrilović Lj. (2008). Hidrologija, Zavod za udžbenike, Beograd.

Hrvačević S. (2004). Resurski površinskih voda Crne Gore, EPCG, Nikšić.

Mijušković. V., Ćulafić G. (2007). Slivski ponor, 6. Simpozijum o zaštiti karsta, ASAK, Beograd. pp. 21–26. UDK 551.435.82(49716).

Milanović S. (2012). Speleologija i speleoronjenje u hidrogeologiji karsta, Departman za hidrogeologiju, Rudarsko-geološki fakultet, Univerzitet u Beogradu, Beograd. ISBN 978-86-7352-262-3.

Milosavljević, M. (1990). Climatology (in Serbian), Scientific book, Belgrade.

Radojičić B. (2005) Geografija Crne Gore - prirodna osnova, DANU, Podgorica.

Radulović M. (2000). Hidrogeologija karsta Crne Gore, JU Republički zavod za geološka istraživanja Podgorica, Posebna izdanja Geološkog Glasnika, Knjiga XVIII, Podgorica.

Stevanović Z., and Blagojević M. (2021). Hydrogeology and Climate Changes Impact on Aquifer System of Drina River basin, Ministry of Agriculture, Foreeestry and Water Management of Montenegro, Podgorica. ISBN 978-86-85799-22-8.

Vlahović, M. (2019). Površinske akumulacije u karstu Nikšićkog polja – hidrogeološki i inženjersko-geološki aspekti, Elektorprivreda Crne Gore Nikšić, Nikšić.

Geomorphological Characteristics of Montenegro

Gojko Nikolić, Filip Vujović, Goran Grozdanić and Aleksandar Valjarević

Abstract

The geodynamics of the region of the southern Dinarides and the southern Adriatic is in many ways specific and has essentially influenced and is influencing the geomorphology of Montenegro. The rhythm of exogenous dynamic relief modification, with different morphosculptural processes, is primarily conditioned by endogenous kinematics, especially neotectonic movements predisposed to the kinematics of older structures. According to the influence on the development of geomorphological processes and the formation of relief types, lithological units are excavated, which include coarse and fine clasts (breccias, conglomerates and sandstones, siltstones and clays), carbonates (limestones, dolomites and dolomitic limestones), igneous rocks, coarse detritus (gravel and sand) and mud and silt. The relief is a consequence of lithofacial composition and tectonics, it is their morphogenetic, dynamic and temporal expression, whose form and visual integrity are reflected and complemented by the following relief units: Montenegrin coast; Central Montenegro; Central high mountains, areas and canyon valleys; Valleys, areas and high mountains of northeastern Montenegro. Depending on the role of geomorphological processes in Montenegro, it is possible to distinguish between the genetic morphostructural type of relief under the influence of endogenous forces (tectonics, neotectonics and volcanism), and exogenous genetic type of relief under the influence of exogenous processes (glacial, glaciofluvial and glaciolimnic, karst, fluvial, hillslope deposit, lake-marsh, marine and anthropogenic). It is important to point out that more than 60% of the terrain is built of carbonate formations on which the karst type of relief is predominantly developed. By analyzing the primary morphometric parameters of the relief (hypsometry, slope, aspect and vertical dissection) using the Digital Relief Model (DMR) in the GIS environment, more detailed geomorphological quantitative characteristics of endogenous and exogenous processes were obtained.

Keywords

Montenegro · Relief · Morphometry · Karst · GIS

1 Basic Physical-Geographical Characteristics of Montenegro

Montenegro is a country located in southeastern Europe, i.e. on the Balkan Peninsula, between latitudes 41°52′–43°32′ N and longitudes 18°26′–20°21′ E. It covers a total area of 13 883 km^2 (MONSTAT 2022) with a population of about 620 000 inhabitants. To the southwest, it relies widely on the Adriatic Sea with a rugged coastline of 293.5 km. The orthodrome distance between the northernmost and southernmost points is 192 km and between the westernmost and easternmost 163 km (Radojičić 2015).

The complex lithofacial composition, structural and physiographic structure are the result of a turbulent and long geoecological evolution that shaped the geospace of Montenegro. Geological history has created, in the lithostratigraphic sense, a complex structure, composed of segments of different geological composition and evolution (Pajović and Radusinović 2010). As a percentage, almost 90% of the rock mass was formed from marine sediments, and significantly less from lake and/or rocks formed on land.

G. Nikolić (✉) · F. Vujović · G. Grozdanić
Faculty of Philosophy, Department of Geography, University of Montenegro, Danila Bojovića Bb, 81400 Nikšić, Montenegro
e-mail: gojkorn@ucg.ac.me

A. Valjarević
Faculty of Geography, University of Belgrade, Studentski Trg 3/III, 11000 Belgrade, Serbia

G. Barovic (ed.), *Speleology of Montenegro*, Cave and Karst Systems of the World,
https://doi.org/10.1007/978-3-031-49375-1_5

Previous geological research shows that the geological history of the territory of Montenegro can be traced back to 400 million years (Mirković et al. 1985). This is indicated by different types of sedimentary, igneous and metamorphic rocks that formed in geological eras: Paleozoic, Mesozoic and Cenozoic. The rocks of the younger Paleozoic occur in shale, marly-clay-sandy layers, with various shales and rare interlayers of limestone and conglomerates. Mesozoic rocks occur in the form of carbonate, igneous, volcanogenic-sedimentary, diabase-horny and flysch rocks. Cenozoic rocks are represented by carbonate and Paleogene flysch; Neogene marine and freshwater sediments and Quaternary unbound sand, gravel, larger blocks with and without clay, glacial, fluvioglacial, limnoglacial, deluvial and alluvial origin. According to the influence on the development of geomorphological processes and the formation of relief types, lithological units are excavated, which include coarse and fine clasts (breccias, conglomerates and sandstones, siltstones and clays), carbonates (limestones, dolomites and dolomitic limestones), igneous rocks, coarse detritus (gravel and sand) and mud and silt.

In the geotectonic sense, four structural-tectonic units can be clearly seen in the geospace of Montenegro, namely Paraautochthon (Adriatic Pine System), Budva–Cukali zone, High Karst zone and Durmitor tectonic unit (Bešić 1969). Systematic research has established that they are extensions of regional geotectonic units of the southeastern Dinarides. They are separated from each other by regional reverse dislocations with the direction of pulling from the northeast and in the territory of Montenegro they are characterized by ridges, faults and folds of considerable dimensions.

Based on the data for the reference climate period, two climates according to Köppen are represented: moderately warm (C) and moderately cold (D). Moderately warm climate is widespread in hypsometrically lower areas, while in the higher mountainous areas in the interior of Montenegro, mainly in the hypsometric zone above 1000 m, climate D is represented (Burić et al. 2014).

The hydrology of Montenegro is largely determined by its hydrogeological structure (Radulović 2000). This is due to the complexity of geological composition and soil structure, relief and climatic conditions. Carbonate rocks, cavernous-fissure and rocks of intergranular porosity make up over 60% of the rock mass of Montenegro (Petrović 1983). Taking into account this fact, it can be assumed that in most of Montenegro underground hydrographic forms (aquifiers with their surface outcrops, springs and karst springs in the form of estavelle, springs, brackish springs, etc.) are more developed than surface ones. The rivers belong to the basins of the Adriatic and the Black Sea, and the watershed consists of the mountain ranges Golija–Vojnik–Lola–Sinjavina–Komovi–Prokletije. In addition to rivers in the territory of Montenegro, there are a large number of lakes of hydrographic forms, among which the Skadar, Zogonjsko, Šasko, Biogradsko, Crno (Black) lakes and numerous smaller lakes on Durmitor and other high mountains were affected by the glacial process during the Quaternary.

The unique multi-component combination of geological background, climate and georelief, hydrography and land cover, has conditioned the emergence of two bio-geographical regions (Alpine and Mediterranean) with exceptional biodiversity in Montenegro's geospace, whose value can be seen in the fact that Montenegro has 3,250 plant species and a floristic diversity index (S / A1) of 0.837, one of the "hot spots" of biodiversity in the Mediterranean, and one of 153 globally important centers of floristic diversity (Pulević 2022).

2 Genetic Types of Relief

The geodynamics of the region of the southern Dinarides and the southern Adriatic is in many ways specific and has primarily influenced and continues to influence the geomorphology of Montenegro. The rhythm of exogenous dynamic relief modification, with different morphosculptural processes, is primarily conditioned by endogenous kinematics, especially neotectonic movements predisposed to the kinematics of older structures. The types of relief are distinguished in relation to the large energy amplitudes of morphogenetic and morphodynamic processes that have conditioned their distinct synergy and diversity. Depending on the role of geomorphological processes in Montenegro, it is possible to distinguish between genetic morphostructural type of relief created under the influence of endogenous forces (tectonics, neotectonics and volcanism) and genetic type of relief created under the influence of exogenous processes (glacial, glaciofluvial and glaciolimnic, fluvial, slope, lake-marsh, marine and anthropogenic).

2.1 Morphostructural Relief Types

In the geotectonic sense, four structural-tectonic units can be clearly seen in the geospace of Montenegro, namely Paraautochthon (Adriatic Pine System), Budva–Cukali zone, High Karst zone and Durmitor tectonic unit (Bešić 1969). These four structural-tectonic units geospatially reflect the regional rupture structure (with three large tectonic blocks limited by systems of regional ruptures of the general NW–SE direction. All three blocks are Dinaric and have a synergistic, very important role in relief formation (Glavatović 1982; Mirković 1997). The first, NE block, includes the mountain ridges and the foothills of Bjelasica,

Sinjajevina and Ljubišnja. The southern boundary of the block is contoured by a system of regional ruptures, the position of which generally coincides with the forehead of the Durmitor overthrust. The second central block includes the zone in the center of which is the Maganik mountain. The northern border of the block consists of a system of regional ruptures that follow the front of the Durmitor overthrust, and from the south it is limited by a system of large faults, which can be traced along the northern edges of the Njegoš mountain and Nikšić polje, then along the northern edge of the Bjelopavlićki depression and further east toward the border with Albania. The third, SW block includes the area from the Njegoš mountain, Nikšić polje and the lower course of the Zeta River to the Adriatic coast. The following morphostructural elements dominate within structural-tectonic units with large tectonic blocks (i.e. the effects of endogenous forces and the processes they cause): negative structures, positive structures, regional covers, deep regional faults and the thickness of the Earth's crust (Mirković 1997).

Negative structures are primarily singled out as tectonic depressions. In the relief, they morphologically reflect deep sea bays (Boka Bay with internal and external bays) and valleys (Sutorina and Grbaljska), lakes (Skadar Lake as the lowest part of the large Ćemovsko polje tectonically lowered above the level of aquifier), karst poljes (Nikšićko, Grahovsko, Brezansko, Ćemovsko and others) and other depressions that are more or less filled with Quaternary deposits (glaciofluvial and glaciolimnic).

Positive structures are morphologically recognizable as relatively elevated forms above the surrounding terrain. The characteristic circular structure is separated in the zone from Lovćen to Danilovgrad with two inaccessible rings that partially intersect. And the second circular structure is in the Bjelasica zone. The origin of the isolated circular structures in the area from Cetinje to Danilovgrad has not been clarified, while the circular structure of Bjelasica is related to magmatic and tectonic activity.

Regional thrust faults as large units primarily determine the relief and its characteristics, especially in mountain systems, river valleys (canyons and gorges) and valleys. The faces of thrust faults, with the general extension in the NW–SE direction, are traced along the Adriatic coast (Budva thrust fault and Visoki krš thrust fault), and in the central part of the terrain, along the SW edges of Durmitor, Sinjajevina, Bjelasica and Komovi (Durmitor thrust fault). Tectonic contact of two environments of different lithological composition, different erodibility and solubility, causes the appearance of a pronounced fracture in the slope, giving a specific morphological appearance of the relief.

The thickness of the earth's crust is variable and shows a regional character. It is the largest in the area of high mountains (maximum thickness is in the zone of Maganik, Golija, Njegoš, from 49 to 54 km), which are tectonically elevated, and the smallest in places of relative tectonic sinking. Analysis of changes in the thickness of the Earth's crust is one of the criteria for the analysis of neotectonic activity and morphological appearance of the terrain in isolated zones of the regional rupture assembly.

Deep regional faults represent morphostructural elements that form the boundaries of the area of relative elevations, i.e. sinking, i.e. restricted tectonic blocks. The northern fault separates the Bjelasica block from the Maganik block. There are two deep faults located within the southwestern tectonic block: a fault from the northern rim of Skadar Lake to the rim of the Orjen mountain and a fault located in the seabed along the Adriatic Montenegrin coast. However, there are no fully confirmed relevant results for these two faults.

2.2 Exogenous Relief Types

Morphogenetic factors form several types of exogenous relief (Fig. 1), which are expressed through glacial, glaciofluvial and glaciolimnic, karst, fluvial, slope, lake-marsh, marine and anthropogenic genetic forms (Fig. 2). The dominance of the processes changed cyclically, they were repeatedly restored, they are very important for the intensity of exodynamic processes, especially erosion-denudation processes and inundation processes that lead to gravitational leveling of the relief surface.

It is assumed that the mean height of the equilibrium line (ELA) during glaciation in the Pleistocene was above 1750 alt. in the mountains of Montenegro, which makes up about 8% of the territory of Montenegro (Cvijić 1899; Djurović 1996, 2009; Stepišnik 2011). The glacial type of relief is mainly related to morphological markers and shapes belonging to the Virmic glaciation. Whether they are defined as erosive (cirques, waves, glacial shoulders, nunats, mutated rocks) and accumulative (frontal, side and bottom moraines and loose moraine material, etc.), these forms are preserved on all high and higher mountains in Montenegro: Volujak, Durmitor, Sinjajevina, Bjelasica, Maganik, Prekornica and Prokletije. In the coastal area, distinct traces of glacial relief have been isolated on the naked karst of Orjen and Lovćen. According to morphological features, four types of glaciers have been reconstructed: valley, plateau (piedmont), karst and regenerated.

According to the methodological framework of Mirković and Pavlović (2002), glaciofluvial and glaciolimnic reliefs stand out as a special genetic type. According to their energy, watercourses transported large quantities of unbound glacial material, which, depending on the length of transport and other factors, was processed, rounded and classified by size. Since the area affected by glaciation has

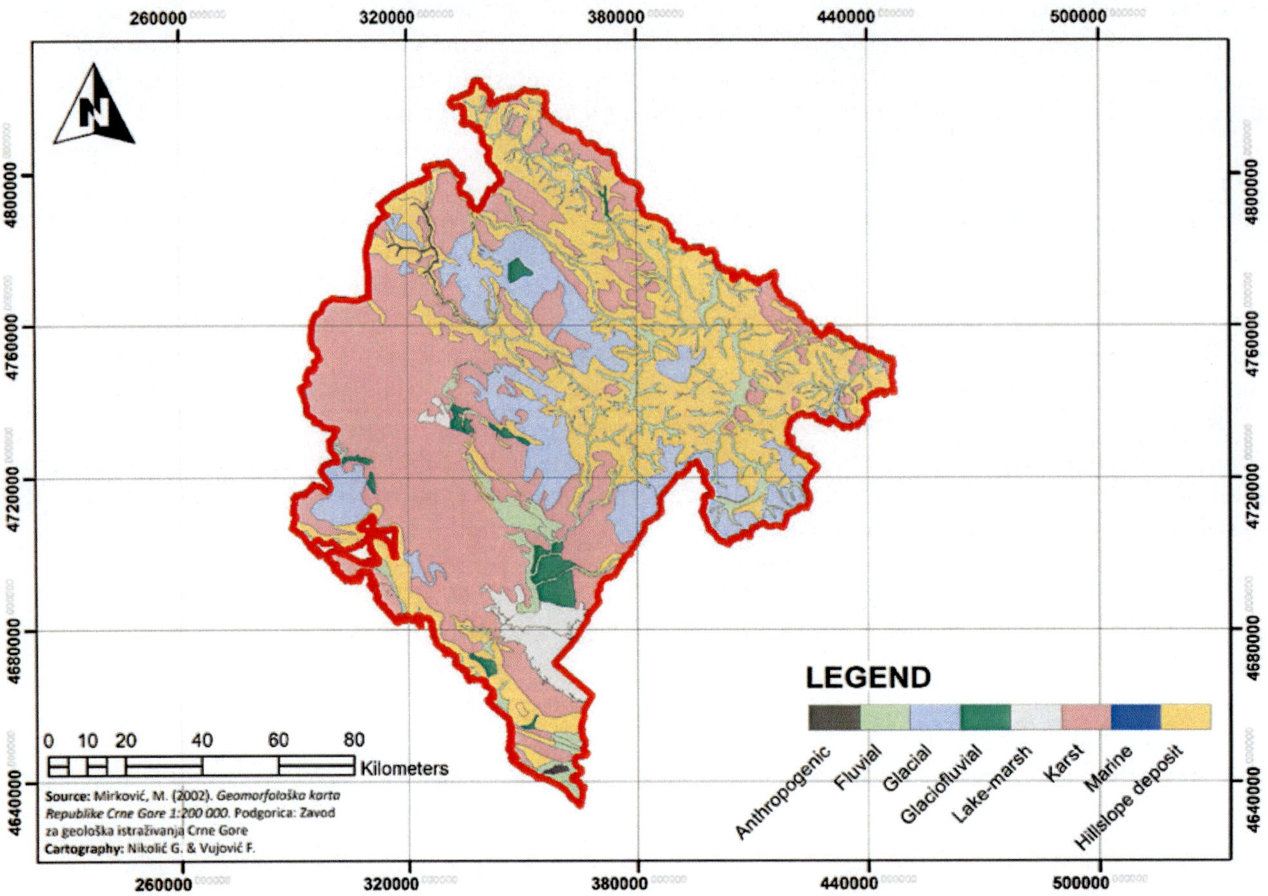

Fig. 1 Overview map of exogenous relief types

a karst character, tectonic depressions in these areas were suitable places for the disposal of this material. These karst depressions are covered with thick layers of material, their bottoms are flattened by accumulation and today they represent karst polje (Nikšićko, Grahovsko, Dragaljsko and Cetinjsko), and the material that covers them is classified as glaciolimnic type of relief. Where material was deposited along the bottom of river valleys, glaciofluvial relief was formed. In addition to glaciofluvial sediments deposited at the bottom of river valleys and glaciolimic sediments deposited in karst polje, these two types of relief are also characterized by shapes: terminal basin (Lake Biograd and Lake Plav) and glaciofluvial fans (such as fans in Grahovo, Dragalj and the largest in Ćemovsko polje, several tens of meters thick).

Karst relief type is the most common genetic type of relief in Montenegro with all surface and underground forms and specific hydrogeological properties. The karst type of relief is a combination of recent karst and stratigraphically different paleocarps formed during numerous regressions from the beginning of the Mesozoic to the present day. Boginjavi karst are characteristic of the western, and nival for the high mountainous part (Bešić 1969; Vlahović 1975; Djurović 1996; Lješević 2005). The following are very important for the rhythm of morphogenetic processes and karst development: conditions and forms of karst, hydrogeology and circulation of water in karst, the relationship between river erosion and karst process and especially the role of neotectonics. Positive vertical movements lead to the lowering of karst outcrops and karstification zones, due to which developed holokarsts with meso and micro forms (bays, sinkholes, uvala, cliffs, caves and shafts) appear in the elevated zone, especially in pure batches of powerful Mesozoic limestones. Carbonate rocks, cavernous-fissure, and rocks of intergranular porosity make up over 60% of the rock mass. Taking into account this fact, it can be assumed that underground hydrographic forms are more developed than surface ones. Karst relief was developed in the western, southwestern, central and northwestern parts (Golija, Njegoš, Orjen, Lovćen, Sinjajevina, Durmitor, Piva Mountains, Vojnik, Maganik, Prekornica, Maglić and the southwestern hillslopes of Prokletije). The karst relief is much less developed northwest of the Tara River, in the east and southeast. Karst poljes are the most prominent forms in the surface karst relief, they are predisposed to negative tectonic structures, and shaped by karst

Fig. 2 Genetic types of relief: **a** Grebaje Valley, pearl of Prokletije, view from another meadow, NE are Karanfili and NW Talijanka, Valušnica and Popadija—glacial relief, **b** Captain's Lake below the top of Stožac with guttered grooves and taluses—glaciolimnic relief, **c** Škrka Valley from Durmitor's Planinica—glaciofluvial relief **d** Nikšićko polje, the largest polje in the karst of Montenegro, from Ostroške grede, in the background Vojnik and Zla gora—karst relief, **e** Morača Valley at the exit near Bioče, view from Vežešnik—fluvial and glaciofluvial relief, **f** Pestingrad, to the west is Vrmac, frontally in front is the Bay of Kotor and the Orjen massif—profile and relief of the coastal slope **g** Barno lake on Durmitor overgrown with dense marsh vegetation—lake-marsh relief, **h** Veslo bay rocky shore on the west side relief, **i** Red mud basins of The Aluminium Plant Podgorica, A and B—7.5 million tons—anthropogenic relief (Authors of images: Vujović Drago, Nikolić Gojko, Vujović Filip)

and other processes that have changed. They are filled with glaciofluvial and glaciolimnic sediments that fossilized the earlier karst relief. Highland relief contains 14 of the 16 karst polje in Montenegro, which are distributed in the hypsometry belt from 600 to 1450 m. Karst poljes are relatively small areas, from 1 to 3 km^2, with a maximum of about 5 km^2, with the exception of the Nikšić polje as the largest (66.5 km^2) and Brezna (24 km^2). The most numerous forms of surface karst relief are sinkholes, while the uvalas are mostly present in the southwestern and western parts of Montenegro (Čevo, Ćeklići, Ceklinska, Broćanac). In the prepolje zone of karst polje, as well as throughout the karst relief, other phenomena of the karst process occur, such as cave springs (Bjelopoljska Bistrica from Djalovića cave, Obodska cave behind part of the underground flow of the Crnojević River, Djatlo Petnjik near Berane, Koritnik in Njeguši), hidden springs (Ljuta, Perućica, Dubrovska, Manastirsko near Berane), fissure springs (Studenac, Vidrovan, Vukova, Čelina pod Medjedom), spring eyes (Pivsko, Oboštica, Slansko, sublacustrine in Skadar Lake), spring systems (Glava Zete, Gurdić, Pivsko oko), karst springs (Gurdić, Ljuta, Sopot, Spila), brackish springs (Kotorska, Morinjska), lake eye springs (Murići, Krnjice, Karuč, Bolje sestre, Veliko i malo oko) estavelles (inversacs) (Gornjepoljski vir, Slanska jama, Ajdarov ponor, Diljino i Manito oko, Čevska estavela, Moračke estavelles), intermittent springs (Vidov potok, Zaslapnica, Šavnička glava), as well as shafts (Gvozdena na Maganiku, Duboki do na Njegušimi, Vjetrena brda na Durmitoru, Pala rock on Lovćen) and caves (Djalovića, Lipska, Novakova, Patalina, Studendula, Osoja, Ledena).

Under the fluvial genetic type, the relief created under the influence of constant line watercourses is mostly isolated (Petrović 1983). Drainage systems belonging to the Black Sea (Ibar, Lim, Ćehotina, Tara and Piva) and the Adriatic basin (Zeta, Morača, Bojana, smaller rivers in the coastal area) have a dominant role in the formation of relief characteristics of the studied area. According to the methodological framework of Mirković and Pavlović (2002), the valley sides are classified as a slope relief type, regardless of the undoubted and significant influence of rivers on the development of slope processes. The fluvial type of relief includes alluvial plains and river terraces, and in Montenegro it is developed in the valleys of Morača, Lim, Tara, Piva (lower course) and Ćehotina. River valleys as a complex erosion-accumulation form and alluvial plains as an accumulation form have a dominant place within the fluvial type of relief. River valleys are a complex and dynamic erosion–accumulation form of fluvial relief. Following their morphology along the transverse profile in the upper reaches of the valley, they usually have a "V" shape. According to the longitudinal profile and lithological basis, further down the river valley they change morphology and pass either into wide river valleys with flat bottoms on terrains of insoluble rocks, or into deeply incised gorges and canyon valleys on terrains of soluble limestone rocks. Especially in terrains with pronounced elevation relief and height differences in a small space. The same conditions lead to the creation of recognizable composite valleys (Tara, Lim, Ćehotina). The valley of the river Tara is characteristic, which builds a composite valley from its source to Djurdjević Tara, changing its transverse profile several times (gorge—a normal river valley with a wide bottom cut in the alluvium). From the profile of Djurdjevića Tara to Šćepan polje, where it merges with Piva, it builds a deep (over 1000 m) and long magnificent canyon. Drainage systems Piva, Komarnica and Morača in their middle part of the flow build a canyon or gorge type of river valley. A number of valleys are partly cut into the accumulation material in which the terraces have been preserved as an indicator of morphological evolution or cutting of the river valley. Bojana was cut into the accumulating river material at the mouth of the Adriatic Sea, as well as some of its tributaries, then Tara through Kolašin, Ćehotina around Pljevlja, Tušina, Zeta through Nikšićko polje and Zeta plain, Lim around Bijelo Polje. In their lower streams Morača (from Bioče to Cijevna), Cijevna (from Dinoša to the mouth of Morača) and Lim (from Plav to Berane) built on several levels of characteristic glaciofluvial terraces that cut into the glacial material and processed it in fluvial conditions. The appearance of epigeny was found on Ćehotina south of Pljevlja, while the most significant occurrence of river piracy is on Kovren between Ljubovidja and Ćehotina and on the left tributary of the Komarnica.

The hillslope deposit (derasion) type of relief includes forms that are formed under the influence of deluvial, proluvial and colluvial processes. In the morphogenetic sense, the slope is of endogenous origin and plays a crucial role in the formation of tectonics. The most exposed is the relief element, so that the exodynamic processes on them are pronounced. Inclination geometry, planimetric relationships and climatic conditions are also important, because they directly affect the intensity of hillslope processes. In areas where we have a full slope profile, their vertical zonation is observed. The processes of falling, collapsing, rinsing and dredging are some of the most pronounced slope destruction processes represented in the formation of the relief (Lazarević 1970). The largest number of landslides was registered in coastal Neogene sediments, as well as in Paleozoic and Paleogene sediments in the east and northeast of Montenegro. Taluses are extremely common forms on slopes built of limestone. Landslides have been found on very steep rocky slopes and valley sides of rivers that have a canyon or gorge type of valley.

The lake-marsh basins are of very diverse origin because they were formed by the strong interaction of various processes (glacial, karst, colluvial, cryo-leveling, etc.) that have taken place from the Pleistocene to the present day (Mirković and Pavlović 2002). This type of relief includes the lake-marsh plains on the northern and western edges of Skadar Lake near Podhum, in the western part of Nikšić polje (immediately next to the Krupac, Vrtac and Slano reservoirs), Barno Lake on Durmitor and along the edge of Šasko Lake and Lake Plav. The plains are periodically flooded, often with swampy parts in which organogenic-marsh sediments are deposited, which provided material for the formation of primary deposits.

Marine relief was created by the action of abrasion and accumulation processes at the contact of the sea and the land (Gluščević and Djurović 1980). Abrasive forms of relief, characteristic of rocky shores on the high seas, built of clastic rocks of Tertiary flysch and carbonate sediments in certain sections create cliffs. Typical cliffs have been singled out on the coast of Luštica and Donji Grbalj, in the bay of Jaz, around Budva, between Bečići and Sutomore, as well as on the stretch from Bar to Ulcinj. In addition to marine erosion, the creation of abrasive shapes was influenced by the movement of masses and fault neotectonics, which shows that the predominant part of the coastal relief is of polymorphic genesis. Accumulation forms are represented by sandy pebble beaches and the St. Stefan tombolo. Larger sandy and pebble beaches are located in Trašte Bay, Jaz Bay, Budva, Bečići, Petrovac, Buljarica, Sutomore and Bar, while smaller ones are located between Bečići and Petrovac, as well as between Cape Volujica and Cape Mendra. The large beach near Ulcinj, with a fluvial plain in the hinterland, is built of fine-grained sand that originates from the ophiolite belt in the tributary of the tributaries of Skadar Lake. This material, brought by the river Bojana to the littoral part of the sea, was retransported and accumulated by the sea water energy on the low shore as a beach.

The anthropogenic type of relief arises as a result of human activity in the geoenvironment (Nikolić 2010). It is recognizable by its characteristic forms such as surface mines, tailings, accumulations and man-made deposits. As a result of economic activity in the relief of Montenegro there are characteristic forms: saltworks near Ulcinj, bauxite mines near Nikšić and coal near Pljevlja and Berane, with accompanying tailings, artificial reservoirs (Krupac, Slano, Vratčko, Pivsko, Otilovići and Liverovići), surface quarries of decorative and technical-construction stone, landfills and interventions along major infrastructure facilities.

2.2.1 A Brief Overview of Relief and Speleological Objects

The appearance of speleological objects is causal and naturally correlates with mineralogical-petrographic, stratigraphic, tectonic and neotectonic, climate and hydrogeological conditions, as well as the intensity and depth of karstification. The types and position of these objects (water and waterless) mark and indicate the evolution of the karstification process, its relationship with the fluvial process and the glaciation process, surface and groundwater and dependence on neotectonics. These objects have been and are the subject of study of speleologists, bio-speleologists, hydrogeologists, physical geographers, as well as archaeologists, ethnographers. The data indicate that in various forms of documentation, more than 3000 speleological objects have been recorded, mapped or their existence is known (Barović et al. 2018; Djurović and Djurović 2021).

However, it is believed that the number of speleological objects is much higher, especially for the karst area of Montenegro. Shafts, abyss and caves are key speleological objects. Deep shafts usually occur in zones with a strong carbonate base and a large relief dissection. Caves are formed in zones with less pronounced dissection of relief and in somewhat shallower karst. The Montenegrin holokarst, in its occurrence and extent, is geo-genetically more predisposed to shafts of great depths (e.g. Željezna jama) which occur in gates, on slopes, in cliffs, below and in the cone of mountain peaks, and especially in karst valleys. Caves, especially those with more kilometers of cave systems, are quite rare. Their appearance in the relief is related to old river valleys, barriers that morphologically separate uvalas and poljes or to zones with different lithological contacts (limestone, dolomites and hornstones). This is confirmed by numerous examples of speleological objects in the relief of Orjen, Lovćen, Rumija, Zeta-Skadar valley, Bjelopavlićka plain, Krivošija, Grahovo, Katun plateau, Garča, pripolje zone of Nikšić polje, Banjani, Rudina, Golija, Piva, Durmitor, northern and eastern Montenegro and drainage systems of Ćehotina, Lim and Ibar.

3 Basic Characteristics of Relief Units

Although Frankl et al. (2016) proposed a somewhat different methodological geomorphological regionalization, the basic geomorphological characteristics were processed according to the relief units singled out by Radojičić (2015). The relief of Montenegro is a consequence of lithofacial composition and tectonics, it is their morphogenetic, dynamic and temporal expression, whose form and visual integrity are reflected and complemented by four distinct relief units:

- Montenegrin coast,
- Central Montenegro
  - *Plateau of deep karst: Krivošije, Grahovski kraj, Bijele Rudine, Banjani and Oputne Rudine,*
  - *Valley of Central Montenegro,*
- Central high mountains, areas and canyon valleys,
- Valleys, areas and high mountains of northeastern Montenegro.

The Montenegrin coast is a morphologically separated narrow belt along the coast, which is separated from the continental hinterland by steep slopes, Orjen, Lovćen and Rumija (Glusčević and Djurović 1980). The coefficient of divergence is 3.5, and the coastline is mostly jagged in the area of the Bay of Kotor. The Bay of Kotor consists of several bay morphological forms connected by straits. From the entrance gate of Boka Kotorska, between the Luštica peninsula and Cape Oštro, 1500 to 2950 m wide, to Kotor, there are: Herceg Novi and Tivat bays, which are connected as external bays by the Kumbor Strait, deeper into the mainland is the Verige Strait (340 m wide), which connects these two outer bays with the inner Bay of Kotor and Morinj-Risan bay. The average depth of the bay is from 35 to 45 m. From the Luštica peninsula to the Volujica peninsula and southeast to Bar, the coast consists of numerous cliff-shaped capes, among which the most famous coves and beaches are Jaz, Budva, Bečići, Miločer, Petrovac, Sutomore and Bar. From Bar to the Albanian border on the Bojana River, the coastal part consists of four limestone ridges, between which are bays and plains. A number of coves and pebble-sandy beaches were formed there, as well as the 12 km long Ulcinj beach. Plains were formed in this part of the coast. The largest are Ulcinjsko polje 72 km^2 and Barsko polje of 10 km^2.

Central Montenegro is a recognizable geospatial unit of the central part of Montenegro, with two characteristic relief units:

The deep karst plateau is one of the most typical karst regions in the world. It consists of Katunski krš, Krivošije,

Grahovski kraj, Bijele Rudine, Banjani and Oputne Rudine (Krivokapić 1975). It is separated from the Montenegrin coast by the mountains Orjen (1,894 m above sea level), Lovćen (1,749 m above sea level), Sutorman (1,185 m above sea level) and Rumija (1,595 m above sea level), and from the central valley of Montenegro by the mountains Somina (1,586 m above sea level), Njegoš (1,761 m above sea level), Zla gora (1,459 m above sea level), Pusti lisac (1,476 m above sea level), Budoš (1,217 m above sea level) and Garač (1,436 m above sea level). Between them is a plateau with an average height of 800–900 m, about 90 km long and 40 km wide. The capacity of carbonate rocks is over 4000 m, the process of scaling goes to the deepest layers. It contains the main resources of groundwater. This area is a mosaic of micro and macro forms of karst relief with sinkholes, karrens (limestone pavements), crevices, shafts, caves, valleys, uvalas and poljes. The largest karst poljes in this zone are Dragaljsko, Grahovsko, Njeguško and Cetinjsko.

The central valley of Montenegro has a Dinaric direction of extension and gradually descends from northwest to southeast, and is located between the Plateau of Deep Karst and the Central area of high mountains, surfaces and canyon valleys. It consists of Golija and Duga (800 to 1000 m), Nikšić polje (647–660 m), Bjelopavlićka plain (40 to 56 m) and Podgorica–Skadar valley (6 to 67 m), with mutual structural lines and passes. Groundwater and surface waters of the surrounding areas flow through the valley. The first in a series is an elongated stepped valley from Gata to Duga. It is located between Gatačko and Nikšić polje and the mountains Njegoš and Golija, it is 40 km long and 7 km wide. Morphologically, the largest karst polje in Montenegro–Nikšić polje (66.5 km^2) continues on them. It is a closed karst depression surrounded by terrain with an average height of about 1200 m, formed by complex tectonic processes, and partly by erosion, covered with fluvioglacial gravel and sand, up to 40 m thick. The river Zeta originates from numerous springs, estavelles and short streams in the Nikšić polje. Further across the Planinica pass (685 m above sea level), the Bjelopavlićka plain (72 km^2) continues, which is 28 km long and 8 km wide. It got its shape thanks to the accumulated deposits of clay when a lake was formed here in the Pleistocene (Vlahović 2019). The Podgorica–Skadar valley was formed in a similar way as the Bjelopavlica plain. Between Podgorica and Skadar it is 50 km long, and it is the widest (30 km wide) between Virpazar and Hota, in Albania. The largest part of the structural basin covers Skadar Lake (354 to 506 km^2), while north of Lake Zeta is 240 km^2.

The central high mountains, surfaces and canyon valleys have been shaped by complex tectonic evolution and for the most part glacial erosion. It is the dominant relief unit, consisting of two mountain ranges between which are mountainous areas and deep canyons (Djurović and Petrović 2007). The first chain, in the southwest of this area, consists of the mountains Golija (1,942 m above sea level), Vojnik (1,998 m above sea level), Maganik (2,139 m above sea level), Prekornica (1,927 m above sea level) and Žijeva (2,131 m above sea level). The second mountain range consists of the highest Dinaric mountains Volujak (2,336 m above sea level), Maglić (2,386 m above sea level), Bioč (2,397 m above sea level), Ljubišnja (2,238 m above sea level), Durmitor (2,523 m above sea level), Sinjajevina (2,277 m above sea level), Bjelasica (2,139 m above sea level), Visitor (2,211 m above sea level), Kučki Komovi (2,487 m above sea level) and Vasojevićki Komovi (2,461 m above sea level) and Prokletije (2,530 m above sea level). Glacial erosion has created numerous cirques, waves and moraine forms. This is an area with a pronounced elevation relief and height differences in a small area, because the rivers cut deep canyons against the high mountains. The most famous is the canyon of the river Tara (80 km), then the canyons of the rivers Piva, Komarnica, Bukovica, Morača and its tributaries Mrtvica and Mala rijeka. The height of the canyon sides of the river Tara reaches 1300 m, and the canyons of Piva, Komarnica and the river Morača up to 1000 m. Numerous mountain lakes that were created in cirques and waves of former glaciers should be pointed out here, such as the Black Lake under Durmitor and the Biograd Lake on Bjelasica.

The valleys, surfaces and high mountains of northeastern Montenegro differ in relief from the rest of Montenegro. The northeastern area stretches in the direction northwest–southeast in the length of 140 km and a width of 35 km. The most prominent geomorphological features are the valleys of the rivers Ćehotina, Lim, Ibar and their tributaries (Knežević and Kićović 2004). The river Ćehotina has a narrow valley with sudden widenings, the largest of which is the Pljevlja valley (16 km^2), at an average height of 770 m. The composite valley of the Lim River consists of valleys, narrow valleys and gorges. In the source part is the Plav–Gusinje basin, with Lake Plav, at an altitude of 900 to 1000 m. It is 25 km long and up to 5 km wide. It is a glacial valley, into which rivers flow, whose valleys represent glacial waves. The Berane valley (24 km^2) is the largest extension in the Lim valley. In the middle course of the Lim is the Bijelo Polje valley, 11.5 km long and up to 3 km wide. Between the valleys of the rivers Lim and Ibar, the terrain has a distinct highland character, divided into branches and ridges of the Hajla mountain (2,403 m), intersected by tributaries of Lim and Ibar. The Ibar Valley, in the area of Montenegro, is an extended crest, with the characteristics of a high valley from 900 to 1000 m, between the Pešterska plateau, Turjak and the branches of the Hajla mountain. Its reflection is the river network with numerous small rivers, from which the river Ibar is formed.

4 Analysis of Quantitative Morphometric Characteristics of Relief in GIS Environment

Quantitative geomorphological analysis involves the expression of morphological processes, shapes and reliefs using numerical parameters. However, it is important to emphasize that the number of these parameters is unlimited and that they can be grouped into three groups: (1) morphometric, (2) hydrological and (3) climatological (Marković 1983; Hengl et al. 2003, Valjarević et al. 2022, 2023).

Morphometric parameters (also called primary digital geomorphological parameters) define morphology, e.g. hypsometry, slope, aspect and vertical dissention (Šiljeg et al. 2015, 2018). For the purpose of mapping and analyzing the morphometric characteristics of the relief of Montenegro, the Digital Relief Model (DMR) of medium resolution (25 m) EU-DEM in GIS (Geographic Information Systems) environment was used. This model, developed by the European Environment Agency, according to studies, shows a squared error value for a vertical accuracy of about 7 m (Mouratidis and Ampatzidis 2019). So for more relevant analyses, the use of DMR of better resolution and quality would be significant. ESRI ArcGIS 10.5 was used as a software for the analysis of morphometric parameters and the production of overview maps with Spatial Analyst modules.

Hypsometry is the basis of all geomorphological research and gives an idea of the characteristics of the relief that we analyze. The hypsometric distribution is a numerical parameter and refers to the amount of area of a certain space or the entire territory of a country or area, which belongs to a certain height category. With this we come to know whether it is a plain, hilly or mountainous relief. Depending on that, possibilities for planning and proper land use are imposed (Manojlović et al., 2004). According to the previous analysis of Radojičić (2015) the average height value is 1050 m; the hypsometric category of up to 200 m in relation to the area it occupies participates with 10%, the category between 200–1000 m with 35%, the category between 1000–1500 m participates with 40% and over 1500 m makes 15% of the relief. However, this analysis has slightly different results shown according to the hypsometric categories that are standardized for the area of Yugoslavia according to Mladenović (1984) in Table 1.

Table 1 Hypsometry category

Hypsometry category	Coverage (%)
0–200 m Lowland relief	10.30
200–500 m Hilly relief	5.70
500–1000 m Low mountainous relief	27.06
1000–1500 m Mid-mountainous relief	39.09
1500–2000 m Mountainous relief	16.87
>2000 m High-mountainous relief	0.98
Total	100.00

An overview hypsometric map is shown in Fig. 3a. Based on this analysis and previous research, we conclude that in the hypsometric sense, Montenegro has a predominantly highland relief, whose disintegration character is especially recognizable in the area of surfaces, plateaus and mountains. In the relief of Montenegro, three hypsographic levels or steps are recognized: the first, which includes a narrow coastal belt, the Podgorica–Skadar valley and the Bjelopavlićka plain, from 0 to 100 m; second, which forms a plateau of deep karst: Katunski krš, Krivošije, Grahovski kraj, Rudine and Banjani, from 800 to 900 m. On this plateau, above 600 m, depressions of karst poljes were formed: Nikšić, Grahovo, Cetinje and Njeguši with their prepolje zones; the third refers to the zone with high plateaus and surfaces, from 1200 to 1600 m, intersected by deep canyons of Piva, Tara, Morača and Ibar, from which rise mountain massifs with peaks over 2000 m.

The relief can be presented as a combination of slopes of different inclination categories (Bognar 1990). The slope is often considered the most important morphometric parameter used for more efficient relief analysis and description. Slope analysis is widely used in hydrological modeling, analysis of slope processes, erosion studies, as well as land use planning (Vujačić et al. 2023). The analysis of the spatial distribution and coverage of certain categories of slope hillslopes is important because they are an important indicator of the scope and intensity of morphostructural and exogeomorphological processes that influenced the morphogenesis of slopes during the paleogeomorphological period, but also as an indicator that is, denudations and accumulations (Radoš et al., 2012). In order to simplify the analysis of the spatial distribution and coverage, an overview map of the slope (Fig. 3b) and interpretation (Table 2) was made, according to Šiljeg et al. (2018) geomorphological classification of the inclination of slopes.

Aspect is a morphometric parameter that represents the orientation of the slope with respect to the sides of the world (Burrough and McDonell 1998). In the northern hemisphere, the southern slopes receive the most solar radiation, and the northern ones the least. The eastern and western aspect slopes receive less radiation than the southern and more than the northern; the eastern ones receive the most radiation in the morning and the western ones in the early evening (Nikolić et al. 2023). The influence of aspects on geomorphological processes is very significant because it differently affects the characteristics of climatic elements (temperature, precipitation, etc.) as exogenous-geomorphological agents (Šegota and Filipčić 1996). Therefore, aspects indirectly affect changes related to slope processes. In order to simplify the analysis of the spatial distribution

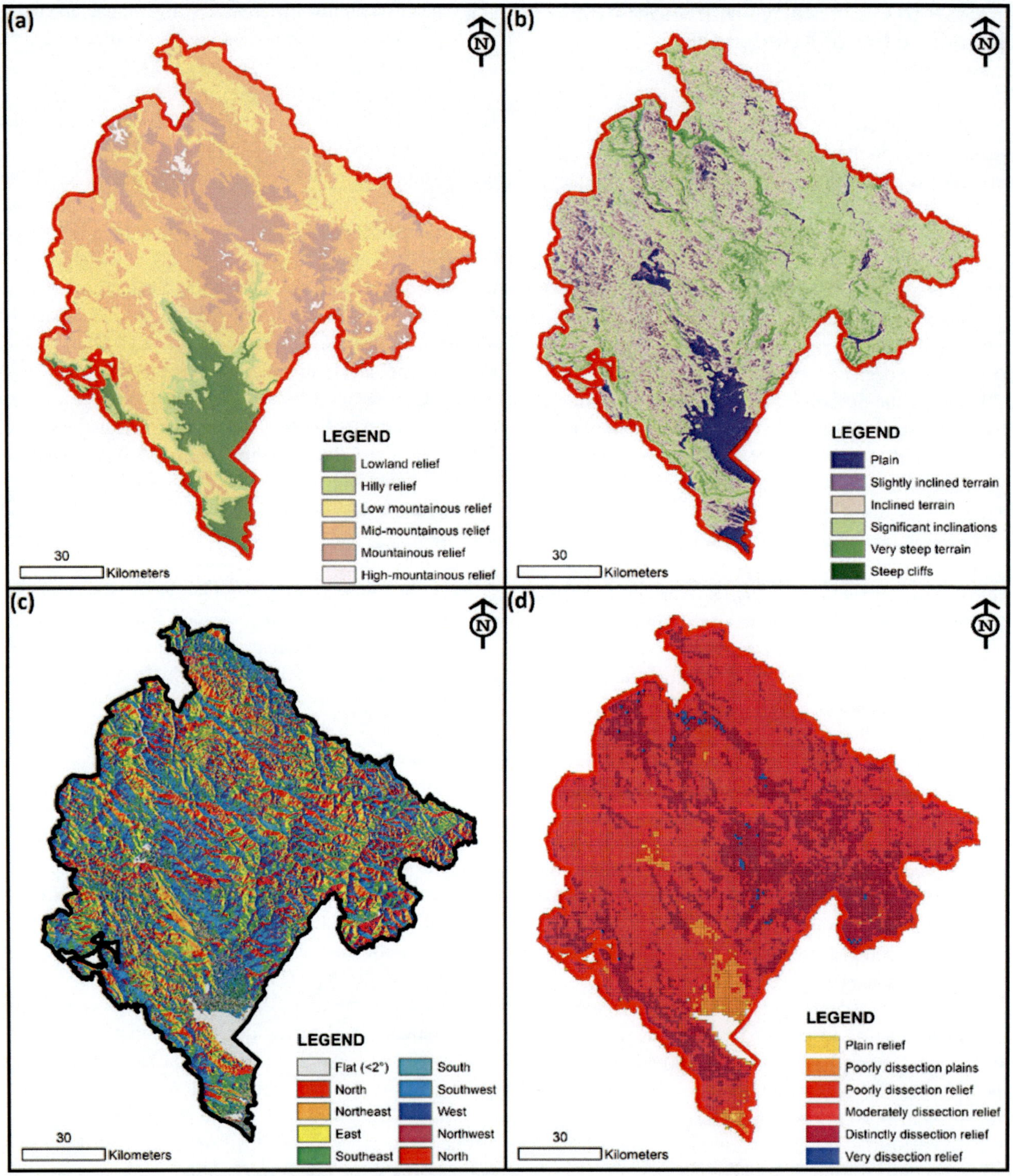

Fig. 3 Overview map of primary digital geomorphological parameters. **a** Hypsometry, **b** Slope, **c** Aspect, **d** Vertical dissection

and coverage, an overview map of the aspect was made (Fig. 3c), and the surface and percentage values of the analysis are shown in Table 3.

Vertical dissection is a morphometric relief parameter that represents the height difference between the lowest and highest point within the observed area of 1 km^2 (Lozić, 1995). Regarding local characteristics, the vertical dissection of the relief is due to (lithological composition and distribution, the amount of water in the terrain) and is a parameter of the intensity of the development of exogenous processes. In areas with higher vertical dismemberment, the intensity of erosion is higher, and in areas with less dismemberment, there is an increased accumulation. The results obtained on the basis of this parameter are

Table 2 Slope categories and their characteristics

Inclination (°)	Characteristics of geomorphological forms and processes	Participation (%)
<2	Plain. Mass movements are not observed. Denudation intensity is minimal	9.11
2–5	Slightly inclined terrain. Denudation poorly expressed. Soil denudation and slippage can be significant	10.82
5–12	Inclined terrain. Increased denudation and mass movement. Sliding and flow of material come to the fore. Terrain endangered by slope processes	28.58
12–32	Significant inclinations. Denudation is intense. Very strong slope processes. The terrain is very endangered by leaching and the appearance of mass movements	45.13
32–55	Very steep terrain. Material removal dominates. Accumulation material is retained only in certain places (thin cover). The slopes are rocky and mostly bare	6.21
>55	Steep cliffs. Shedding and collapsing dominate	0.15
Total		100.00

Table 3 Aspect categories

Aspect	Participation (%)
Flat (Inclination < 2°)	9.05
North	11.47
Northeast	12.93
East	10.42
Southeast	9.61
South	12.04
Southwest	13.89
West	11.05
Northwest	9.53
Total	100.00

Table 4 Categories of vertical relief dissection

Category m/km^2	Description	Participation (%)
0–5	Plain relief	0.32
5–30	Poorly dissection plains	3.25
30–100	Poorly dissection relief	7.87
100–300	Moderately dissection relief	54.85
300–800	Distinctly dissection relief	33.13
>800	Very dissection relief	0.59
Total		100

significant in engineering geology and geomorphology (slope stability), in the research of mineral resources, pedology (soil utilization). Regionally speaking, the vertical dissection is a reflection of the youngest tectonic movements. The values obtained in this analysis show the position of neotectonic structures, intensity and sign of movement (Šiljeg et al. 2018). The results of the analysis (without Lake Skadar) are shown in Table 4 and on the overview map (Fig. 3d).

References

Barović, G., Kicińska, D., Mandić, M., Mulaomerović, J. (2018). Icecaves of Montenegro and Bosnia and Hercegovina. In A. Persoiu & S. E. Lauritzen (Ed.), *Ice caves* (pp. 263-283). Amstedrdam: Elsevier

Bešić, Z. (1969). *Geologija Crne Gore: Karst Crne Gore*. Titograd: Zavoda za geološka istraživanja Crne Gore

Bešić, Z. (1983). *Geologija Crne Gore: Geotektonika i Paleografija Crne Gore*. Titograd: Društvo za nauku i umjetnost Crne Gore

Bognar, A. (1990). Geomorfološke i inženjersko-geomorfološke osobine otoka Hvara i ekološko vrednovanje reljefa. *Hrvatski geografski glasnik*, 52(1), 49-64

Burić, D., Ducić, V., Mihajlović, J. (2014). The climate of Montenegro: Modificators and types-part two. *Glasnik Srpskog geografskog društva*, 94(1), 73-90. https://doi.org/10.2298/GSGD1401073B

Burrough, P. A., McDonell, R. A. (1998). *Principles of Geographical Information Systems*. New York: Oxford University Press

Cvijić, J. (1899). *Glacijalne i morfološke studije Bosne, Hercegovine i Crne Gore*. Beograd: Glas Srpske kraljevske akademije

Djurović, M., Djurović, P. (2021). Review of the most significant caves in Montenegro. *Acta Carsologica*, 50(1), 49-65. doi: https://doi.org/10.3986/ac.vi.7182

Djurović, P. (1996). *Visokoplaninski kras Durmitora - geomorfološka studija (doktorska disertacija)*. Beograd: Univerzitet u Beogradu, Geografski fakultet

Djurović, P. (2009). Reconstruction of the Pleistocene glaciers of mount Durmitor in Motenegro. *Acta Geographica Slovenica*, 49(2):263-289. https://doi.org/10.3986/AGS49202

Djurović, P., Petrović, A. (2007). Large canyons in Dinaric and Prokletije mountains region of Montenegro. *Geographica Pannonica*, (11), 14-18. https://doi.org/10.5937/GeoPan0711014D

Frankl, A., Lenaerts, T., Radusinović, S., Spalevic, V., Nyssen, J. (2016). The regional geomorphology of Montenegro mapped using Land Surface Parameters. *Zeitschrift für Geomorphologie*, 60(1), 21-34. https://doi.org/10.1127/zfg/2016/0221

Glavatović, B., Starčević, M., Dubak, M. (1982). *Analiza i sistematizacija gravimetrijskih, geoelektričkih i seizmičkih podataka na teritoriji SR Crne Gore*. Titograd: Zavod za geološka istraživanja

Glusččević, B., Djurović, M. (1970). *Potencijalni uspona priobalnog područja Crne Gore*. Titograd: Instiut za društveno ekonomska istraživanja

Hengl, T., Gruber, S., Shrestha, D. P. (2003). *Digital terrain analysis in ILWIS: lecture notes and user guide*. Enschede: International Institute for Geo-information Science and Earth Observation (ITC)

Knežević, M., Kićović, D. (2004). Natural features of the Polimske Prokletije mountains from the point of their active protection. *Bulletin of the Serbian geographical society,* 84(1), 89-100

Krivokapić, B. (1975). *Katunski krš*. Cetinje: Obod

Lazarević, R. (1970). *Stanje, problemi i savremeni metodi za borbu protiv erozije i bujica (SR Crna Gora)*. Beograd: Institut za šumarstvo i drvnu industriju

Lozić, S. (1995). Vertical relief dissection of the land part of the Republic of Croatia. *Acta Geographica Croatica*, 30(1), 17-26

Lješević, M. (2005). *Karst Pive*. Podgorica: Crnogorska akademija nauka i umjetnosti

Manojlović, P., Dragićević, S., Mustafić, S. (2004). Osnovne morfometrijske karakteristike reljefa Srbije. *Glasnik Srpskog geografskog društva*, 84(2), 11-20

Marković, M. (1983). *Osnovi primljene geomorfologije*. Beograd: Geoinstitut

Mirković, M. (1997). *Tumač za strukturno-tektonsku kartu Crne Gore 1:200.000*. Titograd: Zavoda za geološka istraživanja

Mirković, M., Pavlović, R. (2002). *Tumač geomorfološke karte Republike Crne Gore 1:200.000*. Podgorica: Zavod za geološka istraživanja Crne Gore

Mirković, M., Živaljevic M., Djukić, V., Petrović, Z., Kalezić, M., Pajović, M. (1985). *Tumač Geološke Karte SR Crne Gore 1:200.000*. Titograd: Zavod za geološka istraživanja

Mladenović, T. (1984). Visinska struktura reljefa zemljišta u SFR Jugoslaviji. Beograd: Vojnogeografski institut

MONSTAT. (2022). *Statistički godišnjak 2022*. Podgorica: MONSTAT. http://www.monstat.org/uploads/files/publikacije/godisnjak%202022/Godisnjak%202022_za%20web_print.pdf

Mouratidis, A., Ampatzidis, D. (2019). European digital elevation model validation against extensive global navigation satellite systems data and comparison with SRTM DEM and ASTER GDEM in Central Macedonia (Greece). *ISPRS International Journal of Geo-Information*, 8(3), 108. https://doi.org/10.3390/ijgi8030108

Nikolić, G. (2010). *Metodologija geoekološkog istraživanja na primjeru zaštićenih područja Crne Gore (doktorska disertacija)*. Beograd: Univerzitet u Beogradu, Geografski fakultet

Nikolić, G., Vujović, F., Golijanin, J., Šiljeg, A., Valjarević, A. (2023). Modelling of Wildfire Susceptibility in Different Climate Zones in Montenegro Using GIS-MCDA. *Atmosphere*, 14(6), 929. https://doi.org/10.3390/atmos14060929

Pajović, M., Radusinović, S. (2010). Mineralne sirovine Crne Gore; Crna Gora u XXI stoljeću u eri kompetitivnosti. In M. Burić (Ed.), *Crna Gora u XXI stoljeću — u eri kompetitivnosti: Životna sredina i održivi razvoj* (pp. 237–282). Podgorica: Crnogorska akademije nauka i umjetnosti (CANU)

Petrović, J. (1983). *Kraške vode Crne Gore*. Novi Sad: Univerzitet u Novom Sadu, Prirodno matematički fakultet, Institut za geografiju

Pulević, V. (2022). *Botanički leksikon Crne Gore*. Podgorica: Crnogorska akademija nauka i umjetnosti

Radojičić, B. (2015). *Crna Gora: Geografski Encikopedijski Leksikon*. Nikšić: Univerzitet Crne Gore, Filozofski fakultet

Radoš, D., Lozić, S., Šiljeg, A. (2012). Morfometrijske značajke šireg područja Duvanjskog polja, Bosna i Hercegovina. *Geoadria*, 17(2), 177-207. https://doi.org/10.15291/geoadria.352

Radulović, M. (2000). *Hidrogeologija karsta Crne Gore*. Podgorica: Zavod za geološka istraživanja Crne Gore

Stepišnik, U., Žebre, M. (2011). *Glaciokras Lovčena*. Ljubljana: Univerza v Ljubljana, Filozofske fakultete, Oddelek za geografijo

Šegota, T., Filipčić, A. (1996). *Klimatologija za geografe*. Zagreb: Školska knjiga

Šiljeg, A., Barada, M., Marić, I. (2018). *Digitalno modeliranje reljefa*. Zagreb: Alfa

Šiljeg, A., Lozić, S., Šiljeg, S. (2015). The Accuracy of Deterministic Models of Interpolation in the Process of Generating a Digital Terrain Model—The Example of the Vrana Lake Nature Park. *Technical Gazette*, 22, 853-863. https://doi.org/10.17559/TV-20131010210942

Valjarević, A., Algarni, S., Morar, C., Grama, V., Stupariu, M., Tiba, A., Lukić, T. (2023). The coastal fog and ecological balance for plants in the Jizan region, Saudi Arabia. *Saudi Journal of Biological Sciences*, 30(1), 103494. https://doi.org/10.1016/j.sjbs.2022.103494

Valjarević, A., Popovici, C., Djekić, T., Morar, C., Filipović, D., Lukić, T. (2022). Long-term monitoring of high optical imagery of the stratospheric clouds and their properties new approaches and conclusions. *The Egyptian Journal of Remote Sensing and Space Science*, 25(4), 1037-1043. https://doi.org/10.1016/j.ejrs.2022.11.006

Vlahović, M. (2019). *Površinske akumulacije u karstu Nikšićkog polja: hidrogeološki i inženjersko-geološki aspekti*. Nikšić: Elektroprivreda Crne Gore (EPCG)

Vlahović, V. (1975). *Kras Nikšićkog polja i njegova hidrogeologija*. Titograd: Društvo za nauku i umjetnost Crne Gore

Vujačić, D., Milevski, I., Mijanović, D., Vujović, F., Lukić, T. (2023). Initial results of comparative assessment of soil erosion intensity using Wintero model: A Case Study Of Polimlje And Shirindareh Drainage Basins, *Carpathian Journal of Earth and Environmental Sciences*, 18(2), 385 – 404. https://doi.org/10.26471/cjees/2023/018/267

Biodiversity of Montenegrin Caves

Lidija Polović

Abstract

The territory of Montenegro represents one of the impressive hotspots of cave-dwelling fauna diversity in the world. As many as 1600 speleological objects are registered in Montenegro, but only 5–10% of them have been explored in terms of biodiversity. An overview of literature data on cave biodiversity in Montenegro is given here. A total of 72 species of micromycetes were registered in 7 caves and the most abundant species were *Acrodontium salmoneum, Aureobasidium pullulans, Cladosporium oxysporum, Mucor racemosus f. racemosus, Penicillium chrysogenum,* and *Rhizopus stolonifer*. Regarding dwelling invertebrates, the following groups are represented in Montenegrin caves: Protozoa—ciliates; Turbelaria—tricladids and temnocephalids; Gastropoda; Oligochaeta and Hirudinea; Cladocera, Copepoda, and Ostracoda; Isopoda; Amphipoda; Aranea; Pseudoscorpiones; Palpigradi; Opiliones; Chilopoda; Diplopoda; Collembola; Coleoptera—over 80 species. According to the non-invasive survey method based on environmental DNA (eDNA), indications of the presence of the olm (*Proteus anguinus* Laurenti 1768) were reconfirmed by finding traces of environmental DNA at several sites in Montenegro.

Keywords

Caves · Biodiversity hotspot · Montenegro

1 Introduction

Many new species, even genera, are being described from biodiversity-rich karst areas of Dinarides, Alps, or Pyrinees. These regions are considered as areas with the highest concentration of subterranean biodiversity hotspots in the world. This was conditioned by various factors such as temperature, precipitation, number and size of caves in the area, connectivity and availability of karst landscapes, and climatic and paleogeographic events. Pleistocene climatic oscillations greatly contributed to the existing patterns. Falling temperatures and glacier coverage, associated with the Permafrost, which extended as far as the southern Alpine valleys, decimated the subterranean fauna north of the Alps (Delić and Reboleira 2021). The extent of the Last Glacial Maximum, approximately 21,000 years ago, limited the distribution of many subterranean taxa (Stoch and Galassi 2010). Analysis of species richness regarding subterranean amphipods and beetles revealed the existence of two peaks of species richness and one of them is located in southeastern Herzegovina, Dalmatia, and Western Montenegro. Endemics known only from one locality have a large proportion of 23 and 31 percent of subterranean amphipods and fauna of beetles in the Dinarides (Delić and Reboleira 2021).

Most of the above-mentioned patterns and processes are related to the origins of subterranean life through physical changes in the environment, however, a large part of these highly specialized taxa arose through processes of ecological speciation, which implies different use of environmental resources, behavioral changes related to predator avoidance, etc. The most representative example of ecological speciation in caves is represented by species-rich communities of closely related subterranean amphipods from the genus Niphargus, so-called hygropetricolous, highly specialized filter-feeding beatles (Delić and Reboleira 2021).

L. Polović (✉)
Natural History Museum of Montenegro, Podgorica, Montenegro
e-mail: lidija.polovic@pmcg.co.me; lidijapolo@t-com.me

G. Barovic (ed.), *Speleology of Montenegro*, Cave and Karst Systems of the World,
https://doi.org/10.1007/978-3-031-49375-1_6

2 Cave Biodiversity

Until the discovery of a blind beetle *Leptodirus hochenwartii* Schmidt, 1832, animal with unusual morphology from Postojnska jama in Slovenia (Schmidt 1832a; Schmidt 1832b) (Fig. 1), there was no awareness that any form of life could live underground. After that, the discoveries of underground animals were greatly increased. It started from the Dinaric Karst, and then it was transferred to other mountain areas in Europe.

Underground ecosystems belong to the biomes with the largest number of narrowly distributed and relict taxa (Culver and Sket 2000; Holsinger 1990; Marmonier et al. 1993). This is conditioned by the geographical isolation of underground habitats, which facilitates evolutionary drift (Gibert and Deharveng 2002; Romero 2014). These well-protected habitats were not affected by climatic fluctuations over a long period of time; this can be explained by the absence of Pleistocene glaciations (Holsinger 1990; Gibert and Deharveng 2002; Howarth 1983). Compared to terrestrial biomes, subterranean habitats were previously considered to have less species richness (Culver and Sket 2000); however, findings from the last few decades and the recently described high incidence of cryptic diversity mainly in invertebrates. (Gibert and Deharveng 2002; Rouch and Danielopol 1997; Lefebure et al. 2006) indicate that this opinion should be revised. While the obligate subterranean fauna is dominated by invertebrates (Gibert and Deharveng 2002; Sket et al. 2004), salamanders and bony fish have been able to successfully colonize these habitats (Culver and Sket 2000; Soares and Niemiller 2013; Bendik et al. 2013; Romero 2009). Underground biodiversity has been considerably less studied in comparison to biodiversity of taxa living on or near the earth's surface, mainly due to physical inaccessibility or inadequate sampling strategies (Gibert et al. 2009).

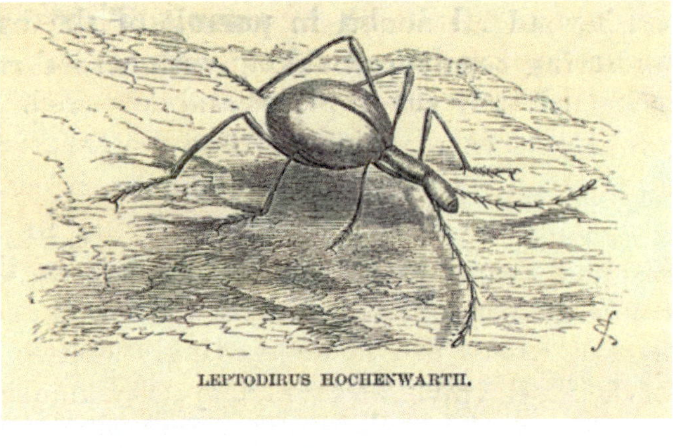

Fig. 1 Drawing of the first discovered and scientifically described subterranean animal, *Leptodirus hochenwartii* Schmidt, 1832

Caves are unique habitats characterized by relatively constant conditions (light, temperature, humidity) in comparison to surface habitats that are characterized by highly variable conditions. The most obvious difference, compared to surface habitats, is the absence of light. Caves represent specific habitats where only specially adapted species can survive. Since autotrophs cannot survive underground, there is no photosynthesis and this leads to a reduction of the available nutrients. As a result we have the absence of species feeding on plant food (Barović et al. 2013). Available nutrients mainly come from the surface by sink rivers, water droplets, and occasional visitors (bats, crickets, etc.). As a result, the combined effects of the aforementioned habitat characteristics are summed up in dramatic morphological changes that are noticeable in subterranean animals.

Underground habitats are characterized by the absence or reduction of light, as well as the specificity of environmental factors such as humidity, temperature, as well as isolation from other environments and spatial limitations. As a result, the combined effects of the above-mentioned habitat characteristics are summed up in morphological changes, behavioral changes, as well as the development of specific physiological mechanisms during evolution.

Cave animals are divided into three categories depending on their ecological features. Trogloxenes are temporary cave dwellers who spend part of their lives in the cave but also outside the cave. The survival of these animals does not depend directly on caves, and they also do not show special adaptations for cave ecosystems. Representatives of trogloxen animals are skunks, moths, centipedes, beetles, butterflies, flies, etc. Troglophiles can live in complete darkness, but they can also survive outside of caves. From time to time they go outside searching for food. The most representative troglophiles are bats. Some bat species prefer the constant temperature of caves to hibernate and raise their young. Also, this group of animals includes certain species of insects, frogs, salamanders, some species of crustaceans, etc.

Troglobionts are true cave dwellers that spend their whole life in a cave. They live in complete darkness, we can only find them in caves and they are not able to survive outside the caves. During their evolution, these organisms have developed numerous morphoanatomical and ecophysiological adaptations that help them to survive in underground habitats. All troglobionts show certain common features, such as lack of pigment, leading to more uniform body color from completely white, to various shades of brown and grayish brown, then absence or atrophy of the sense of sight, which is manifested by the pulling of skin or chitin over the eyes, while sensory cells located on different parts of the body are highly developed. Animals that have completely adapted to cave life are some species of fishes, crustaceans, centipedes, vertebrates, and insects.

2.1 Cave Biodiversity of Montenegro

The Dinaric mountain chain, to which Montenegro belongs, is an impressive hotspot of cave-dwelling fauna biodiversity in the world. Intensive biospeleological research conducted in the Dinaric karst area during the last two decades has led to the discovery of many new subterranean species and genera (Monguzzi 1993; Casale and Guéorguiev 1994; Casale and Jalžić 1999; Quéinnec 2008; Quéinnec and Pavićević 2008; Quéinnec et al. 2008; Lohaj and Jalžić 2009; Lohaj and Lakota 2010; Lakota et al. 2010; Casale et al. 2012; Lohaj and Mlejnek 2012). Despite that, the biodiversity of speleological objects in Montenegro is not researched enough. According to the Register of speleological objects (Environmental Protection Agency), there are more than 1600 speleological objects in Montenegro. One of the first studied caves in Montenegro is Lipa Cave, discovered around the same time as the Postojna Cave in Slovenia, the only cave in Montenegro valorized for tourist purposes (Fig. 2a, b). Biodiversity has been investigated in only 5–10% of the total number of speleological objects, so it is considered that as a result of future research, we can expect a large number of new species, especially invertebrates (Karaman 2018).

2.1.1 Biodiversity of Fungi

The composition of micromycete communities in both photic and aphotic entrance zones was examined for the first time in seven caves in Montenegro: Obod cave, Golubinja cave, Veluštica cave, Vrbačka pit, Pit ER-1, Njegoš cave as well as one unnamed cave (Kozlova and Mazina 2020). Seventy-two species of micromycetes were identified in total: 46 species from illuminated photic zones of caves, including 43 species on phototrophs, 10 species from air, 15 species on substrates, and 50 species from non-illuminated aphotic zones, including 35 species from air and 37 species on substrates. The most abundant species were *Acrodontium salmoneum, Aureobasidium pullulans, Cladosporium oxysporum, Mucor racemosus f. racemosus,*

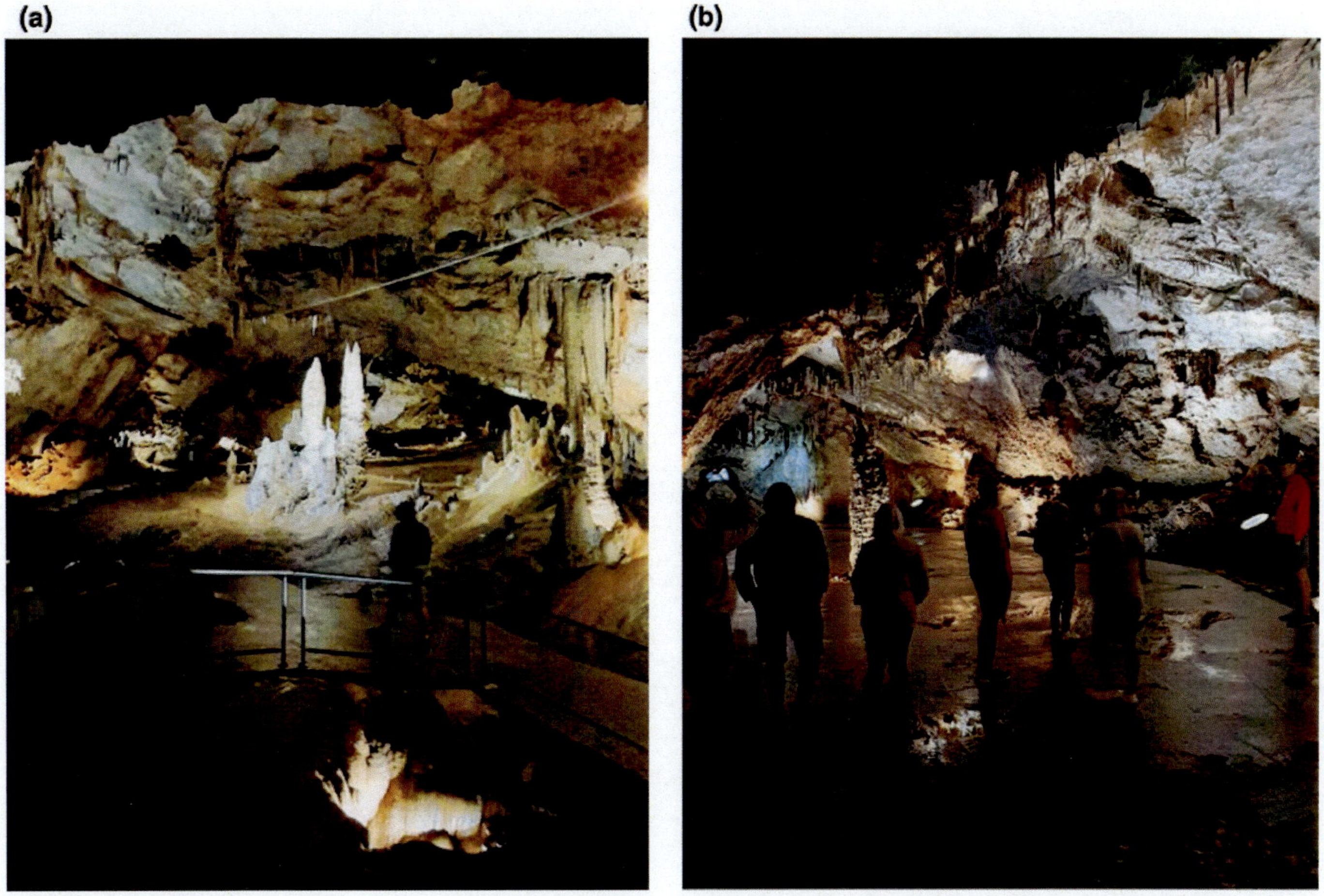

Fig. 2 **a**, **b** Lipa Cave (Montenegrin: Lipska pećina) is one of the largest and most beautiful speleological objects in Montenegro. Lipa Cave is the only valorized cave in Montenegro open to visitors (Photos by L. Polović)

Penicillium chrysogenum, and *Rhizopus stolonifer* (Kozlova and Mazina 2020).

2.1.2 Biodiversity of Invertebrates

There is a general opinion for terrestrial cave organisms in Montenegrin caves that their immediate ancestors are dwelling invertebrates living in the forest, soil, and litter that invaded caves and were isolated there during the Pleistocene interglacials (Barr 1968; Peck 1981; Ćurčić et al. 2008). No direct fossil evidence has been found for this but strong indirect evidence does exist:

1. caves in glaciated areas have very scarce terrestrial fauna or it is not present at all.
2. the closest relatives of many cave-restricted species are found in the leaf litter of boreal forests.
3. the largest number of terrestrial troglobites live in the regions most affected by the Pleistocene climate, except that they are actually covered by ice.
4. regions that lacked forests throughout the Pleistocene have very scarce terrestrial cave fauna.
5. estimates of time since divergence of closely related species, using Nei's (1975) formula and based on electrophoretic data, are consistent with Pleistocene isolation (Delay et al. 1980).

Groups of dwelling invertebrates present in Montenegrin caves (according to Ćurčić et al. 2008):

Protozoa—a certain number of ciliates were found on some cave specimens of Niphargus, although free-living protozoans have not been researched (Ćurčić et al. 2008); Turbelaria—although Tricladids are unexplored, it is known that Tricladids are present in some caves. Some unspecified temnocephalids have been recorded on specimens of four Niphargus species (Ćurčić et al. 2008); Gastropoda living in cave waters are most often represented by hydrobiid and emmericiid snails represented by one species each (*Adriohydrobia gigantinella* and *Emmerici expansilabris*) as well as orientalid snails represented as many as 14 species belonging to genera *Orientalina* (6 species), *Anagastina* (5 species), and *Antibaria*, *Bracenica, Plagigeyeiaria* (1 species each). This group has not been sufficiently researched (Ćurčić et al. 2008); Oligochaeta and Hirudinea are represented by some depigmented species from different caves not studied in detail. Cladocera, Copepoda, and Ostracoda are common in both underground lakes and streams (Ćurčić et al. 2008); Isopoda—several terrestrial isopod families are represented in Montenegrin caves by a number of endemic species and genera. Aquatic isopods are represented as well (Ćurčić et al. 2008); Amphipoda includes a high number of endemic lacustrine and brackish *Niphargus* species, as well as some relict interstitial forms (Ćurčić et al. 2008); Aranea (Fig. 3a)—cave component of spider fauna includes several genera and species of different origin: *Folkia mrazeki*, *Stalagtia hercegovinensis*, *Stalitella noseki*, *Troglohyphantes lesserti*, *Troglohyphantes pretneri*, *Typhlonesticus absoloni* (Ćurčić et al. 2008); Pseudoscorpiones—numerous species (24 species, according to list by Ćurčić et al. 2008) found in many Montenegrin caves belong to the genera *Chthonius, Neobisium,* and *Roncus* (Ćurčić et al. 2008); Palpigradi—numerous undetermined species belonging to genera *Eukenenia* and *Pauropus* are occasionally found in cave habitats, mostly under rotten wood or under litter of organic origin (Ćurčić et al. 2008); Opiliones—Apart from a presently known troglophilous species, it is possible that some Montenegrin caves are inhabited by several species (like the travuniids) from the neighboring Herzegovina

a Aranea **b** Diplopoda

Fig. 3 **a, b** Some groups of dwelling invertebrates present in Montenegrin caves. *Photos* Natural History Museum of Montenegro Photo Archive https://pmcg.co.me/zbirka-beskicmenjaka/

(Ćurčić et al. 2008); Chilopoda—include three species *Eupolybothrus gloriastygis, Lithobius sketi*, and *Lithobius remyi* (Ćurčić et al. 2008); Diplopoda (Fig. 3b)—according to Ćurčić et al. (2008), this group is well represented in Montenegro by 10 species: *Macrochaetosoma bifurcata, Macrochaetosoma troglomontanum, Heterolatzelia nivalis, Typhloglomeris coeca, Typhloglomeris seuti, Polydesmus gradjensis, Polydesmus jugoslavicus, Brachydesmus herzegowinensis, Brachydesmus stygivagus*, and *Brachydesmus subterraneus* (Ćurčić et al. 2008). Collembola—seven species of springtails (belonging to five genera and three families) are known to occur in Montenegrin caves: *Heteromurus gradgensis, Heteromurus absoloni, Heteromurus anagastumensis, Lepidocyrtus serbicus, Pseudosinella joupani, Tomocerus gradjackae*, and *Oncopodura jugoslavica* (Ćurčić et al. 2008). Coleoptera—the most important terrestrial group in the caves of Montenegro. This old, geographically limited, and varied fauna includes over 80 species representatives of carabids, leiodids, and staphylinids in particular (Ćurčić et al. 2008).

2.1.3 Potential Presence of *Proteus anguinus* in Caves of Montenegro

The olm, *Proteus anguinus* Laurenti 1768 has always attracted the attention of many researchers due to its troglomorphic characteristics (Dumas and Chris 1998; Schlegel and Bulog 1997), longevity (Voituron et al. 2011), ecology (Durand and Delay 1981), and behavior (Uiblein et al. 1992). It prefers calm waters rich in oxygen and low temperatures that vary from 6 to 15 °C. It feeds on water shrimps, insect larvae, and molluscs. Lives up to 100 years. The olm is the only European troglobiont chordate species that inhabits the underground waters of the Dinaric Karst, ranging from Trieste in Italy, Slovenia, Croatia to Bosnia, and Herzegovina (Kletečki et al. 1999; Sket 1997). There are indications of its presence in Montenegro as well (Ćurčić et al. 2008).

Traditional techniques used for epigean species are often not feasible in the surveillance of subterranean fauna, especially in the case of rare or elusive aquatic species. A non-invasive survey method based on environmental DNA (eDNA) was developed to detect the presence of the red-listed cave dwelling amphibian, *Proteus anguinus*, in the caves of the Dinaric Karst (Voros et al. 2017). Using this method, it was successfully confirmed the presence of *P. anguinus* in ten caves in Croatia and detected the species for the first time in five others (Voros et al. 2017). This method was recently applied in the research of Montenegrin caves as part of the project named "Capacity building for conservation of the subterranean biodiversity of the Skadar/Shkodra Lake basin (Montenegro and Albania)". The results of this project reconfirmed the presence of proteus environmental DNA trace at several sites in Montenegro, pointing at the extension of its range as far as the NW edge of Skadar Lake (Năpăruş-Aljančič et al. 2018).

References

Barr, T.C. (1968). Cave Ecology and the Evolution of Troglobites. In: Dobzhansky, T., Hecht, M.K., Steere, W.C. (eds) Evolutionary Biology, pp. 35–102. Springer, Boston, MA. https://doi.org/10.1007/978-1-4684-8094-8_2

Bendik, N.F., Meik, J.M., Gluesenkamp, A.G., Roelke, C.E., Chippindale, P.T. (2013). Biogeography, phylogeny, and morphological evolution of central Texas cave and spring salamanders. *BMC Evolutionary Biology* 13, 201 (2013). https://doi.org/10.1186/1471-2148-13-201

Casale, A., Guéorguiev, V.B. (1994). *Albanotrechus beroni*, nuovo genere, nuove specie di Trechini cavernicoli di Albania (Coleoptera, Carabidae). *Bolletino del Museo regionale di Scienze naturali di Torino* 12(2), 413–423

Casale, A., Jalžić, B. (1999). *Croatotrechus* (new genus) *tvrtkovici* n.sp., a new species of eyeless trechine beetle from Gorski Kotar (Coleoptera, Carabidae, Trechini). *Natura Croatica* 8(2), 137–145

Casale, A., Jalžić, B., Lohaj, R., Mlejnek, R. (2012). Two new highly specialised subterranean beetles from the Velebit Massif (Croatia): *Velebitaphaenops* (new genus) *giganteus* Casale y jalžić, new species and *Velebitodromus ozrenlukici* Lohaj, Mlejnek & Jalžić, new species (Coleoptera: Cholevidae: Leptodirini). *Natura Croatica* 21(1), 129–153

Culver, D.C., Sket, B. (2000). Hotspots of subterranean biodiversity in caves and wells. *Journal of Cave and Karst Studies*, 62, 11–17

Ćurčić, S.B., Brajković, M.M. (2003). *Curcicia*, a new genus of endemic ground beetles (Trechini, Carabidae, Coleoptera) based on *Duvalius bolei* Pretner. *Archives of Biological Sciences*, 55(3/4), 27–28. https://doi.org/10.2298/ABS030427PC

Ćurčić, S.B., Brajković, M.M., Ćurčić, B.P.M., Mitić, B.M. (2003a). *Javorella*, a new genus of endemic ground beetles (Trechini, Carabidae, Coleoptera) from west and southwest Serbia. *Archives of Biological Sciences*, 55(1/2), 15–22. https://doi.org/10.2478/s11756-006-0082-0

Ćurčić, S.B., Brajković, M.M., Mitić, B.M., Ćurčić, B.P.M., (2003b). *Rascioduvalius*, a new genus of cave and endogean Trechines (Trechini, Carabidae, Coleoptera) from the Balkan Peninsula. *Periodicum Biologorum*, 105(4), 483–486

Ćurčić, S.B., Pavićević, D., Ćurčić, B.P.M. (2001). *Serboduvalius dragacevensis*, a new genus and a new species of Trechines from caves in southwestern Serbia (Trechinae, Carabidae, Coleoptera). *Archives of Biological Sciences*, 53(1/2), 51–56. https://doi.org/10.2298/ABS1004947B

Ćurčić, B.P.M., Decu, V., Juberthie, C. (2008). Cave-dwelling invertebrates in Montenegro. In: Makarov, S.E., & Dimitrijević, R. N. (eds.). Advances in Arachnology and Developmental Biology. Papers dedicated to Prof. Dr. Božidar Ćurčić, pp. 35–55. Inst. Zool., Belgrade; BAS, Sofia; Fac. Life Sci., Vienna; SASA, Belgrade & UNESCO MAB Serbia

Delić, T., Reboleira, A.S. (2021). Underground life and ecology of subterranean habitats. Speleomedit cave biology. https://repositorio.ul.pt/bitstream/10451/50837/1/SPELEOMEDIT%20Cave%20Biology.pdf. Accessed 29 July 2022

Dumas, P., Chris, B. (1998). The olfaction in *Proteus anguinus*: a behavioural and cytological study. *Behavioural Processes*, 43, 107-113. https://doi.org/10.1016/S0376-6357(98)00002-3

Durand, J.P., Delay, B. (1981). Influence of temperature on the development of *Proteus anguinus* (caudata: Proteidae) and relation with its habitat in the subterranean world. *Journal of Thermal Biology*, 6, 53-57. https://doi.org/10.1016/0306-4565(81)90044-9

Gibert, J., Culver, D.C., Dole-Olivier, M-J., Malard, F., Christman, M.C., Deharveng, L. (2009). Assessing and conserving groundwater biodiversity: synthesis and perspectives. *Freshwater Biology*, 54, 930–941. https://doi.org/10.1111/j.1365-2427.2009.02201

Gibert, J., Deharveng, L. (2002). Subterranean ecosystems: a truncated functional biodiversity. *BioScience*, 52, 473–481. https://doi.org/10.1641/0006-3568(2002)052[0473:SEATFB]2.0.CO;2

Holsinger, J.R. (1993). Biodiversity of subterranean amphipod crustaceans: global patterns and zoogeographic implications. *Journal of Natural History*, 27, 821–835. https://doi.org/10.1080/00222939300770501

Howarth, F.G. (1983). Ecology of Cave Arthropods. *Annual Review of Entomology*, 28, 365–389. https://doi.org/10.1146/annurev.en.28.010183.002053

Karaman, M. (2018). General report on cave habitat type within the project „Establishment of Natura 2000 network in Montenegro". Part of the IPA Project "Establishment of NATURA 2000 network" in Montenegro. Natura 2000, Standard Data Form. Explanatory Notes

Kletečki, E., Jalzic, B., Rada, T. (1996). Distribution of the olm (*Proteus anguinus*, Laur.) in Croatia. Mémoires de Biospéologie, 23, 227-231. https://doi.org/10.1371/journal.pone.0170945

Kozlova, E.V., Mazina, S.E. (2020). Biodiversity of Fungi in the photic and aphotic zones of Montenegro caves. *Aerobiologia* 36, 589–604. https://doi.org/10.1007/s10453-020-09654-8

Barović, G., Krivokapić, M., Nikolić, G., Mitrović, L., Vujanović, M. (2013). Plan upravljanja za zaštićeno prirodno dobro "Lipska pećina" u Prijestonici Cetinje, (Management plan for the protected natural resource "Lipa Cave" in the royal capital of Cetinje), Cetinje Capital

Lakota, J., Lohaj, R., Dunay, G. (2010). Taxonomical and ecological notes on the genus *Scotoplanetes* Absolon, with the description of a new species from Montenegro (Coleoptera: Carabidae: *Trechini*). *Natura Croatica*, 18(2), 229–242

Lohaj, R. (2013). A new species of the genus Duvalius sg. Neoduvalius from Montenegro with taxonomical remarks on the genus Duvalius (Coleoptera, Carabidae, Trechini). *ZooKeys,* 278, 91–104. https://doi.org/10.3897/zookeys.278.4650

Lohaj, R., Jalžić, B. (2009). *Minosaphaenops croaticus*, a new species of the cave dwelling Trechinae beetle from Croatia, with description of the male specimen of *Derossiella nonveilleri* Quéinnec (Coleoptera: Carabidae: Trechinae). *Natura Croatica*, 18(2), 229–242

Lohaj, R., Lakota, J. (2010). Two new genera and species of aphaenopsoid cave-dwelling Trechini beetles from Croatia and Montenegro (Coleoptera: Carabidae: Trechinae). *Natura Croatica,* 19(1), 77–97

Lohaj, R., Mlejnek, R. (2012). A new species of the genus *Acheroniotes* Lohaj & Lakota from Ljubišnja Mts., Bosnia & Herzegovina (Coleoptera: Carabidae: Trechini). *Natura Croatica*, 21(1), 155–163

Lefebure, T., Douady, C.J., Gouy, M., Trontelj, P., Briolay, J., Gibert, J. (2006). Phylogeography of a subterranean amphipod reveals cryptic diversity and dynamic evolution in extreme environments. Molecular Ecology, 15, 797–1806. https://doi.org/10.1111/j.1365-294X.2006.02888

Marmonier, P., Vervier, P., Giber, J., Dole-Olivier, M.J. (1993). Biodiversity in ground waters. Trends in Ecology and Evolution, 8, 392–395. https://doi.org/10.1016/0169-5347(93)90039-R

Monguzzi, R. (1993). *Dalmataphaenops* (n.gen.) *chiarae* (n.sp.) nuovo eccezionale Trechino troglobio della regione Dinarica e considerazioni sul genere *Aphaenopsis* G. Müller, 1913 *Natura Bresciana*, 28, 231–242

Năpăruş-Aljančič, M., Pavićević, M., Merzlyakov, L., Turk, T., Théou, P., Ulqini, D., Shumka, S., Aljančič, G. (2018). Capacity building for conservation of the subterranean biodiversity of the Skadar/Shkodra Lake basin (Montenegro and Albania), 20, 69-72

Peck, S.B. (1981). Evolution of cave Cholevinae in North America (Coleoptera: Leiodidae). *Proceedings of the Eighth International Congress of Speleology*, 2, 503-505

Quéinnec, E. (2008). Two new genera and species of ´aphaenopsoid´ cave beetles from the Balkan Peninsula and considerations about the evolutionary trends of the so-called „hyper- -specialised" trechines. In: Pavićević D., Perreau M. (Eds) Advances in the studies of the fauna of the Balkan Peninsula. Papers dedicated to the memory of Guido Nonveiller, pp. 157–176. Institute for Nature Conservation of Serbia, Belgrade

Quéinnec, E., Pavićević, D. (2008). A new species of cave-dwelling Trechine from the Eastern Herzegovina karst (Coleoptera, Carabidae, Trechinae). In: Pavićević D., Perreau M. (Eds) Advances in the studies of the fauna of the Balkan Peninsula. Papers dedicated to the memory of Guido Nonveiller, pp. 157–176. Institute for Nature Conservation of Serbia, Belgrade, Monographs

Quéinnec, E., Pavićević, D., Ollivier, E. (2008). Preliminary description of a new *Adriaphaenops* from Lebršnik Nountain (Herzegovina, BiH) (Coleoptera, Carabidae, Trechinae). In: Pavićević D., Perreau M. (Eds) Advances in the studies of the fauna of the Balkan Peninsula. Papers dedicated to the memory of Guido Nonveiller, pp. 154–156. Institute for Nature Conservation of Serbia, Belgrade, Monographs

Romero, A. (2011). The evolution of cave life. *American Scientist*, 9, 144–151

Rouch, R., Danielopol, D.L. (1997). Species richness of microcrustacea in subterranean freshwater habitats. Comparative analysis and approximate evaluation. *Internationale Revue der gesamten Hydrobiologie und Hydrographie*, 82, 121–145

Schlegel, P., Bulog, B. (1997). Population-specific behavioral electrosensitivity of the European blind cave salamander, *Proteus anguinus*. *Journal of Physiology*, 91, 75-79. https://doi.org/10.1016/S0928-4257(97)88941-3

Schmidt, F. (1832a). Beitrag zu Krain's Fauna. Leptodirus Hochenwartii, n. g., n. sp. *Illyrisches Blatt,* 3, 9-10

Schmidt, F. (1832b). Leptodirus Hohenwartii und Elater Grafii. In: Gistl. J. (eds.), (Faunus: Zeitschrift für Zoologie und vergleichende Anatomie.Vol. 1. pp. 83–84. - M. Lindauer, München

Sket B., Paragamian, K., Trontelj, P.A. (2004). Census of the obligate subterranean fauna of the Balkan Peninsula. In: Sket, B., Paragamian, K., Trontelj, P., (eds.). Balkan Biodiversity: Pattern and Process in the European Hotspot. pp. 309–322. Dordrecht, Springer. https://doi.org/10.1007/978-1-4020-2854-0_18

Soares, D., Niemiller, M.L. (2013). Sensory adaptations of fishes to subterranean environments. *BioScience,* 63, 274–283. https://doi.org/10.1525/bio.2013.63.4.7

Stoch, F., Galassi, D.M.P. (2010). Stygobiotic crustacean species richness: a question of numbers, a matter of scale. *Hydrobiologia,* 653, 217–234. https://doi.org/10.1007/s10750-010-0356-y

Uiblein, F., Durand, J.P., Juberthie, C., Parzefall, J. (1992). Predation in caves: the effects of prey immobility and darkness on the foraging behaviour of two salamanders, *Euproctus asper* and *Proteus anguinus*. *Behavijoural Processes*, 28, 33-40. https://doi.org/10.1016/0376-6357(92)90046-G

Voituron, Y., de Fraipont, M., Issartel, J., Guillaume, O., Clobert, J. (2011). Extreme lifespan of the human fish (*Proteus anguinus*): a challenge for ageing mechanisms. *Evolutionary biology*, 7, 105-107. https://doi.org/10.1098/rsbl.2010.0539

VoÈroÈs, J., MaÂrton, O., Schmidt, B.R., Gal, J.T., Jelić, D. (2017). Surveying Europe's Only Cave-Dwelling Chordate Species (*Proteus anguinus*) Using Environmental DNA. *PLoS ONE* 12 (1): https://doi.org/10.1371/journal.pone.0170945

Pedological Characteristics of Montenegro

Velibor Spalevic

Abstract

Speleology is a cross-disciplinary field that studies caves and other karst features, as well as their structure, physical properties, history, life forms, and the processes by which they form and change over time. The understanding of geology and pedology, combined with the knowledge of chemistry, biology, physics, meteorology, cartography, and related fields is also significant for the understanding of caves as complex, evolving systems. Pedology (from Greek: pedon, "soil"; and logos, "study"), defined by Amundsen, is a discipline within soil science which focuses on understanding and characterizing soil formation, evolution, often in the context of the natural environment. With the idea of contributing to a wider understanding of speleology in Montenegro, an analysis and cartographic presentation of the soils of Montenegro was made including physical–geographical background, mapping of soils and interrelationships between geomorphic processes, soil formation, biodiversity, land use, including soil erosion. The author's idea with this presentation was to contribute to the completion of a mosaic about Montenegro, with all the research reports of the related disciplines, to provide an integral presentation of the Speleology of Montenegro.

Keywords

Pedology · Geomorphology · Soils · Soil Erosion · Montenegro

V. Spalevic (✉)
Biotechnical Faculty, University of Montenegro, Mihaila Lalica 15, 81000 Podgorica, Montenegro
e-mail: velibor.spalevic@ucg.ac.me

V. Spalevic
Faculty of Philosophy, Geography, University of Montenegro, 81400 Niksic, Montenegro

1 Introduction

As defined in Britannica, soil is the biologically active, porous medium developed in the upper layer of earth in which plants grow, consisting of a mixture of organic remains, clay, and rock particles and serving as a reservoir of water and nutrients, as a medium for the filtration and breakdown of injurious wastes, and as a participant in the cycling of carbon and other elements through the global ecosystem. It has evolved through weathering processes driven by biological, climatic, geologic, and topographic influences. Soils differ widely in their properties because of geologic and climatic variations over distance and time. Even a simple property, such as the soil thickness, can range from a few centimetres to many metres, depending on the intensity and duration of weathering, episodes of soil deposition and erosion, and the patterns of landscape evolution. Soils have a unique structural characteristic that distinguishes them from mere earth materials and serves as a basis for their classification: a vertical sequence of layers produced by the combined actions of percolating waters and living organisms (Sposito 2022).

The most important pedological characteristics, including land degradation of Montenegro have been presented in the book Soils of Montenegro by Fustic and Djuretic (2000), but also, with regional analyses in the papers of Spalevic et al. (2021, 2020, 2017, 2016), Topalovic et al. (2018), Knezevic et al. (2018), Topalovic and Knezevic (2016), Salkovic et al. (2017), Skataric et al. (2021), Zejak et al. (2021), Jovovic et al. (2021), and Nyssen et al. (2014).

From a geographical point of view, Montenegro is generally divided into three regions which share climatic, lithologic, hydrographic, and vegetation characteristics: Coastal Montenegro, Central Montenegro North(east)ern Montenegro. Most of the soils in Montenegro are shallow, with low plant and nutritional potential. The lands of Montenegro are classified into five categories of effective fertility. Lands of high production capacity of I and II

G. Barovic (ed.), *Speleology of Montenegro*, Cave and Karst Systems of the World,
https://doi.org/10.1007/978-3-031-49375-1_7

categories cover only 1.5% of the total area of Montenegro; Land of medium fertility of III and IV category, 3%. Significant areas are lands with limited fertility, of V and VI categories, and they are represented by a quarter of the territory of Montenegro. Most of the lands are with low fertility, of VII and VIII categories, which cover approximately half of the territory of our Montenegro (49%). There are significant areas with infertile land, which is not classified in any rating class. Such areas cover 23% of the territory of Montenegro.

In the specific Montenegrin topographic conditions of high relief dynamics, under the combined influences of surface and groundwater, precipitation and temperature, vegetation that differently protects land from degradation in different regions of Montenegro, and under anthropogenic influence, many types of soils have developed. The most dominant are the following soil types: (1) Litosol and Regosol; (2) Calcomelansol; (3) Rendzina; (4) Humus silicate soil—Ranker; (5) District Cambisol (Brown acid soils); (6) Eutric Cambisol (Brown Eutric soil); (7) Kalko cambisol (Brown soil on limestone); (8) Terra Rossa; (9) Deluvial and alluvial soils.

This chapter summarizes information on the physical–geographical background of Montenegro with a short presentation on geomorphology, biodiversity, and land use. The main messages are in relation to the Soils of Montenegro including the land degradation processes, with special attention to the Soil Erosion in Montenegro.

2 Physical–Geographical Background of Montenegro

In terms of landforms, Montenegro (Crna Gora) shows great variability, ranging from sand and rock coasts (some corresponding to a ria coast), karst plateaus, large (intramontane) polja, high mountains holding a (peri) glacial imprint, canyons, and more; all of it comprised in only 13 812 km^2 and within an elevation range of 2535 m. It is characterized by a Mediterranean climate, with warm and dry summers and autumns, and relatively cold winters with heavy snowfall in the continent. The coast is very indented with a narrow coastal belt that is in the hinterland with rugged high limestone mountains and plateaus. The average altitude is 1,086 m; lowest point: Adriatic Sea 0 m; highest points: Prokletije—Zla kolata, 2534 m.asl., Dobra kolata, 2528 m; Durmitor Kuk—Bobotov kuk, 2,522 m (Spalevic et al. 2020; Nyssen et al. 2014).

The structure of land use is as follows: agricultural land: 38.2% (estimated in 2018); arable land: 12.9% (estimated in 2018); crops: 1.2% (estimated in 2018); pasture: 24.1% (estimated in 2018); forest: 40.4% (estimated in 2018); other: 21.4% (estimated 2018); Irrigated land 24 km^2 (2012) (Spalevic et al. 2020).

3 Geomorphology of Montenegro

From a geographical point of view, Montenegro is generally divided into three regions which share climatic, lithologic, hydrographic, and vegetation characteristics: Coastal Montenegro, Central Montenegro North(east)ern Montenegro.

Coastal Montenegro. This region consists of a narrow zone along the coastline, mainly characterized by steep limestone mountains rising from the sea to an elevation of ca. 800 m a.s.l. and representing a ria coast. The climate is Mediterranean: wet and mild winters and long, hot and dry summers. However, it has to be noted that the climate depends strongly on elevation; the Krivošije area above Kotor is among the wettest of Europe, receiving on average 4600 mm y^{-1} due to orographic precipitation. The Bojana on the boundary of Albania is the largest river in the coastal part, draining Skadar Lake (basin of 530 km^2 shared by Montenegro and Albania) into the Adriatic Sea. Vegetation in Coastal Montenegro is dominated by Mediterranean shrubs (Wraber 1983; Nyssen et al. 2014).

Central Montenegro. Central Montenegro consists of a karst plateau dissected by a depression filled with fine-grained sediments. Because of its fertility, it is densely populated and economically very important; and houses the country's two largest cities Podgorica and Niksic. In the lower parts of Central Montenegro, the climate is still Mediterranean, which becomes more continental in inland valleys. To the North, the continental character of the climate increases as the Dinaric Mountains form a barrier between the Mediterranean and continental air currents. As it is mainly a karst plateau, the hydrography in this region is very complex and numerous basins drain to subsurface networks. Most rivers within this territory drain to Skadar Lake, with the Moraca and the Zeta being the two main rivers. Concerning vegetation, Mediterranean forests are dominant (Wraber 1983; Nyssen et al. 2014).

Northern and North-Eastern Montenegro. The N and NE parts of Montenegro are very mountainous, with the presence of deep incised valleys (in Limestone Mountains) but also a rather hilly, quite densely populated part with the towns of Pljevlja and Bijelo Polje. In this region, the highest peaks of Montenegro are found, with Bobotov Kuk in the Durmitor Mountains (2522 m a.s.l.) and Zla Kolata in Prokletije Mountains (2535 m a.s.l.). The climate is continental, i.e. with cold winters and warm, wet summers.

Rivers in this region drain to the Black Sea. Some of these rivers (Tara, Piva) form deep canyons in limestone formations, but further downstream they form broader valleys flowing through softer Paleozoic material. This region represents different vegetation types such as subalpine deciduous forests, pine forests in higher areas, and alpine herbs on the high plains (Wraber 1983; Nyssen et al. 2014).

Starting from the Adriatic coast, seven more or less parallel NW–SE geomorphological regions could be distinguished: Coastal Montenegro, High Karst, Inland Depression, Durmitor Flysch, Northwestern Highlands, Prokletije, and Northern Crystalline Hills. Geomorphological regions of Montenegro are presented in Fig. 7.1.

Coastal Montenegro. Two NW–SE oriented units are aligned in Coastal Montenegro: de Budva zone consists of Triassic limestone and the Dalmatian zone represents different linear structures consisting of Cretan-Eocene limestone (anticlines) and flysch deposits (synclines), also called the Para-autochthonous zone (Zivaljevic 1989).

The High Karst Zone. The High Karst zone is the largest and most homogenous geomorphological region of Montenegro. Geologically it consists of thick-bedded and heavily fractured Cretaceous limestone (Zivaljevic 1989).

Inland Depression. In the Inland Depression, four different zones are discerned: the plain of Podgorica on the debris fan of the Moraca River, the Zeta-valley, and the high polje of Nikšić and Skadar Lake.

Durmitor Flysch. The lithology here is relatively uniform and consists of flysch; a sequence of conglomerates, sandstones, siltstones, marls, and limestone sediments dated to the Cretaceous and Paleogene.

Northwestern Highlands. These highlands are part of the Durmitor geotectonic unit or Durmitor nappe, that is overthrusted on the Kuci geotectonic unit (Morley 2007). The lithology of this region is complex but mainly

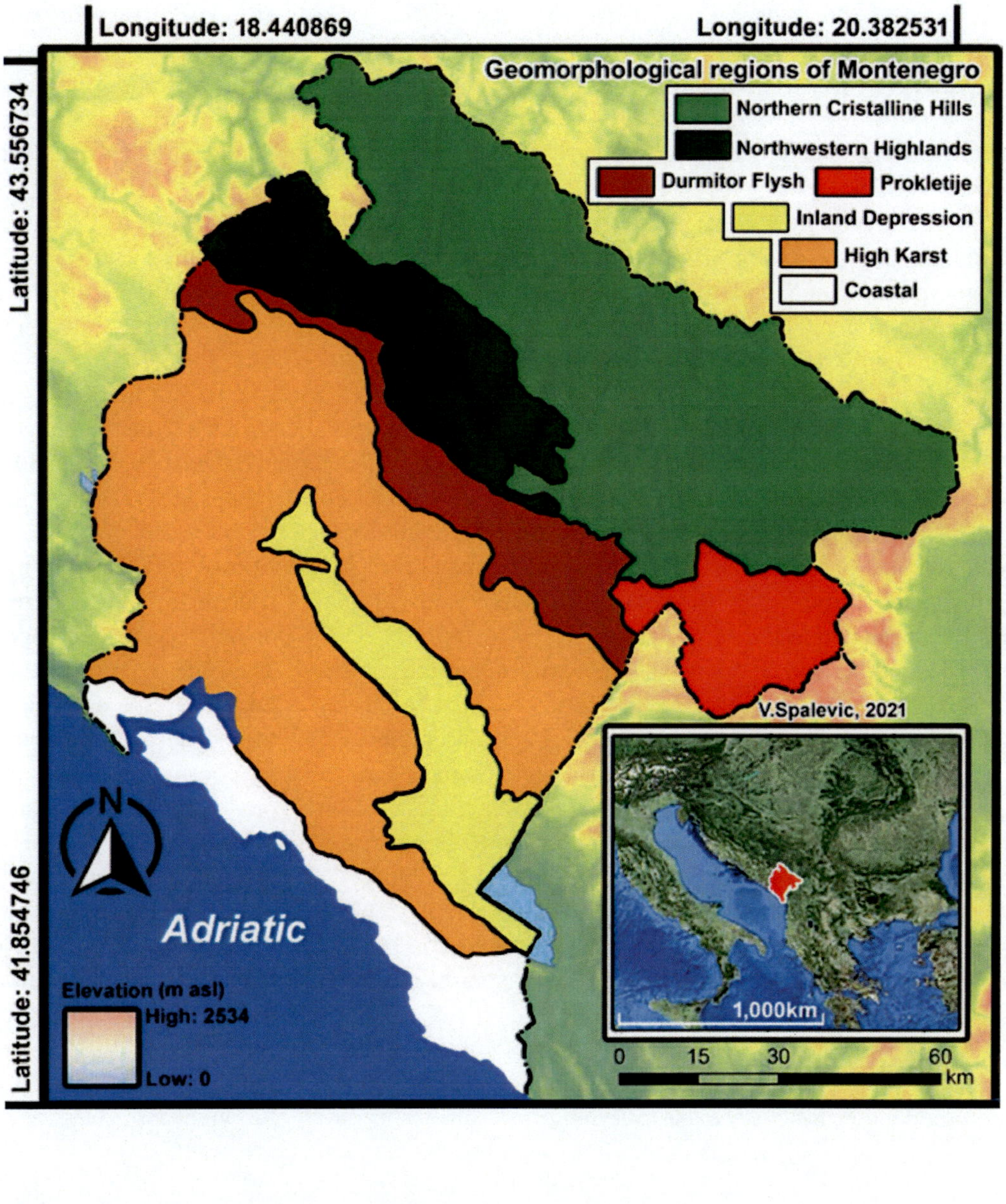

Fig. 7.1 Geomorphological regions of Montenegro. *Source* Original, Velibor Spalevic (2021), based on own research and research completed with the team from the Gent University: Frankl et al. (2016)

represents Triassic limestone in the Durmitor and Piva area and Jurassic limestone to the east of Durmitor.

Prokletije. Geologically, the geomorphological region of Prokletije is also part of the Durmitor geotectonic unit. The lithology of the area is very complex with the delineation of three parts: the western part of Prokletije region (around Komovi mountains) consists of Tertiary sandstones and limestones and additionally volcanic outcrops identified as "keratophyre and quartz-keratophyre" on the geological map (Mirkovic et al. 1985). The South part, close to the village Vusanje, consists of Mesozoic limestones and dolomites. The East part, to the Southeast of Plav, is more complex with Paleozoic (Devonian and Carboniferous) phyllites and schists and Permian–Triassic conglomerates.

The Northern Crystalline Hills. This region includes Biogradska Gora National Park, the Bjelasica mountains, and the North zone forming the border with Serbia around Pljevlja and Bijelo Polje. Flysch and sandstone sediments are dominant, although carbonate rocks are still abundantly present (Spalevic et al. 2020; Nyssen et al. 2014).

4 Soils of Montenegro

Soil is a physical space—pedosphere, which was formed under certain topographic conditions (relief) and conditioned by: geology (parent substrate and hydrogeological conditions), hydrosphere (surface and/or groundwater), climate (precipitation and temperature), vegetation (natural and anthropogenic biocenoses), and the results of anthropogenic activities such as hydromelioration, terracing, fertilization, tillage, and similar.

Agricultural land, with a total area of 515,740 ha, covers 37.4% of the total territory of Montenegro. In relation to the number of inhabitants, it amounts to 0.79 ha per inhabitant, which at first sight represents an important resource for the development of agriculture. After Northern Ireland (1.36 ha/inhabitant), according to this data, Montenegro is ahead of all European countries.

The type of use of the land is of great importance for agricultural production, the resistance of the land to degradation and its consequent hydrological responses, about which numerous local and international authors have written about the territory of Montenegro. However, the structure of agricultural land use is unfavourable, because pastures (323,953 ha) and natural meadows (126,990 ha) have a dominant share (87% in total). The share of arable land, orchards, and vineyards with an area of 62,154 ha or 0.095 ha/inhabitant is below the European average and the average of all neighbouring countries.

Due to the specific orography-dynamics of the relief, geological composition, and other related conditions, agricultural land is used very extensively.

Demographic discharge with depopulation of the rural area of Montenegro has a negative impact on valorization of pastures and meadows, which are gradually overgrown with forests, and these areas are converted into forest land.

Most of the soils in Montenegro are shallow, with low plant and nutritional potential. The lands of Montenegro are classified by five categories of effective fertility. Lands of high production capacity of I and II categories cover an area of 20,000 ha, or 200 km^2, which represents only 1.5% of the total area of Montenegro. Land of medium fertility of III and IV category covers an area of 40,000 ha and 400 km^2, which is 3% of the total area. Significant areas are lands with limited fertility, of V and VI categories, and they are represented in areas of about 350,000 ha, or 3500 km^2, which is a quarter of the territory of Montenegro. The majority of the lands have low fertility, of VII and VIII categories, which cover 640,000 ha, or 6400 km^2, which is approximately half of the territory of Montenegro (49%). There are significant areas with infertile land, which is not classified in any rating class. Such areas cover 312,000 ha, or 3120 km^2, which represents about 23% of the territory of Montenegro.

In the specific Montenegrin topographic conditions with high relief dynamics, under the combined influences of surface and groundwater, precipitation and temperature, vegetation that differently protects land from degradation in different regions of Montenegro, and under anthropogenic influence, many types of soils have developed, of which the most dominant are the following soil types: (1) Litosol and Regosol; (2) Calcomelansol; (3) Rendzina; (4) Humus silicate soil—Ranker; (5) District Cambisol (Brown acid soils); (6) Eutric Cambisol (Brown Eutric soil); (7) Kalko cambisol (Brown soil on limestone); (8) Terra Rossa; (9) Deluvial and alluvial soils. A map of basic soil types of Montenegro is presented in Fig. 7.2.

1. Litosol and Regosol have developed on igneous, compact rocks and scattered regolith, which in the process of mechanical decomposition give the stone fragments. They are represented in an area of 38,470 ha, or 385 km^2, which is an area of about 3% of the total area of Montenegro. They belong to the group of undeveloped or poorly developed soils in which skeletal fractions predominate (stone and gravels). The vertical profile of these soils is not more than 20 cm. They have a very low natural production capacity. Litosol using modern agricultural machines (milling machines) that can crush such stones, i.e. rocks, can be used, with implemented irrigation and the addition of more fertile soil in planting holes, for growing olives, vines, and so-called stone fruits. Regosol is formed by erosion of flysch and marly sediments. It is also an undeveloped soil primarily used for vineyards, but very good olive groves can also be

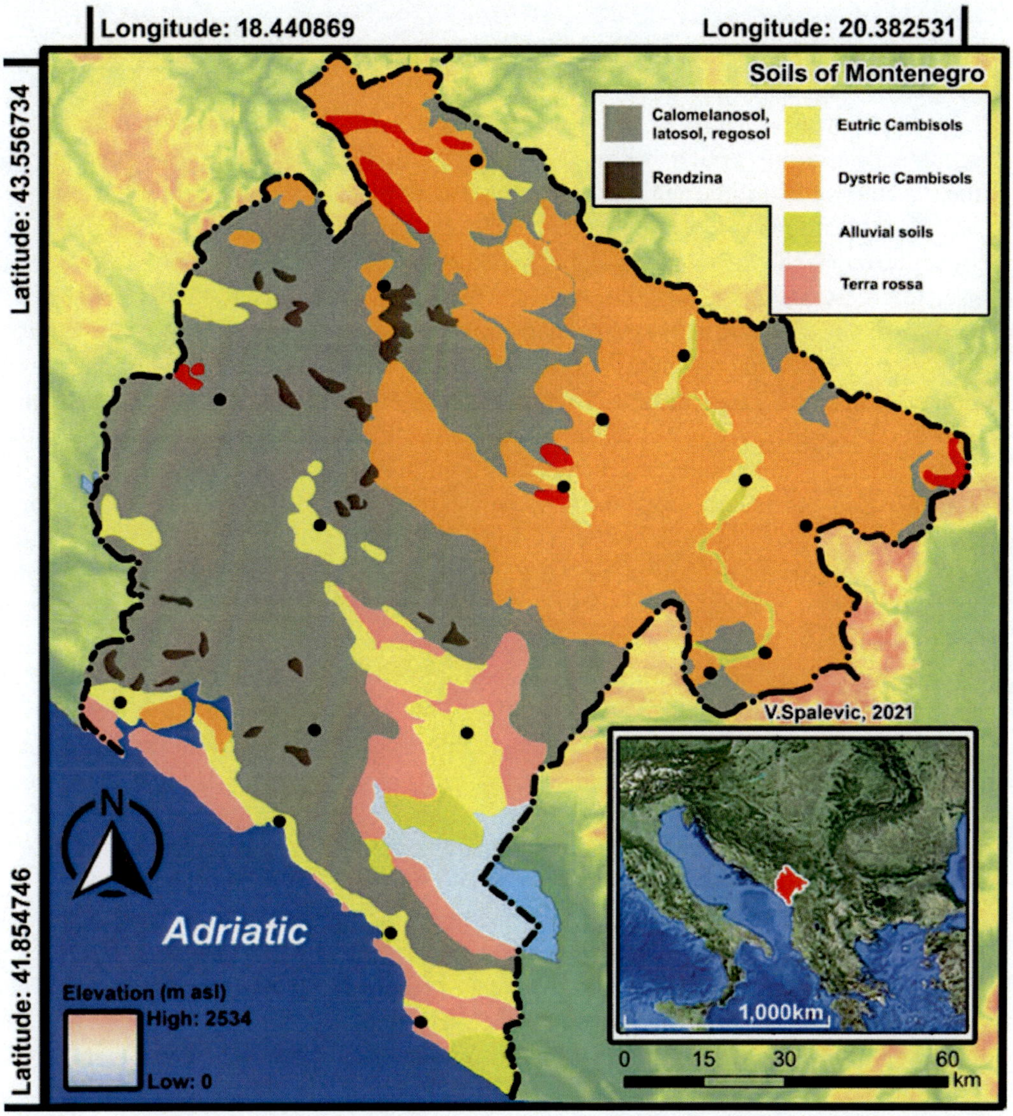

Fig. 7.2 Map of basic soil types of Montenegro. *Source* Original, Velibor Spalevic; based on collected data of previous field research of Djuretic and Fustic, 1964–1988 and Spalevic 1995–2021

found on them. Limitations in relief, soil erosion, and nutrient deficiencies, as well as a surplus of carbonates are the main limitations of these soils.

2. Calkomelansols (Limestone-dolomite) is the most widespread soil type in Montenegro and covers an area of about 660,000 ha, which is 6600 km^2 or 50% of the territory of the country. They are formed on hard and compact limestones and dolomites and belong to dry and warm, very porous soils with a high content of supracapillary pores. They are very permeable to water and well aerated, which often leads to a deficit of water in the soil for plants. These lands are mainly dominated by xerophytic vegetation. Plants on these soils often suffer from drought due to strong water permeability and shallow soil depth.
The reactions of this soil are neutral to slightly acidic. In the higher mountains, they are more acidic because the bases are washed away, and at lower altitudes, they are neutral to slightly acidic because there is no leaching of organic matter. They are rich in humus (10–25%), which also causes a high adsorption capacity. Most of these lands are under meadows and pastures, and a smaller part is used for growing vegetable and field crops (potatoes, rye, barley, wheat). These soils are not exposed to water erosion due to high water permeability. However, they are often exposed to aeolian erosion, especially if the plant cover is destroyed.
3. Rendzina (or rendsina) is a soil type recognized in various soil classification systems, including those of Britain and Germany as well as some obsolete systems. The term rendzina originated via Russian from the Polish rędzina, of unknown origin. They are humus-rich shallow soils that are usually formed from carbonate- or occasionally sulphate-rich parent material. In the World Reference Base for Soil Resources, Rendzina soils would be classified as leptosols, chernozems, kastanozems, or phaeozems, depending on their specific

characteristics. Rendzina soils are often found in karst and mountainous regions of Montenegro and cover insignificant areas (31,205 ha or 312 km^2), which represents 2.5% of the territory of Montenegro. This humus-accumulating soil is similar to limestone black soil, with the structure of its profile and properties, but it is formed on a loose carbonate substrate: marl, flysch, soft limestone, carbonate deluvium, and dolomite and had a lot of skeleton in its profile.

4. Ranker (Humus silicate soil) covers insignificant areas in Montenegro (6825 ha, less than 70 km^2, 0.5% of the territory). It is formed on silicate substrates above 1500 m above sea level, on non-carbonate, solid rocks. It is characterized by a very acidic reaction and high humus content. Rankers are the soils with moderate to large restrictions.
5. District Cambisol (Acid brown soil), covers an area of 394,825 ha (about 4000 km^2, or 30% of the territory of Montenegro). It is most widespread in north-eastern Montenegro. It is limited by poor pedochemical properties, primarily with high acidity, which is a limitation for many cultivated plants. Due to its geographical position, relief, erosion, and acid reaction, but also having a significant skeletoid profile, it shows high fertility only in some smaller, flat localities. The large distribution of District Cambisols is influenced by the large share of silicate rocks in the geological structure of Montenegro. Among them are significantly represented sandstones and shales of Paleozoic and Mesozoic age, eruptive rocks of the Triassic and diabase hornblende formations, then hornblende, quartzites, breccias, and conglomerates as well as Upper Cretaceous or the so-called Durmitor Flysch. It is spread mosaically on the flattened foothills and terraces of the River Lim, Tara River, and Cehotina River basin.
6. Eutric Cambisol (Brown Eutric soil), covers an area of 118,275 ha (about 1200 km^2, or about 9% of the territory of Montenegro), covers the lowest parts of river valleys (old river terraces), valleys, and karst fields. Eutric brown soils are soils with average or good physical and chemical properties. The variability of fertility and usability of these soils depends mostly on the parent substrate and relief. Loose parent substrates deepen the ecological depth of this type, so these soils are a very good habitat for growing olives and similar crops. The most fertile soils of this type are developed on marly flysch sediments, which often need to be deepened by rigging.
7. Calkocambisol (Brown soil on limestone) is formed on karstified pure limestones, dolomitic limestones, and dolomites. In Montenegro, it covers an area of 35,000 ha (350 km^2, 2.5% of the territory of Montenegro). It is formed mainly on milder forms of relief where the natural vegetation is deciduous forest shrubs and grass communities in the lower zones, and in the higher mixed coniferous forests and grasslands. There is very little arable land on this type of land, so it is often called forest land. The profile of Calcocambisol is mostly with medium depth, less often shallow or deep. It is mostly with clayey and some skeletons in the profile. Calcocambisol is poor in phosphorus (rarely with more than 10 mg per 100 g of soil). Potassium content is medium. On the vertical soil profile, the potassium content gradually decreases. The varieties of deeper soil in sinkholes have the highest production value and fertility.
8. Terra Rossa is from the Mediterranean karst areas, i.e. from the terrains built of limestone that receive a significant amount of precipitation during the year. It covers an area of about 84,000 ha (840 km^2, which represents 6.5% of the total territory of Montenegro). It is distributed on the Montenegrin coast in the Basin of Skadar Lake up to about 500–600 m above sea level. Terra Rossa is heavy soil, compacted, with little humus content. It easily absorbs and retains water for a long time, which allows the plants to survive on it even during the dry, hot, and long Mediterranean summer. This type of soil is formed by dissolving limestone and dolomite and represents its non-degradable residue. Terra Rossa occurs in Montenegro from Herceg Novi to the mouth of the Bojana to Adriatic Sea but also on the slopes of the southern shore of Skadar Lake, in the Zeta and Bjelopavlici plains. The humus content is about 1–3%. It contains oxides of iron and aluminium, which give it a red colour. At higher altitudes, it turns into groves and Podzols thanks to the increase in the amount of precipitation. Terra Rossa is suitable for growing vines and fruits (figs, olives).
9. Deluvial and Alluvial Soils are represented by about 43,500 ha (435 km^2, 3.5% of the total territory of Montenegro). Alluvial soil (aluvion, lat. Alluvius) is porous soil of fluvial origin. The process of its formation begins with erosion, continues with the reshaping of water and transport on the slope, and ends with the deposition or formation of alluvial sediments. Alluvium usually consists of various materials such as small particles of silt and clay, or larger particles such as sand and gravel. Deluvial and Alluvial lands occupy the lowest terrains, foothills, and alluvial plains along watercourses and on the shores of Šasko, Plav, and Skadar lakes.

Given that Montenegro has limited resources in terms of areas of high fertility land (200 km^2, 1.5%) and medium fertility land (400 km^2, 3% of the total area), the society must oppose land degradation in Montenegro. Pressures by sectors are from (1) agriculture (now less than in previous periods) and (2) inappropriate forest management.

The pressures come from (3) industry (now less than in earlier periods) and (4) energy sector. Recently (5) traffic; (6) urbanization; (7) the impact of the mines; and (8) the impact of landfills, significantly affect the degradation of land in Montenegro. The scientists at the global level consider (9) land degradation caused by climate change. This problem should be studied more seriously and this problem should be addressed in an organized manner, including multidisciplinary teams of experts, using the experiences of advanced developed countries.

5 Biodiversity and Land Use in Montenegro

The diversity of the geological base, landscape, physical–geographical characteristics, climate, and soil, and the position of Montenegro on the Balkan Peninsula and Adriatic Sea, created the conditions for high biological diversity. The number of species per area unit index in Montenegro is 0.837, which is the highest index recorded in the European Continent. Biological estimates suggest that over 1,200 species of freshwater algae, 300 species of marine algae, 589 species of moss, 7,000–8,000 species of vascular plants, 2,000 species of fungi, 16,000–20,000 species of insects, 407 species of marine fish, 56 species of reptile, 333 species of regularly visiting birds and a high species diversity of mammals are found in Montenegro.

Montenegro can be divided into two main biogeographic regions, which include the Mediterranean Biogeographic Region and the Alpine Biogeographic Region. It is also home to three terrestrial ecoregions: Balkan mixed forests, Dinaric Mountains mixed forests, and Illyrian deciduous forests. The total share of protected areas in Montenegro is 9.05% of the country's area, which mainly comes from the five national parks of Montenegro.

Map of the Land use of Montenegro is presented in Fig. 7.3.

6 Soil Erosion in Montenegro

Soil erosion is a natural process in which water and wind transfer soil material from one place to another. Water and wind in their movement have a certain mechanical energy with which they move the components of the soil and carry them downstream. In doing so, depending on a number of factors, erosion can occur normal (geological) or accelerated (excessive).

Very intensive soil erosion processes are taking place in Montenegro. These processes are not of the same intensity in certain narrower areas and regions. Some areas are, thanks to the prevailing natural (different soils, topography, geology, vegetation, land use, and climate) and anthropogenic factors in them, far more endangered by these processes, and thus the occurrence of far greater damage.

Montenegro consists of three clearly separated regions: coastal, central, and northern, and we can conclude that erosion processes are far more pronounced in the coastal and northern regions and much less in the central. An extremely important factor in the development of soil erosion is the condition of the vegetation cover, especially in areas extremely susceptible to these processes. This protective vegetation cover consists of numerous species of tall and low trees, shrubs, grasses, and other plants, whose roots bind loose soil material and with their developed canopies and leaves protect the soil from the destructive influence of raindrops, which greatly prevents or mitigates erosion processes.

The greatest kinetic energy that water receives in its circulation is that which raindrops have when they reach the ground because then the water has the highest speed. This energy must be destroyed by contact with the soil surface, which leads to a weakening of the cohesion of the soil and its preparation for further strong erosion. The plant cover destroys this energy in a very simple way, receiving the impact of raindrops on its above-ground organs and leaves and destroying their kinetic energy by friction. If we analyse this form of destruction of the kinetic energy of water, we will see that the role of plants in this regard is useful if they have more developed and lower crowns and if there are more leaves on the soil. Hence, young deciduous forests, and even grasses until they are mown, are the best protectors of soil from erosion. Rare large trees or forests with decorated crowns represent almost no protection of the soil from erosion.

Underground organs of plants, roots, and veins are another factor that opposes erosion. These organs provide the veins at a certain depth and thus perform its reinforcement, creating cohesion, so that the soil is no longer opposed to erosion by individual particles, but by a mass of connected particles. This factor is also stronger in the protection of soil from erosion if the network of veins is denser, smaller, and closer to the surface. On this basis, too, young forests and uncut grass represent the best protection of the soil from erosion.

The two factors described are related, as we have seen, to the destruction or reduction of the kinetic energy of water droplets and the increase of soil cohesion. After the water reaches the soil, a new movement of water occurs, depending on the slope of the land, which the first two factors have not changed. But the plant cover still acts strongly against erosion, creating greater roughness and greater porosity of the soil. The water slows down the movement and sinks, continuing further underground, very slow movement, harmless to the ground. This happens in two ways: first, the

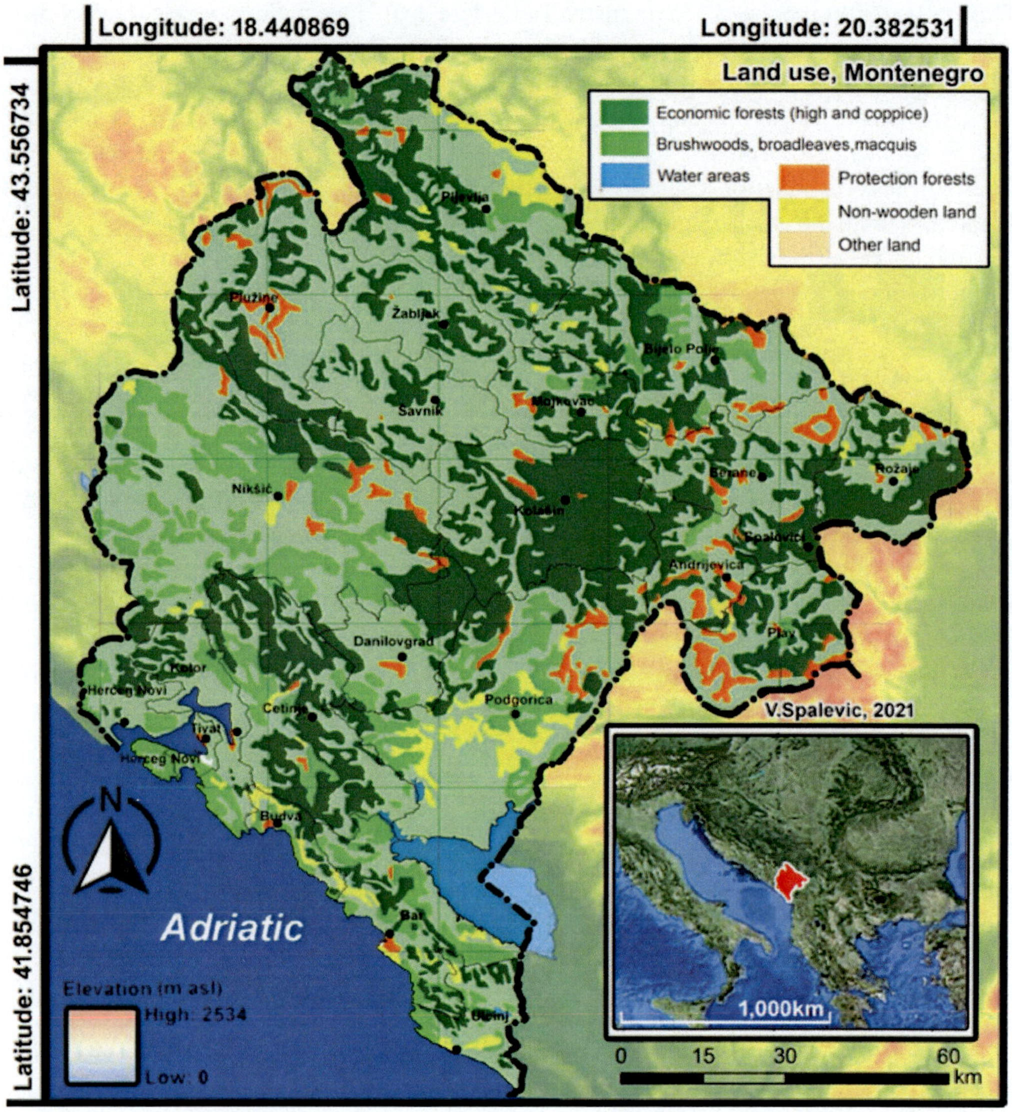

Fig. 7.3 Land use of Montenegro. *Source* Original, Velibor Spalevic

plants, developing a network of underground organs, create a significant porosity of the soil for the purpose of obtaining air and water from the surface; secondly, the plant leaves form a porous and rough humus cover. This type of soil is able to strongly absorb water, so there is no movement on the surface, and thus no erosion processes. Vegetation has a special beneficial effect on water runoff and water regime, and the coefficient of runoff from land areas covered with vegetation, due to stronger evaporation and plant nutrition, is lower than the coefficient of runoff from bare surfaces. Swelling occurs in a completely different, mild form, uniformly, without particularly pronounced highs and lows. Watercourses abundantly fed with spring and spring water never dry up, destructive water does not occur at all, and swelling is contained mainly in the form of useful water.

The Map of Soil erosion of Montenegro is presented in Fig. 7.4.

The most important processes of land degradation in Montenegro are the degradation caused by water erosion. Various types of erosion occur across Montenegro. According to the Spatial Plan of Montenegro (2008), the dominant hydrogeological characteristics of Montenegro are manifested in processes and occurrences of karst erosion (holokarst), which are followed by occurrences and processes of river and glacier erosion and coastal abrasion. Additionally, fluvial erosion processes are significant in river beds of the Piva, Gornja Moraca, Tara, Lim, and Ibar rivers. Erosion of agricultural land is also a problem across the country.

The Degradation category and Erosion process intensity in Montenegro are presented in Table 7.1 and some examples of land degradation and soil erosion processes on photos 5–10 (Figs. 7.5, 7.6, 7.7, 7.8, 7.9 and 7.10).

The processes of land degradation caused by soil erosion in Montenegro can be decreased by terracing the slopes in

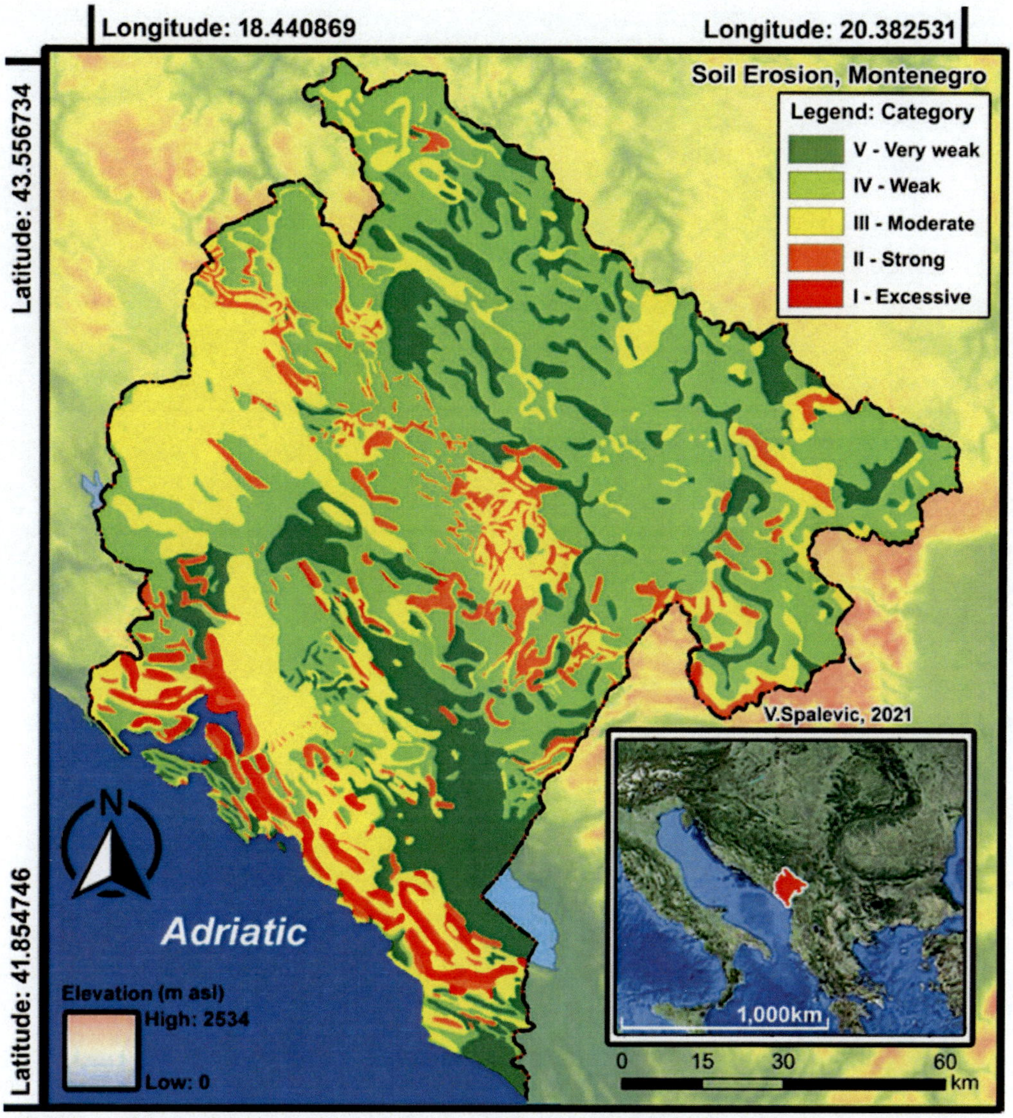

Fig. 7.4 Map of Soil erosion of Montenegro. *Source* Velibor Spalevic (2021)

the river basins; by introducing contour tillage and planting, and by selecting the proper crop rotation.

In relation to land degradation, control of the use of mineral fertilizers is needed, and the control of soil fertility but also on rational use of pesticides. The intensification of agricultural production will require the modernization of irrigation systems. It is necessary to work on soil conservation in terms of acidity and soil structure and to stimulate the introduction of a larger amount of organic and lime fertilizers. Education of farmers and agricultural producers plays a significant role, according to the experiences of other developed countries, where the application of anti-erosion measures and sustainable management of agricultural land and environmental protection play a significant role. One of the first steps in the fight against land degradation is the establishment of a database on agricultural land, using previous pedological research (1964–1988), by conducting monitoring and quality control.

The problem in relation to the land degradation Montenegro may face in the future is soil compaction due to the use of heavy tractors and machines during intensive land exploitation. By establishing permanent soil monitoring, the condition and loss of organic matter, i.e. depletion of soil in nutrients, should be closely monitored, and at the same time, biological degradation and soil pollution should be taken into account.

7 Conclusion

In the specific Montenegrin topographic conditions of high relief dynamics, under the combined influences of surface and groundwater, precipitation and temperature, vegetation that differently protects land from degradation in different regions of Montenegro, and under anthropogenic influence, many types of soils have developed, of which

Fig. 7.5 Sand pyramids in Miljevina near Foča, on the border between Bosna i Herzegovina and Montenegro. *Photo* Velibor Spalevic (2021)

Fig. 7.6 Đavolja varoš in Serbia. *Photo* Velibor Spalevic (2021)

Fig. 7.7 Scree slopes, Komovi, Montenegro. *Photo* Velibor Spalevic (2021)

Fig. 7.8 The first sediment deposits close to the mountain tops, Hridsko jezero, Bogićevica, Crna Gora. *Photo* Velibor Spalevic (2021)

Fig. 7.9 Small torrents easily transport sediments, Ravnjak, Montenegro. *Photo* Velibor Spalevic

Fig. 7.10 Exploitation of sediments from riverbeds, Tara, Montenegro. Photo Velibor Spalevic

Table 7.1 Degradation category and Erosion process intensity in Montenegro

Degradation category	Erosion process intensity	Erosion intensity	Area endangered	
		($m^3\ km^2\ y-^1$)	km^2	%
I	Excessive	>3000	180	1.3%
II	High	1500–3000	1,354	9.8%
III	Moderate	1000–1500	5,027	36.4%
IV	Weak	500–1000	6,367	46.1%
V	Very weak	70–500	884	6.4%
	Total		13,812	100

Source Water Resources Management Basic Plan of Montenegro (2001)

the most dominant are the following soil types: (1) Litosol and Regosol; (2) Calcomelansol; (3) Rendzina; (4) Humus silicate soil—Ranker; (5) District Cambisol (Brown acid soils); (6) Eutric Cambisol (Brown Eutric soil); (7) Kalko cambisol (Brown soil on limestone); (8) Terra Rossa; (9) Deluvial and alluvial soils. The most important processes of land degradation in Montenegro are the degradation caused by water erosion. Karst erosion (holokarst) is followed by occurrences and processes of river and glacier erosion and coastal abrasion. Fluvial erosion processes are significant in river beds of the Piva, Gornja Moraca, Tara, Lim, and Ibar rivers. Erosion of agricultural land is also a problem across the country. Establishing permanent soil monitoring should be taken into account.

References

Fustic, B.; Djuretic, G. (2000). Soils of Montenegro. University of Montenegro, Podgorica

Jovovic, Z.; Dolijanovic, Z.; Spalevic, V.; Dudic, B.; Przulj, N.; Velimirovic, A.; Popovic, V. (2021). Effects of Liming and Nutrient Management on Yield and Other Parameters of Potato Productivity on Acid Soils in Montenegro. Agronomy 2021, 11, 980

Knezevic, Mirko; Zivotic, Ljubomir; Cerekovic, Natasa; Topalovic, Ana; Kokovic, Nikola; Todorovic, Mladen. (2018). Impact of climate change on water requirements and growth of potato in different climatic zones of Montenegro. Journal of Water and Climate Change. 9. jwc2018211. https://doi.org/10.2166/wcc.2018.211

Nyssen, J.; Van Den Branden, J.; Spalevic, V.; Frankl, A.; Van De Velde, L.; Curovic, M.; Billi, P. (2014). Twentieth century land resilience in Montenegro and consequent hydrological response; Land Degradation and Development. 25 (4): 336–349. https://doi.org/10.1002/ldr.2143

Salkovic, E.; Djurovic, Igor; Knezevic, Mirko; Vesna, Popovic-Bugarin; Topalovic, Ana. (2017). Digitization and mapping of national legacy soil data of Montenegro. Soil and Water Research. 13. https://doi.org/10.17221/81/2017-SWR

Skataric, G.; Spalevic, V.; Popovic, S.; Perosevic, N.; Novicevic, R. (2021). The Vernacular and Rural Houses of Agrarian Areas in the Zeta Region, Montenegro. Agriculture 2021, 11, 717. https://doi.org/10.3390/agriculture11080717

Spalevic, V., Barovic, G., Fikfak, A., Kosanovic, S., Djurovic, M., Popovic, S. (2016). Sediment yield and Land use changes in the Northern Montenegrin Watersheds: Case study of Seocki Potok of the Polimlje Region. Journal of Environmental Protection and Ecology: 17 (3): 990–1002

Spalevic, V., Lakicevic, M., Radanovic, D., Billi, P., Barovic, G., Vujacic, D., Sestras, P., Khaledi Darvishan, A. (2017). Ecological-Economic (Eco-Eco) modelling in the river basins of Mountainous regions: Impact of land cover changes on sediment yield in the Velicka Rijeka in Montenegro. Notulae Botanicae Horti Agrobotanici Cluj-Napoca: 45(2):602–610. https://doi.org/10.15835/nbha45210695

Spalevic, V., Zejak, D., Curovic, M., Glisic, I., Radovic, A. (2021). Analysis of the impact of fruit growing development on the intensity of soil erosion and runoff: Case study of Krusevo, Bijelo Polje, Montenegro. Agriculture and Forestry, 67 (2): 37–51. https://doi.org/10.17707/AgricultForest.67.2.03

Spalevic, V.; Barovic, G.; Vujacic, D.; Curovic, M.; Behzadfar, M.; Djurovic, N.; Dudic, B.; Billi, P. (2020). The Impact of Land Use Changes on Soil Erosion in the River Basin of Miocki Potok, Montenegro. Water 2020, 12, 2973. https://doi.org/10.3390/w12112973

Sposito, Garrison. "soil". Encyclopedia Britannica, Invalid Date, https://www.britannica.com/science/soil. Accessed 3 July 2022

Topalovic, Ana; Knezevic, Mirko. (2016). Status of Nutrients in Vineyards of Cemovsko Polje. Agriculture and Forestry, 62. 137–143

Topalovic, Ana; Knezevic, Mirko; Trifunovic, S.; Novakovic, Miroslav; Pesic, Milica; Djurovic, Dijana. (2018). Effects of soil properties and fertilization on quality and biological activity of Swiss chard. European Journal of Horticultural Science. 83. 374–381. https://doi.org/10.17660/eJHS.2018/83.6.5

Wraber, T. (1983). The chart of national potential vegetation of SFR Yugoslavia, Scale 1:1000000 – Skopje: Jovan Hadji, Biological Institute ZRC SAZU

Zejak, D.; Radovic, A.; Spalevic, V.; Glisic, I. (2021). Production of planting material of raspberry variety "Glen Ample" in the North Montenegro. Agriculture and Forestry, 67 (2): 245-259

Zivaljevic, M. (1989). Geological Map of Montenegro, 1:200000. Titograd, The Institute for Geological exploration of SR Montenegro

Protected Speleological Objects

Duško Vujačić, Đurđica Perović, Goran Grozdanić
and Filip Vujović

Abstract

Speleological objects, in the narrower sense, are considered to be caves and pits as underground forms of relief created by the karst process. Horizontal and subhorizontal underground channels are called caves. Pits are called vertical and subvertical channels. In the world (English-speaking world), the single term "caves" has long been accepted for both pits and caves. In addition to the morphology of the underground channels, these objects are particularly interesting, attractive, intriguing and, certainly, scientifically significant, due to the presence of water, accumulative forms—cave jewelry and flora and fauna. Montenegro has a great potential for speleological objects on its territory, but it has not yet been used in the right way. In recent years, works on the valorization of protected objects in Montenegro have become more intense. One part of Lipska cave was valorized for tourism a few years ago, and construction works are underway to valorize the tourist potential of Đalovića Cave.

Keywords

Caves · Pits · Canals

D. Vujačić (✉) · Đ. Perović
Faculty of Tourism and Hospitality Management, University of Montenegro, Kotor, Montenegro
e-mail: duskov@ucg.ac.me

Đ. Perović
e-mail: duda@ucg.ac.me

G. Grozdanić · F. Vujović
Faculty of Philosophy, Department of Geography, University of Montenegro, Nikšić, Montenegro
e-mail: gorangr@ucg.ac.me

F. Vujović
e-mail: filip.v@ucg.ac.me

1 Lipska Cave

The Lipska cave is the most famous Montenegrin cave. It is partially arranged as a tourist object. So far, 1200 m of it has been examined, and 450 m of it has been geodetically recorded. It is rich in cave jewelry that is partially damaged in the part up to 300 m from the entrance. Research established that it belongs to the unique system of Cetinje caves together with the Manastir Cave and the Obod Cave.

The Lipska Cave is located in the village of Lipa in Dobrski Selo, between Cetinjsko field and Dobrska Uvala. It is 7 km away from Cetinje and was created by the action of underground waters that flow from the Cetinje field towards Rijeka Crnojevića. The entrance to the cave is at an altitude of 490 m asl. It was formed in Cretaceous limestone at its contact with dolomites. This is a very developed speleological object, the length of the cave channels explored so far is about 3500 m, while the vertical distance between the highest and the lowest point in the object is about 300 m. The basic direction of the entire object is southeast–northwest. There are several units in the cave: Entrance channels, Njegoševa Hall, Crystal Hall and the corridor with stone blocks. Further research on this object should be aimed at widening several "bottlenecks", especially in the extreme part of the cave, in order to pass into the lower levels of the object. About 120 m from the entrance is a large three-pronged hall, called Njegoševa Hall, which is also the largest room in Lipska cave (Radojičić 2015).

The Lipska cave, like all the caves belonging to the Cetinje cave system (Cetinjska, Obodska and Lipska), was created by the water of the Cetinje river. The Cetinje River moved its course to lower levels by intensive dissolution of the substrate. Every year in the spring and autumn, a small underground lake with several tributaries and a drain appears in Njegoševa hall. The present-day channels of Lipska cave are flooded only on occasions when the inflow of water is very large, so the underground course of

G. Barovic (ed.), *Speleology of Montenegro*, Cave and Karst Systems of the World,
https://doi.org/10.1007/978-3-031-49375-1_8

the river cannot receive all the amount of water that comes into it. In the cave, there is a lot of jewelry represented by basins, large tubs, ponds, etc.

Proclaimed protected on 28/12/1968, Number of official list Sl. newspaper SRCG 30/68.

Republic Institute for Nature Protection.

Proclamation Act Decision no. 01–959 from 12.12.1968.

Category of protected area III—Monument of nature (Environmental Protection Agency).

2 Jama Duboki Do—Njeguši

It is located in Njeguši near Cetinje. The depth of the pit is 310 m. It has a periodic lake at 120 m and a siphon lake at the bottom, as well as cave jewelry in several places. It is significant as a geomorphological and hydrological object because of its connection with the springs near Kotor that were established by coloring.

The history of exploration of deep pits in Montenegro takes us back to the beginning of the twentieth century when the Austro-Hungarian government hired G. Lahner in 1916 to explore a pit in Duboka dol on the edge of the Krstac field near Njeguš for the purpose of water supply for the army and the population. At that time, researchers reached an unreal depth of 340 m and the Duboki cave was officially the deepest in the world at the time.

Proclaimed protected on 28/12/1968, Number of official list Sl. newspaper SRCG 30/68.

Republic Institute for Nature Protection.

Proclamation Act Decision no. 01–959 from 12.12.1968.

Category of protected area III—Monument of nature **(Environmental Protection Agency).**

3 Grbocica

The cave as a whole is rich in well-preserved jewelry. It was researched in a length of 1000 m. Several small permanent ponds have been established in it. For now, it is the most interesting cave in the system of Trnov caves.

It is the oldest in terms of geological origin, the highest in altitude and the most beautiful in terms of cave jewelry. The name Grbočica comes from the adjective deep, i.e. the Proto-Slavic form, which means hollow. The entrance to the cave faces northeast and is horizontal in shape, 2.7 m high and 9.5 m wide. It is overgrown with plants and oak trees, which survived the big fire in 2013. The entrance is located at about 80 m relative height, above the bottom of the bay where the village is located. The absolute height is 560 m asl. The entrance part of the cave is fossilized and divided into several levels, which represent a network of cave channels and corridors. The upper and lower levels of the cave are in different degrees of morphological evolution. The entrance is located on the highest level and opens into a large entrance hall, from which channels separate in three directions. The right channel is about 250 m long and has approximately the same level along its entire length. The middle channel is about 80 m long and ends with collapsed blocks. The left channel is actually the main channel of the cave. It maintains the entrance level at a height of about 600 m asl. The higher level is in the fossil phase, that is, the phase of creation of different cave forms. The Main Channel is particularly interesting, which after a few hundred meters turns into a canyon and goes deep into the cave. The canyon channel is completely covered with thick silt, although in some places it is more than 10 m deep. That part of Grbočica abounds with numerous speleological jewelry of various shapes, which could only have been created by an artist, i.e. the nature. The researchers reached a narrowing along the steep edge in the middle of the canal, which lead to a larger hall that did not have a name, so the researchers called it the Temple of Veles, after the Slavic god of the underworld. The lower levels of the cave, after passing through the hall of the Veles temple, represent a series of smaller and larger halls arranged in a cascade, some of which have a length of about 100 m. At the end of the third hall, there is a passage to the fourth hall at the very bottom. A slit-shaped channel separates above it. The lower level of the cave is younger and assumes a hydrological function (Kovačević 2018) (Fig. 1).

In the back hall, there is a visible cave water flow. During two speleologists' expeditions, minor leakage and pouring of water from the upper part of the cave through a narrow channel was visible, which slowly filled the siphon lake at the end of the hall.

During the expeditions, 795 m of the Grbočica channel were topographically recorded. A few years ago, paint was thrown into the siphon lake in order to trace the underground path of the water and to see where it appears—sources. Color appeared at several springs on Lake Skadar in Caterpillar Bay.

It should be noted that in the literature and according to all available information it is stated that Grbočica is 2,650 m long. However, it is necessary to add other channels, so Grbočica cave deserves further detailed research (Kovačević 2018).

Proclaimed protected on 28/12/1968, Number of official list Sl. newspaper SRCG 30/68.

Republic Institute for Nature Protection.

Proclamation Act Decision no. 01–959 from 12.12.1968.

Category of protected area III—Monument of nature **(Environmental Protection Agency).**

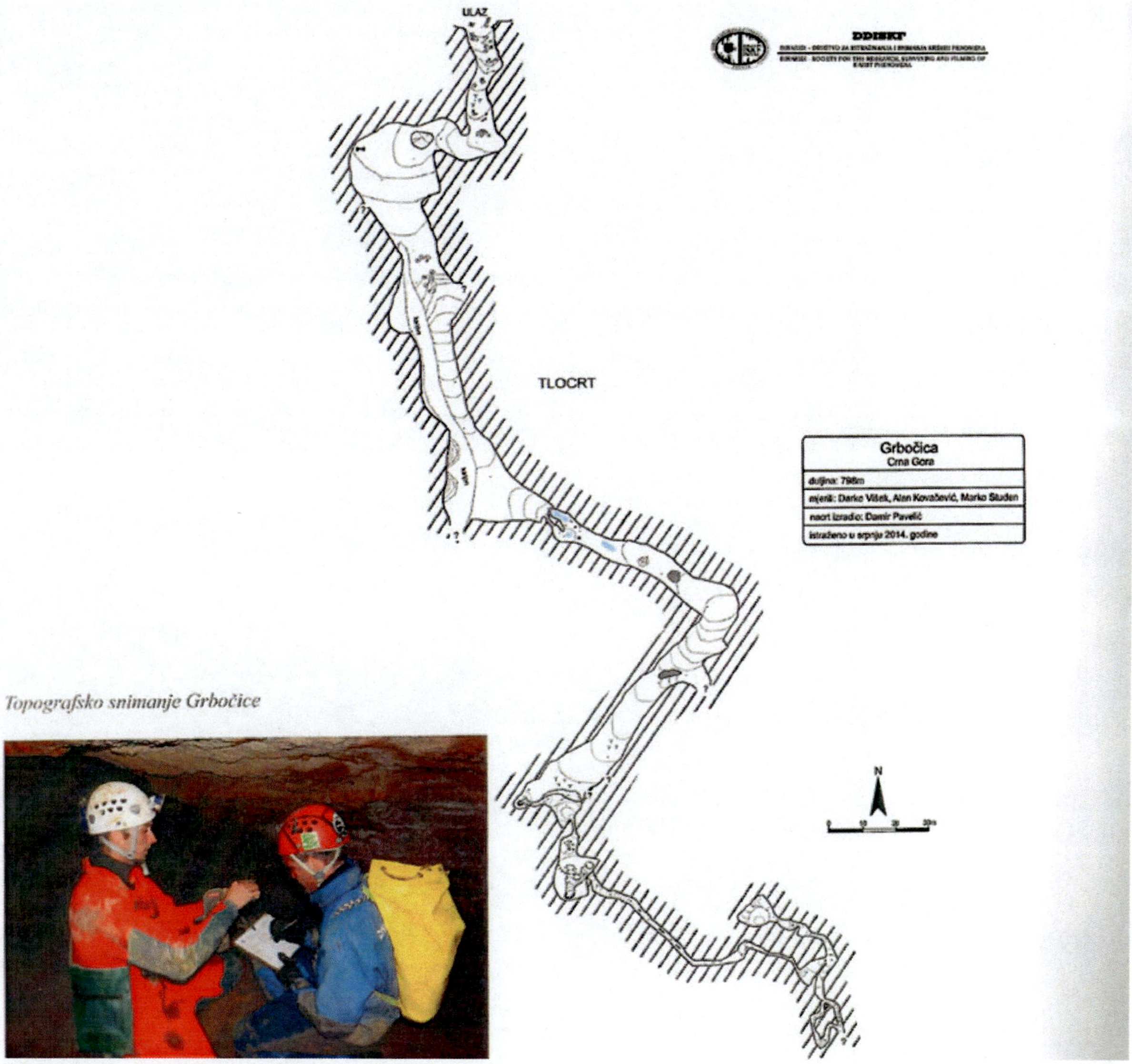

Fig. 1 Grbocica cave—draw. *Source* Kovacevic (2018)

4 Babatusa

It is located in the village of Trnovo near Virpazar. It is a low, short and quite dry cave. The hall is filled with clay deposits up to 10 m thick. It also belongs to the system of Trnov caves, Municipality of Bar.

Babatuša is actually a cave that connects the "supreme beauty" Grbočica with the lowest and "always working, active" Spila, which throws out water as a natural vent, most likely of all three caves. When Spila cannot expel all the accumulated water from the surrounding terrain and the water level in it suddenly rises, then Babatusa starts working. On several occasions, according to the story of the locals, it flooded Trnovac Bay together with Spila. The reason is that the surface sinkholes are still not large enough to receive such amounts of water, which comes up from the caves after heavy rains (Kovačević 2018) (Fig. 2).

Babatuša is in a transitional phase of development. In fact, it turns from a semi-active cave into a stable fossil beauty. Of course, this will happen after the Spila canals are developed and expanded enough to absorb all the water that will flow underground to the source on Lake Skadar or discharge it through the inlet.

According to the geological origin, the sequence of descent of the channel from Grbočica to Spila, via Babatuša, is logical. However, that connection has not been revealed yet, but we assume it will be revealed in the near future. The altitude of the entrance is 381 m asl. The entrance can be reached by a gentle climb along a path with stone markings, which leads towards Grbočica, through a thickly overgrown hornbeam grove. The entrance is a horizontal crack about ten meters wide and about two meters high.

Fig. 2 Babatusa cave. *Source* Kovacevic (2018)

Through fundamental paleontological and archaeological research, many moments from prehistory and the life of man in these rocks will certainly be revealed—especially in Babatusa. From the entrance hall, you pass by the collapsed blocks and go down the muddy slope to the Great Hall, which captivates with the beauty of the cave decorations in the upper part. The hall is specific in that a thick layer of mud descends in a funnel-like manner from the middle to the bottom. After the slow retreat of the water towards Spila, the mud settles in the Great Hall and other parts of Babatuša. The landslide inside the Great Hall and the mud funnel are also interesting. At the bottom of the funnel is a pool of seepage water, which slowly drains through a narrow canyon towards Spila (Kovačević 2018).

Proclaimed protected on 28/12/1968, Number of official list Sl. newspaper SRCG 30/68.

Republic Institute for Nature Protection.

Proclamation Act Decision no. 01–959 from 12.12.1968.

Category of protected area III—Monument of nature **(Environmental Protection Agency).**

5 Spila—Ispila

It is located in the area of the village of Trnovo near Virpazar and forms the lowest gallery in the system of Trnovo caves. An intermittent flow emerges from it. This cave, unfortunately, has no jewelry.

Spila is the only cave in the system in its active geological development. During high water, when heavy rains fall, or the snow that covers this area melts, a torrential stream flows from it, which after a short cascading flow descends to the bay, and through a series of sieve sinkholes disappears again and joins the underground flow, to the source on Lake Skadar.

It is the youngest of all three caves, but also the most interesting for researchers, due to its narrow crevice channels, filled with water for most of the year. The sudden arrival of water would be quite dangerous for the explorers. There is also a problem with Spila's name, as with Grbočica. Slovenian speleologists call it Ispila.

It is the closest cave to the settlement. It can also be reached by a path from the houses of Perović, along a stone wall, through a dry river bed, to the place where the path continues to Babatuša and Grbočica. The place is marked by a stone pile. With a few small jumps upwards, through an overgrown trough, you can easily reach its entrance. The dimensions of the entrance are 5 m width and about 3.5 m height (Kovačević 2018).

During the dry years, when all the springs would dry up, the inhabitants of Trnovo would go to Špila for water to feed their livestock and in many cases for drinking. During 2015, researchers dived through a siphon in the channel behind the sign for the junction of Špila and Babatuša. After diving through the siphon, they passed a passage to the right and entered a channel about 30 m long. The channel is 2 to 3 m wide, and its height is over 20 m. In essence, it is a crack, which ends in a 3 × 4 m siphon lake (Fig. 3).

It was newly topographically recorded at a length of 290 m. The dived-through siphon and the channel behind it are not drawn (Kovačević 2018).

Proclaimed protected on 28/12/1968, Number of official list Sl. newspaper SRCG 30/68.

Republic Institute for Nature Protection.

Proclamation Act Decision no. 01–959 from 12.12.1968.

Category of protected area III—Monument of nature **(Environmental Protection Agency).**

6 Magara

An attempt was also made to valorise the cave of **Me (a) gara** in Tološi near Podgorica. This object was adapted and its use is planned for the provision of health services;

Fig. 3 Spila—Ispila cave. *Source* Kovacevic (2018)

however, the realization of this project did not occur due to the disagreement of the locals with it (Habe F.1980).

To all this, we should add two objects whose interior has been turned into a place of worship. In the first place, it is the Ostrog monastery on the road Podgorica—Nikšić, and in the second is the Dajbabe monastery near Podgorica.

It is located in Velje brdo near Podgorica. The entrance is about 15 m from the field. It is difficult to pass because of siphon narrowings. In addition to jewelry, in the southern hall, there are many basins filled with water. It is characterized by complex microclimatic elements and high humidity that rises from the entrance to reach its maximum after the second siphon.

The first investigations of the Magara cave were carried out by Prof. Dr. Jovan Petrović with a team of collaborators back in 1965. In the book "Traces of the Cave Man", J. Petrović writes that it is a cave of "magic and terror". In the same book, it is pointed out that the air temperature, pressure and air humidity were so high that such characteristics do not occur in any cave in the world. In that attempt, they did not succeed in photographing the interior of the hall, as well as in several subsequent entries, because their photo equipment failed. In the literature, it is mentioned as Magara (Donkey), because the locals called it that, seeing as they used to imprison donkeys in it.

Megara Cave is located in Tološi near Podgorica. The cave channel is located at the foot of Velje brdo (283 m). The entrance to the cave is about 15 m above the level of the field, which is 40 m above sea level and is clearly visible from the road. The length of the tested channels in the building is about 160 m.

Proclaimed protected on 28/12/1968, Number of official list Sl. newspaper SRCG 30/68.

Republic Institute for Nature Protection.

Proclamation Act Decision no. 01–959 from 12.12.1968.

Category of protected area III—Monument of nature **(Environmental Protection Agency).**

7 Novakovica Cave

The mysterious stories about the Novakovica cave lured people with a passion in their hearts to overcome their fear and step into the underground world. Legend has it that countless treasures are hidden in the dark labyrinths, and the desire to find them was stronger for many than the darkness and the terrible stories about people who never returned.

What is the giant black ram guarding? The mythical creature in the bowels of the earth born previously by the flame of the imaginative storyteller's fire? Perhaps the shiny pearls and gold-bound monastery doors that the monks put away after committing themselves to a vow of silence. Maybe a secret passage to something more precious than gold itself. Or "just" a priceless natural treasure that nature has been growing for millennia. Drop by drop (Fig. 4).

Speleologists illuminated their valuables in the cave. Cave decorations, the creation of which requires so little, water dropping from the stone and so much time. Stalagmites, stalactites, basins, corals and draperies in Novakovica Cave were created in deep prehistoric times when caves were the only shelters for animals (Madžgalj 2011).

Due to these values and its natural features, Novaković Cave was declared a protected natural area by the Republic Institute for Nature Protection.

It is located 28 km from Bijelo Polje in the heart of the Vraneška valley. It is 25 km on the main road to the village of Muslići near Tomasevo and then three kilometers on a gravel road that leads from the village church to the spring of the Vranštica river and the site of Novakovića greda. Above the hot spring, 100 m along the left edge of the large hill, there lies the entrance to the cave.

The cave can also be reached by walking from the village church through the very beautiful canyon of the river Vranštica, which is three kilometers long. The path follows the course of a clear river with rapids and smaller waterfalls and offers nature lovers an unforgettable experience with numerous resting places (Madzgalj 2011) (Fig. 5).

At the end of the river course is the spring of Vranštica, which originates from the river Stožernica, which sinks on the Stožerska plateau. The Stožernica sinkhole is most likely the main "culprit" for the creation of the numerous corridors of the Novakovica Cave.

7.1 Cave Labyrinth

According to the morphological characteristics, Novakovica Cave is a branched speleological object. Due to the numerous intertwined channels, the locals had a special fear of entering the cave, which was reinforced by the stories that many curious people got lost and wandered for hours through the cave labyrinth, unsuccessfully trying to find a way out.

During numerous explorations of the cave, speleologists came across fluorescent markings, marks in the sand or stretched rope that treasure hunters left to help them return to the entrance. Fears were unwarranted as research, detailed measurements and speleological design showed that all channels are circular and rejoin the main channel (Madzgalj 2011).

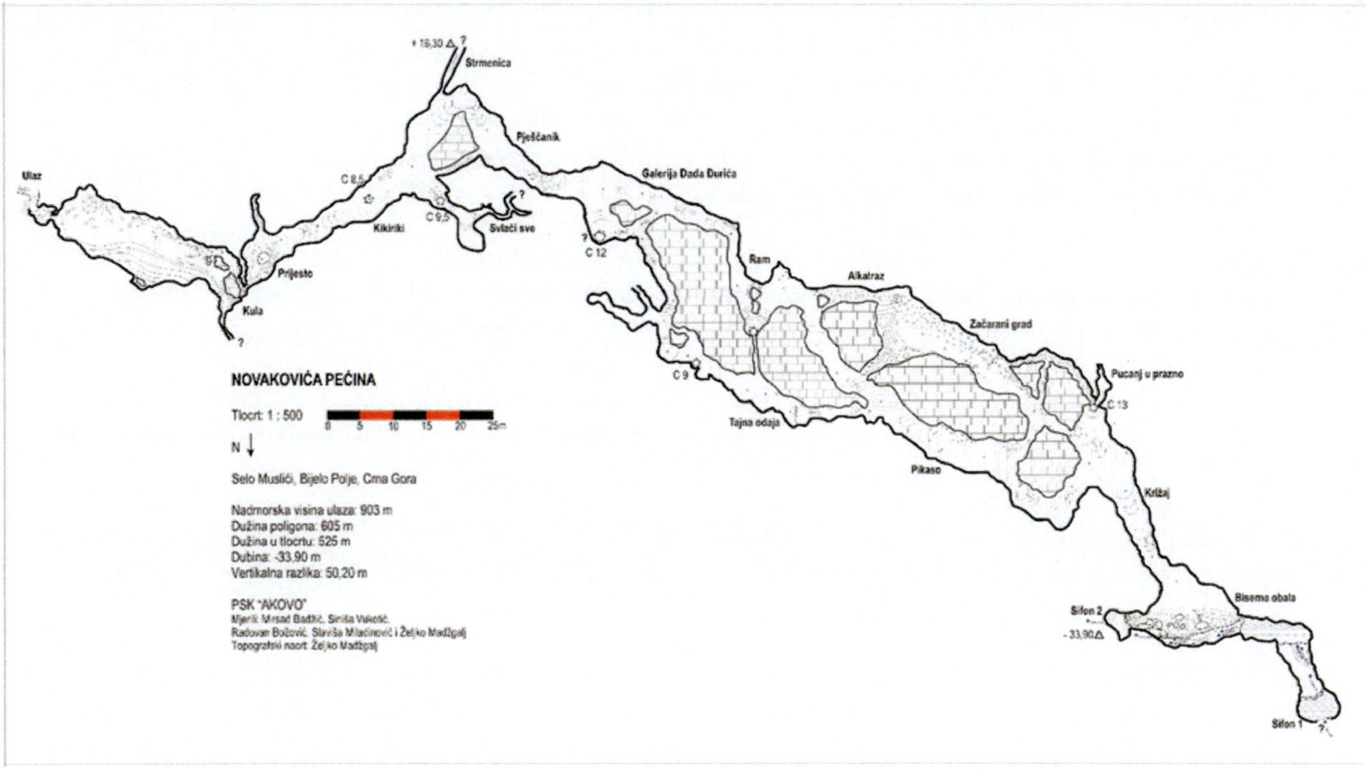

Fig. 4 Novakovica cave—draw. *Source* photo archive of the center for karst research and protection

Fig. 5 Novakovica cave. *Source* photo archive of the center for karst research and protection

7.2 Gallery in Honor of Dado Djuric

The current length of Novaković Cave is 605 m, which makes it one of the longest speleological objects. On the list of the longest caves in the municipality of Bijelo Polje, it is in third place after the extremely long Pećina above Vrazji firovi (about 20 km) and the "Vow of Silence" cave, which is about 2 km long.

The entrance has small dimensions (width—1.80 m, height—0.65 m). After a few meters of the narrow entrance channel, the cave "opens" into a larger room measuring 4 × 20 m. The cave is easy to pass and the only more difficult place is located not far from the entrance at a place called "Kula" where at the end of the narrow passage there is a small jump of three to four meters.

The most beautiful parts of the cave are the "Hidden Chamber" located in the upper part of the cave, as well as the parts called "Castle", "Picasso", "Ram" and "Dado's Gallery".

The highest elevation in the cave is in the "Strmenica" channel, and the lowest is at the end of the underground river course. The total height difference between the highest and the lowest elevation is 50.20 m (Madžgalj, 2011).

7.3 Preko "Tobogana" Do "Biserne Obale"

Novakovića Cave is a structure with a constant water flow, and according to its hydrogeological function, it represents a periodic spring (a jump spring—an intermittent spring).

The most important feature of the soil in the cave is the large deposits of sand brought in by water through numerous chimneys and windows as well as by the underground river flow. There is especially a lot of sand in the part of the cave called "Pješčanik" and "Tobogan". On the "Slide" (where the slope of the channel is greater than 30°) at one point there were so many deposits of sand that speleologists had to remove it in order to open the channel and pass to the "Pearl Coast" and the underground river course. The shorter water course begins and ends with siphons.

The potential for further research of the cave is significant and especially refers to the part behind the inflow siphon, as well as to numerous inaccessible windows and chimneys. Of the six chimneys, two have been expanded. The chimney, which is called "Lasso", was foamed by throwing a rope on a stalagmite located on the edge of the vertical. This chimney extended the cave by about 30 m. The chimney "Shooting into the void" was expanded by "killing" 6 bolts, but this direction quickly ended with an impassable narrowing.

Finally, the problem related to the devastation of the cave should be highlighted. Although the cave has legal protection as a natural monument, its accessibility, passability and poor ecological culture have led to great damage, breaking of cave decorations in the most beautiful parts of the cave (Madžgalj, 2011).

Measurements:

Entrance altitude: 903 m.

Length: 605 m.

Depth: 33.90.

Height difference: 50.20.

Air temperature: 8–8.7 °C.

Air humidity: 100%.

Water temperature in the underground stream: 8.4 °C.

The pH value of the water in the underground stream is 7.65.

7.4 Black Ram—Guardian of the Chamber with Gold

There are several caves in the cliff above the spring of the Vranštica river. Novakovića and Žuta pećina are one above the other, and according to tradition, these two caves are connected by canals. Access to the Yellow Cave is difficult and can only be entered using stairs or scaffolding. Various stories and legends are associated with this cave, and they all say that there is a hidden treasure deep in the dark corridors. Locals often tried to enter it but without success.

Somewhere around 1900, Mušo Muslić, from a nearby village, managed to enter it with the help of several neighbors. They took him up into the cave with the help of scaffolding and he stayed there for a long time. His company got scared, they called him but he didn't answer, in the end, they climbed up after him and pulled him out half dead. They made a stretcher and took him home where he was sick for 90 days. When he regained consciousness and recovered, he said that there were many chambers in the cave and that at the end he came across a lake over which a beam had been passed. There were some incomprehensible inscriptions on the wall and when he crossed the lake he came to a room where he discovered a great treasure guarded by a black ram with big horns. When he started to grab the treasure, the ram rushed at him and hit him hard with its horns. Fleeing through the cave, Muso fainted in front of the entrance. At that moment, his comrades dragged him outside, half dead (Madzhgalj 2011).

Proclaimed protected on 28/12/1968, Number of official list Sl. newspaper SRCG 30/68.

Republic Institute for Nature Protection.

Proclamation Act Decision no. 01–959 from 12.12.1968.

Category of protected area III—Monument of nature (Environmental Protection Agency).

8 Djalovica Cave

Is located in the basin of Bistrica (Bijelo Polje part of Bistrica), a left tributary of the Lim, on the Vražji firovi. It is the longest cave in Montenegro. The entrances to the cave are in the middle part of the Sušice valley, in the Đalovića gorge. The cave is about 15 km away from Bijelo Polje, in the northeast direction. The cave entrances can be reached via the village of Đalovići. The cave was discovered by speleologists of the Speleological Department of the Belgrade Mountaineering Association in 1987. Later, many teams of speleologists, from Montenegro, the Czech Republic and others, explored the cave. The determined length of the discovered channels of the cave is 17.5 km. Research has shown that it is an exceptional speleological object, rich in various speleological forms and hydrological phenomena. In the records of speleologists, it is stated that the speleological complex of Đalovića Cave has three entrances. The main entrance, from the Đalović gorge, is located at the bottom of the northern part of the amphitheater extension, at an altitude of 735 m above sea level. Here, the water gushes out of the cave. The second entrance to the cave, smaller in size, is called the Mala (Little) Cave. At a distance of about 120 m from the entrance, the channels join and in the extension there is a cave with quite rich "jewelry". The third entrance is above the Small Cave, high in the rocks and is difficult to access (Fig. 6).

Certainly the most promising, currently the longest and most beautiful cave in Montenegro is the cave above the Devil's Whirlpools, which is located in Đalovića gorge in the canyon of the river Bistrica in the municipality of Bijelo Polje. Tested so far, the 12 km long cave channel has almost all known types of cave jewelry, water-filled channels, cave lakes, siphons, halls over 30 m high, stalactites over 15 m high and so many other values that we rarely see in our caves that its beauty cast a shadow on the objects explored so far in Montenegro. It is the longest cave in Montenegro. The entrances to the cave are in the middle part of the Sušica valley, in the Đalovića gorge. The cave is about 15 km as the crow flies from Bijelo Polje, in a northeasterly direction. The cave entrances can be reached through the village of Đalovići. The cave was discovered by speleologists of the Speleological Department of the Mountaineering Association of Belgrade in 1987. Later, many teams, speleologists from Montenegro, the Czech Republic and others, explored the cave. The determined length of the discovered canals of the cave is 17.5 km. Research has shown that it is an exceptional speleological object, rich in various speleological forms and hydrological phenomena. The records of speleologists state that the speleological complex of Đalovića cave has three entrances. The main entrance, from the Đalovići gorge, is located at the bottom of the northern part of the amphitheater extension, at an altitude of about 735 m. The second entrance to the cave, smaller in size, is called the Small Cave. At a distance of about 120 m from the entrance—the canals connect and in the extension is a cave with quite rich cave jewelry. The third entrance is above the Small Cave, high in the rocks and difficult to access (Radojičić, 2015). Works on the tourist valorization of this facility are in progress, and soon this cave will also be arranged for tourism. Within the complex, in addition to the beauties inside the cave, the attraction will be the cable car to the entrance to the cave (Fig. 7).

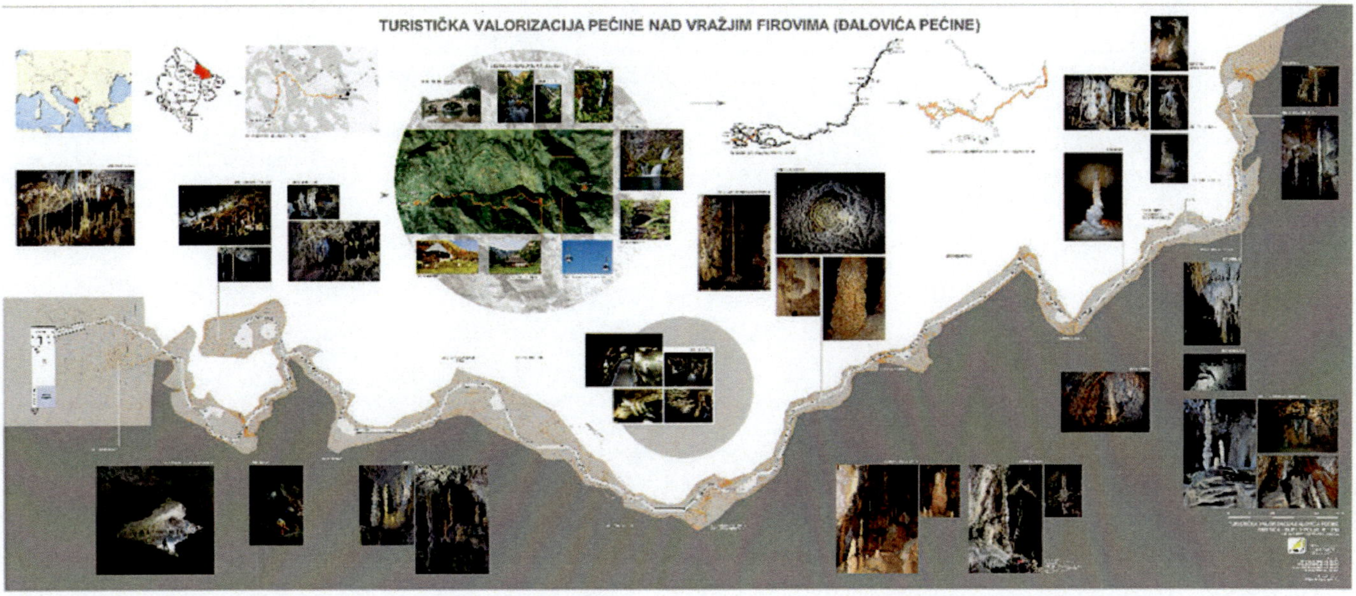

Fig. 6 Valorization map of Djalovica cave. *Source* Duško Vujačić

Fig. 7 Djalovica cave. *Source* Madzgalj (2021)

Literature and Sources

Environmental Protection Agency of Montenegro

Habe F. (1980). Tourist caves in Montenegro and their protection in older literature. VII YUGOSLAV SPELEOLOGICAL CONGRESS, Titograd

Kovačević T. (2018). Stopped time – Pećinski park and ethnic village of Trnovo

Madzhgalj Z. (2011). Speleological record BP—002, Novakovica cave

Radojicic B. (2015). Montenegro—Geographical encyclopedic lexicon, Faculty of Philosophy – Nikšić

Speleological Objects as Tourist Motives

Đurđica Perović, Duško Vujačić and Goran Grozdanić

Abstract

Speleological tourism is a specific type of tourism. This type of tourism, as a segment of the tourist offer, contains unique and unique motives, which are mostly available to a small number of people. Starting from the goal and the number of visitors, we can single out different subtypes of speleotourisms, such as esthetic, sports, educational and scientific research, which can be realized through individual and group visits. Montenegro, with its diverse underground world, which consists of thousands of speleological objects, has great potential for positioning speleotourisms and meeting the needs of this target group of tourists. The interior of speleological objects has always been a strong motive for exploration, but it has long been available to a small number of people who had the courage to step into the earth's interior and embark on the conquest of the unknown. With the progress of human society, there has been a positive shift in this type of human activity, and thus the provision of conditions for acquainting a larger number of people with these natural values. The tourist valorization of these tourist resources is very complex, both in bringing the facilities to the tourist purpose, and in training the staff who will lead the tourists in tours of these magnificent karst underground forms. Probably the only case in the world is Montenegro with its speleological objects. Thanks to the geological structure and the dominant representation of limestone terrains, speleological tourism can in a very high-quality way complement the tourist offer of Montenegro as a tourist destination. During the research of Montenegro so far, several hundred speleological objects have been examined, and several thousands have been recorded. Of all these, two objects deserve more attention for their characteristics.

Keywords

Speleotourism · Speleological objects · Tourists · Karst

1 Tourist Directions

On the main tourist route leading from Serbia to the Montenegrin coast (Bijelo Polje—Kolašin—Podgorica—Cetinje—Budva, and Podgorica—Petrovac) there are several facilities that should be made suitable for tourist use, and thus provide conditions for their wider acquaintance. There are two more tourist destinations that will fit into the already mentioned ones. The first, from Republika Srpska (Trebinje (Srbinje—Foca)—Niksic—and the second Dubrovnik—Herceg Novi—which have yet to gain importance by normalizing relations between the former Yugoslav republics and building a planned Adriatic-Ionian highway that should pass through this area). Another tourist route that has yet to gain importance with the construction of the road Pljevlja—Žabljak—Nikšić—Grahovo—Risan would bring the Durmitor massif, the archeological site Crvena stijena and the Orjen massif under tourist attack.

Caves that stand out in these routes are: The cave above the Devil's Whirlpools, Lipska and Cetinje, Đalovića Cave, Obod Cave, Sopot, Ice Cave, Red Rock Cave and Me(a) gara. Since the mentioned facilities are located on a busy tourist route, their activation would increase the overall tourist offer of Montenegro.

Đ. Perović (✉) · D. Vujačić · G. Grozdanić
Faculty of Tourism and Hospitality Management,
University of Montenegro, Kotor, Montenegro
e-mail: duda@ucg.ac.me

D. Vujačić
e-mail: dule.v@t-com.me

G. Grozdanić
e-mail: gorangr@ucg.ac.me

G. Barovic (ed.), *Speleology of Montenegro*, Cave and Karst Systems of the World,
https://doi.org/10.1007/978-3-031-49375-1_9

Certainly the most promising, currently the longest and most beautiful cave in Montenegro is the cave above the Devil's Whirlpools, which is located in Đalovića gorge in the canyon of the river Bistrica in the municipality of Bijelo Polje (Radojičić 2015).

The speleological map (Fig. 1) should also contain zones with characteristic landscapes, i.e. zones that have been shown in previous research as spaces with a very large number of speleological objects. First of all, there should be the massifs of Durmitor, Moračke planine, Lovćen and Orjen (Lješević et al. 1980).

When training a speleological facility for tourist use, it is necessary to take into account certain conditions that the facility must meet in order to be included in the tourist visit. Those are:

- appearance and size of the cave channel,
- quantitative and qualitative properties of jewelry,
- diversity of flora and fauna in the building,
- archeological finds in the building,
- hydrological characteristics of the facility,
- connections of the building and traces of cultural and historical heritage in it and
- view of the surroundings of the building (2,168).

An important factor is the realization of a possible plan for equipping a given facility, which must be arranged in such a way that in no segment does it disrupt the internal structure of the facility, but to present it to the visitor in the best possible way (Lješević 1980).

To this should certainly be added the fact that for most of the mentioned facilities, there is an already built auxiliary infrastructure that can provide the interested guest with several days of access to the facilities themselves. The current condition, in which most of the mentioned facilities are located, allows the visit of only equipped speleological teams, while others, interested, have to wait for some better times for speleotourism.

Responsible tourist workers should pay attention to some of the mentioned facilities in terms of improving the overall tourist offer because if tourism is one of the main economic branches and directions of development of Montenegro, and almost 80% of its composition is karst, we should use what nature has selflessly given us.

Despite the stated facts about a very large number of researched and recorded research facilities, Montenegro today does have a single cave suitable for tourist visits.

2 The Lipska and the Cetinjska Caves

Lipska and Cetinjska caves experienced a certain level of training for tourist visits, but these activities were interrupted in the 1990s. Today, the only cave in Montenegro that is valorized for tourism is the Lipska Cave. With its landscaped and illuminated trails, it is a safe tourist attraction, which is toured by a tourist train accompanied by trained tourist guides. Its exceptional position 33 km from Budva toward Cetinje, 35 km from Podgorica and 5 km from Cetinje makes it part of the most important tourist destination in Montenegro (Radojičić 2015). The Cetinje cave, in the Cetinje field, is located next to the Cetinje monastery and represents a complex and branched abyss cave system. It consists of four main and several side corridors and canals, on several floors. The total length of the explored part of the cave is 1680 m (Petrović 1982). Before the last great flood of the Cetinje field (1986), 720 m of the most beautiful part of the cave was adapted for tourist visits.

3 The Đalovića Cave

Đalovića Cave is located in the basin of Bistrica (Bijelo Polje part of Bistrica), a left tributary of the Lim, on the Devil's Whirlpools. It is the longest cave in Montenegro. The entrances to the cave are in the middle part of the Sušice valley, in the Đalovića gorge. The cave is about 15 km away from Bijelo Polje, in the northeast direction. The cave entrances can be reached via the village of Đalovići. The cave was discovered by speleologists of the Speleological Department of the Belgrade Mountaineering Association in 1987. Later, many teams of speleologists, from Montenegro, the Czech Republic and others, explored the cave. The determined length of the discovered channels of the cave is 20 km. Research has shown that it is an exceptional speleological object, rich in various speleological forms and hydrological phenomena. In the records of speleologists, it is stated that the speleological complex of Đalovića Cave has three entrances. The main entrance, from the Đalović gorge, is located at the bottom of the northern part of the amphitheater extension, at an altitude of 735 m above sea level. Here, the water gushes out of the cave. The second entrance to the cave, smaller in size, is called the Mala (Little) Cave. At a distance of about 120 m from the entrance, the channels join and in the extension there is a

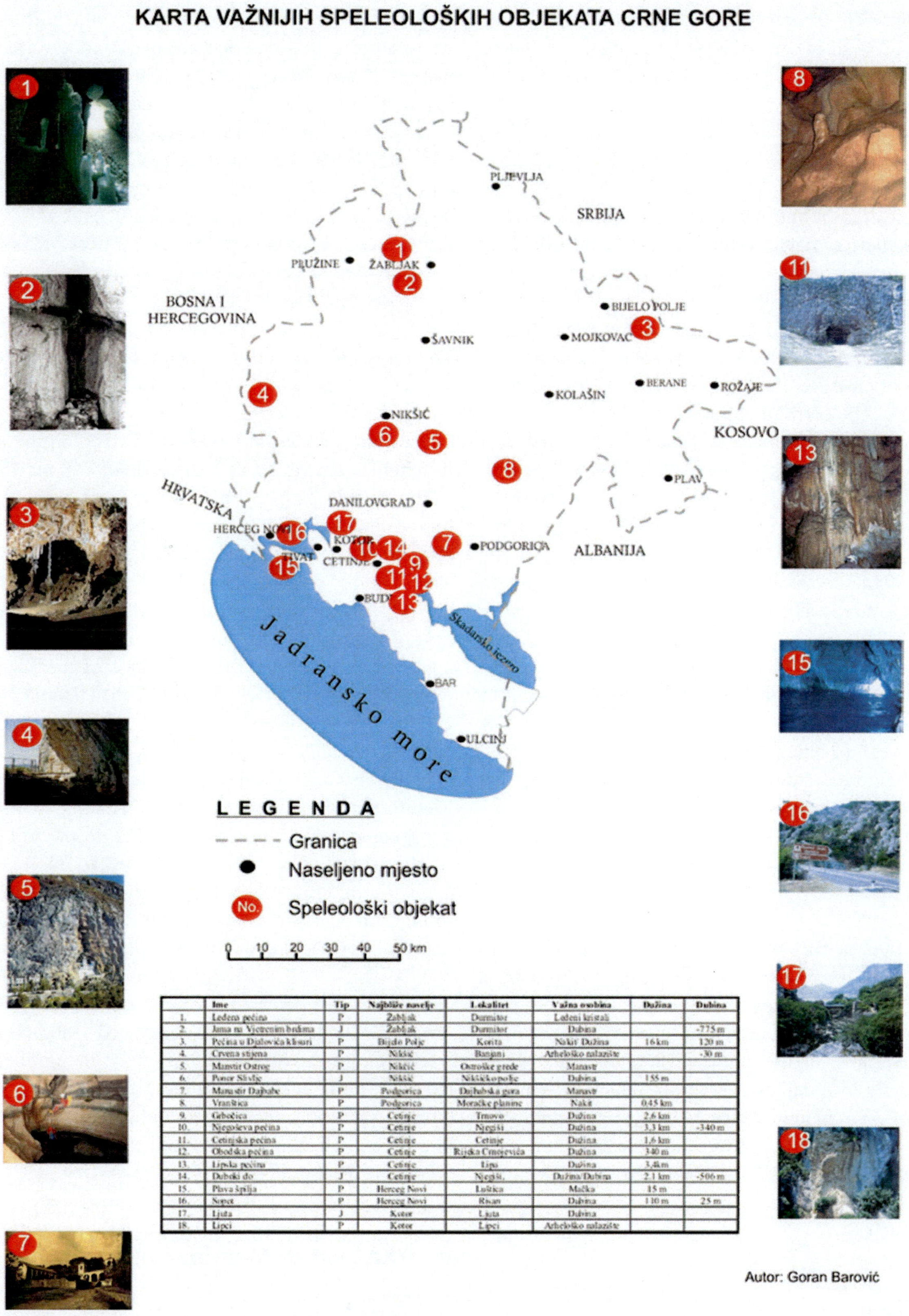

	Ime	Tip	Najbliže naselje	Lokalitet	Važna osobina	Dužina	Dubina
1.	Ledena pećina	P	Žabljak	Durmitor	Ledeni kristali		
2.	Jama na Vjetrenim brdima	J	Žabljak	Durmitor	Dubina		-775 m
3.	Pećina u Djalovića klisuri	P	Bijelo Polje	Korita	Nakit/ Dužina	16 km	120 m
4.	Crvena stijena	P	Nikšić	Banjani	Arheološko nalazište		-30 m
5.	Manastir Ostrog	P	Nikšić	Ostroške grede	Manastir		
6.	Ponor Slivlje	J	Nikšić	Nikšićko polje	Dubina	155 m	
7.	Manastir Dajbabe	P	Podgorica	Dajbabska gora	Manastir		
8.	Vranštica	P	Podgorica	Moračke planine	Nakit	0,45 km	
9.	Grbočica	P	Cetinje	Trnovo	Dužina	2,6 km	
10.	Njegoševa pećina	P	Cetinje	Njeguši	Dužina	3,3 km	-340 m
11.	Cetinjska pećina	P	Cetinje	Cetinje	Dužina	1,6 km	
12.	Obodska pećina	P	Cetinje	Rijeka Crnojevića	Dužina	340 m	
13.	Lipska pećina	P	Cetinje	Lipa	Dužina	3,4km	
14.	Duboki do	J	Cetinje	Njeguši,	Dužina/Dubina	2,1 km	-506 m
15.	Plava špilja	P	Herceg Novi	Luštica	Mačka	15 m	
16.	Sopot	P	Herceg Novi	Risan	Dubina	110 m	25 m
17.	Ljuta	J	Kotor	Ljuta	Dubina		
18.	Lipci	P	Kotor	Lipci	Arheološko nalazište		

Fig. 1 Important speleological objects of Montenegro (*Source* Goran Barović)

cave with quite rich "jewelry". The third entrance is above the Small Cave, high in the rocks and is difficult to access (Radojičić 2015).

4 The Obod Cave

The Obod Cave is at the source of the Obodske vrele, the source of the Crnojević River. At the exit of the cave, there is a narrow valley cut in limestone, with which another deeply cut Suvodolica ravine joins at a length of about 350 m. In the composition, the strongest permanent Obodsko spring springs, which feed Rijeka Crnojević with water. Obodska cave has the appearance of a simple corridor 340 m long. The corridor along its entire length is wider than the entrance door of the cave, which is 12 m wide and 14 m high. After a slight curve, the cave corridor ends with the hall in which Obodsko Lake is located. The water of the Obodska cave comes from the Cetinje surface waters in the rainy period, and in the dry from the groundwater from the Cetinje field, which was also determined by staining. At a time of great influx, water from the cave flows through the riverbed. In recent years, Rijeka has rarely come to flow from the wide opening of the cave. Petrović (1982, 36), states that the overhaul of the ceiling and the shortening of the cave hall took place so quickly that in 10–15 years of the nineteenth century the corridor was shortened by 4–5 m, and during the period from 1935 to 1975 it was shortened by 10–15 m (Radojičić 2015).

5 The Sopot

The Sopot—Such a profile of the cave is maintained, with a gradual fall to 100 m from the entrance, followed by a large cave gallery. After descending about 20 m, between the large blocks is a siphon lake, whose water level is a few meters above sea level in summer. The submerged canal is then gradually lowered, at a length of 200 m, to a maximum depth of 35 m below sea level. The cave channel then climbs steeper for the next 40 m and encounters a cave that is above the siphon lake and above sea level. The total length of the examined submerged Sopot canals is about 380 m (Radojičić 2015).

6 The Ice Cave

In the area of the National Park "Durmitor" there are several facilities that meet most of the conditions for adequate tourist valorization, but a major disadvantage of these facilities is poor traffic connections and high dependence on weather conditions that are not conducive to staying in the mountains for most of the year. These facilities are visited only by mountaineers and speleologists who are willing to make considerable efforts to experience the charms of Durmitor caves. Next to the pit on Vjetrena Brda, the most famous cave in the area of Durmitor is the Ice Cave with beautiful cave jewelry built of ice.

7 The Red Rock Cave

The most famous and most important cave in Montenegro, as far as archeological excavations are concerned, is certainly the Red Rock Cave, which is located in the village of Petrovići (about 45 km from Nikšić), on the left side of the Trebišnjica valley, at an altitude of 710 m. The entrance to the cave is 24 m wide, and the depth from the entrance line to the center of the rock is 15 m. The cave was named after the red color of the steep limestone rock above the cave. The red color of the rocks was formed from the washing of redness or bauxite, as well as from the dry patination of limestone. The opening of the cave faces south and southwest, from where the influence of the Mediterranean warmer climate came through the Trebišnjica valley, and in the Pleistocene (ice age) when Neanderthal man lived in its upper part (Riss and Wirm) and in the interglacials in Crvena stijena. Crvena stijena belongs to the Mediterranean cultural complex. Thanks to its favorable geographical position, primarily due to the milder climate in the Trebišnjica valley, compared to more continental areas, the man of this archeological site managed to adapt to all climate changes and to survive there. This is a building that is comparable in thickness of layers (20–30 m) to the site of EL CASTILLO in Spain. Exploring this cave, J. Basler determined the existence of 31 layers that are believed to have been inhabited 180,000 years ago. Archaeologists and researchers have determined that this cave is one of the largest and most important sites in the world. According to the stated characteristics, this facility is a very interesting tourist motif that should be adequately valorized (Radojičić 2015).

8 The Cave of Me(a)gara

An attempt was also made to arrange the cave of Me(a)gara in Tološi near Podgorica. This facility was adapted and its use is planned for the provision of health services; however, the realization of this project did not occur due to the disagreement of the locals with it (Habe 1980).

References

Habe F. (1980). Tourist caves in Montenegro and their protection in older literature. In: *Seventh Yugoslav Speleological Congress* (pp 211–223). Titograd: Union of speleologists of Yugoslavia.

Lješevic, M. (1980). Geographical space and its structure in the research and environmental protection complex. Proceeding of the Geoographical Institute PMF Belgrade, vol 27.

Lješević, M., Rósler, A. & Belošev, S. (1980). Velike pećine i jame Durmitora. In: *Seventh Yugoslav speleological congress* (pp 271–282). Titograd: Association of speleologists of Yugoslavia.

Petrović, D. (1982). Historical-geographic overview of the knowledge of caves in Serbia before the 20th century. Gazzete of the Serbian Geographical Society, Belgrade, vol 62.

Radojičić, B. (2015). Montenegro–Geographical encyclopedic lexicon. *Faculty of Philosophy, University of Nikšić (in Montenegrin).*

Forests of the Karst Region of Montenegro—Main Characteristics and Vulnerabilities

Milić Čurović

Abstract

This chapter presents the forest ecosystems of the Montenegrin karst region. The Montenegrin karst forests are mainly low forests of thermophilic deciduous trees as well as degradation stages of these forests in the form of shrubs, which have a distinct protective function. The analysis of the condition of the karst forests of Montenegro given in this chapter is certainly one of the contributions to the preservation of these ecosystems. The chapter also points out the importance of forests on the karst, which is often overlooked since the economic value calculated in timber products is small. The significance of these ecosystems is protection of soil, water, habitat and biodiversity. Stability of the karst forests is recognized as an important factor in the prevention of surface and subsurface environmental degradation. Sustainable management and preservation of these forests are of key importance for the nature protection of this area. This chapter also gives an overview of the main risks to these ecosystems, among which the greatest risks are wildfire and erosion processes.

Keywords

Karst forests · Soil erosion · Forest fires · Montenegro

M. Čurović (✉)
Biotechnical Faculty, University of Montenegro, Podgorica, Montenegro
e-mail: curovic@ucg.ac.me

1 Introduction

Forests represent one of the basic natural resources of the karst area of Montenegro. The karst region in Montenegro extends north of the coastal region, that is, from the Orjen-Lovćen-Sutorman-Rumija mountain range. On this surface, there are limestone mountains: Somina, Njegoš, Zla gora, Pusti lisac, Budoš and Garač. The larger plains in this area are the karst fields: Cetinjsko, Njeguško, Dragaljsko, Grahovsko and Nikšićko. In this area, on the horizontal and vertical profile, there are forest communities conditioned by climate (climatogenic forests) and orographic—edaphic conditions, especially altitude and exposure (climate-regional forests), which primarily have a protective function.

Karst forest ecosystems play a key role in preventing surface and subsurface environmental degradation. As the forest stands of this area are mostly located on terrains that are prone to erosion (pluvial and aeolian), their dominant function is precisely to protect those terrains from erosion (Kerckhof et al. 2016). Slopes exposed to the south and west are more subject to runoff and erosion (Bou Kheir et al. 2008). In addition to fulfilling that function, these forest stands are also of key importance for maintaining the karst groundwater regime (Carrière et al. 2017). It was also observed that the increase in forest cover coincided with a significant increase in the growth rate of speleothems—mineral deposits formed from groundwater within underground caverns (Dorale and Liu 2009).

Karst ecosystems are very sensitive areas affected by processes of desertification (Ozgul and Dindaroglu 2021). Different economic reasons over the centuries caused the different ways in which the Montenegrin karst forests were destroyed. Excessive cutting of firewood and felling for the needs of shipbuilding and construction, deforestation in order to expand arable land and pastures, and intensive grazing due to a large number of livestock are the main reasons for the destruction of forests in this part

G. Barovic (ed.), *Speleology of Montenegro*, Cave and Karst Systems of the World,
https://doi.org/10.1007/978-3-031-49375-1_10

of Montenegro. The destruction of forests then led to the intensification of erosion processes, so the recovery of the forests in this area was slow and incomplete, and in some places, the land was permanently stripped.

The need for firewood and wood products, as well as pressures for the expansion of arable land in this area are now decreasing due to population migration to cities. The functions of these forests become primarily protective. The social and landscape functions of these forests have also become more significant. The function of these forests is also to provide a safe habitat—a refuge for many other species that are related to the forest ecosystem and depend on it. The forests in this area do not have a great economic value measured in the production of wood products, but they represent a great potential for secondary forest products such as medicinal herbs, forest fruits (Curovic et al. 2019), game and have an invaluable function in protecting karst areas from numerous degradation processes.

2 Structure of Forest Communities on the Montenegrin Karst

The most common forests of the Montenegrin karst are low thermophilic deciduous forests with South European flowering ash (*Fraxinus ornus*), European hop-hornbeam (*Ostrya carpinifolia*), Turkey oak (*Quercus cerris*), pubescent oak (*Quercus pubescens*), beech (*Fagus moesiacae*) and Montpellier maple (*Acer monspesulanum*) as the main representatives of dendroflora (Fig. 1; Table 1). At higher altitudes, there is a belt of beech forests, as well as the southern border of the fir areal in Montenegro (on the mountains: Bijela gora, Njegoš, Somina, Paklarica and Zla gora). It grows here in mixed forests with beech: *Abieti-Fagetum*. Of the other coniferous species, there are mostly Pine species (Vitali et al. 2019). In the area of Skadar lake, there are hygrophilous forests of willows, poplars and Skadar oak (*Quercus robu*r ssp scutariensis). The largest area in this part of Montenegro (24.2%) is occupied by forests of South European flowering ash (*Fraxinus ornus*), European hop-hornbeam (*Ostrya carpinifolia*) (Fig. 2; Table 1). Apart from these dominant vegetation types are: Pubescent and white oak forest (*Querco-Carpinetum orientalis),* pubescent oak and hop-hornbeam forest (*Ostryo-Quercetum pubescentis) and* Macedonian oak forest (*Quercetum trojanae)* (Blečić and Lakušić 1976; Lakušić 1987).

These forests are mostly of natural origin (95%) and characterized by high biodiversity, a large number of tree and shrub species, which ensures their polyfunctionality and stability. These are mostly low and coppice forests, as well as their degradation stages with thickets, brusch and low forests as part of the dynamics of the process from pioneer species to climax species of vegetation (Lifei et al. 2002).

It should also be mentioned the presence of endemic species found in this region, such as *Acer obtusatum, A. paradoxum, Petteria ramentacea, Dioscorea balcanica, Tulipa grisebachiana, Lilium cattaniae, Helleborus hercegovinus, Fritillaria gracilis, Crocus tommasinia-nus, C. dalmaticus, Colchicum visianii, C. hungaricum, Lonicera formanekiana, Acanthus balcanicus, Eryngium pahnatum, Daphne laureola* etc. (Dees et al. 2013).

Such forest communities represent a significant potential of general beneficial functions of the forest expressed in ecological (protective) and social (social) functions of forests. The production of wood and the use of other forest products in this area is of secondary importance. Existing tree species and ecological conditions on the karst do not ensure the production of more valuable wood assortments. This can also be seen through the structural parameters of these forests, which according to the data of the National Forest Inventory (Dees et al. 2013) are characterized by a low timber volume (V) which is only 41.6 m^3/ha, low wood volume increment (Zv) of 1.1 m^3/ha per year. All this is a result of small average dimensions: medium height (Hs) of 9.9 m and medium stem diameter (Ds) which is 16.9 cm. A characteristic of these forests is their high density, i.e. a large number of trees (N) per unit area, which amounts to an average of 871 trees/ha (Table 1).

Due to the intensive anthropogenic influence over a long period of time, the primeval vegetation of the karst areas of the Mediterranean has been changed, and instead of the forest communities that represent the climax of the vegetation, there are now degradation forms: low forests, thickets, garigs, and rocky fields with rare shrub formations (Gams and Gabrovec 1999; Parise and Pascali 2003; De Waele 2009; Kranjc 2009). In addition to the excessive cutting of forests on the karst, grazing in these areas also had a major impact. The negative impact of excessive grazing is noticeable not only on the existing vegetation but also on the seeds found in the soil, which affects the process of vegetation renewal (Chaideftou et al. 2008; Stefanidis et al. 2022). After the cessation of intense pressure on the forests of this area, the development of progradational forms was also noticeable. This certainly supports the position that low forests and brush of flowering ash, Oriental hornbeam and *European hop-hornbeam* are only a progradation-degradation stadium of forests of Macedonian oak, Turkey oak or pubescent oak.

3 Risks

Forest fires represent a latent danger for the loss of forests and forest lands and are one of the main triggers of erosion processes in the Mediterranean (Stefanidis et al. 2022).

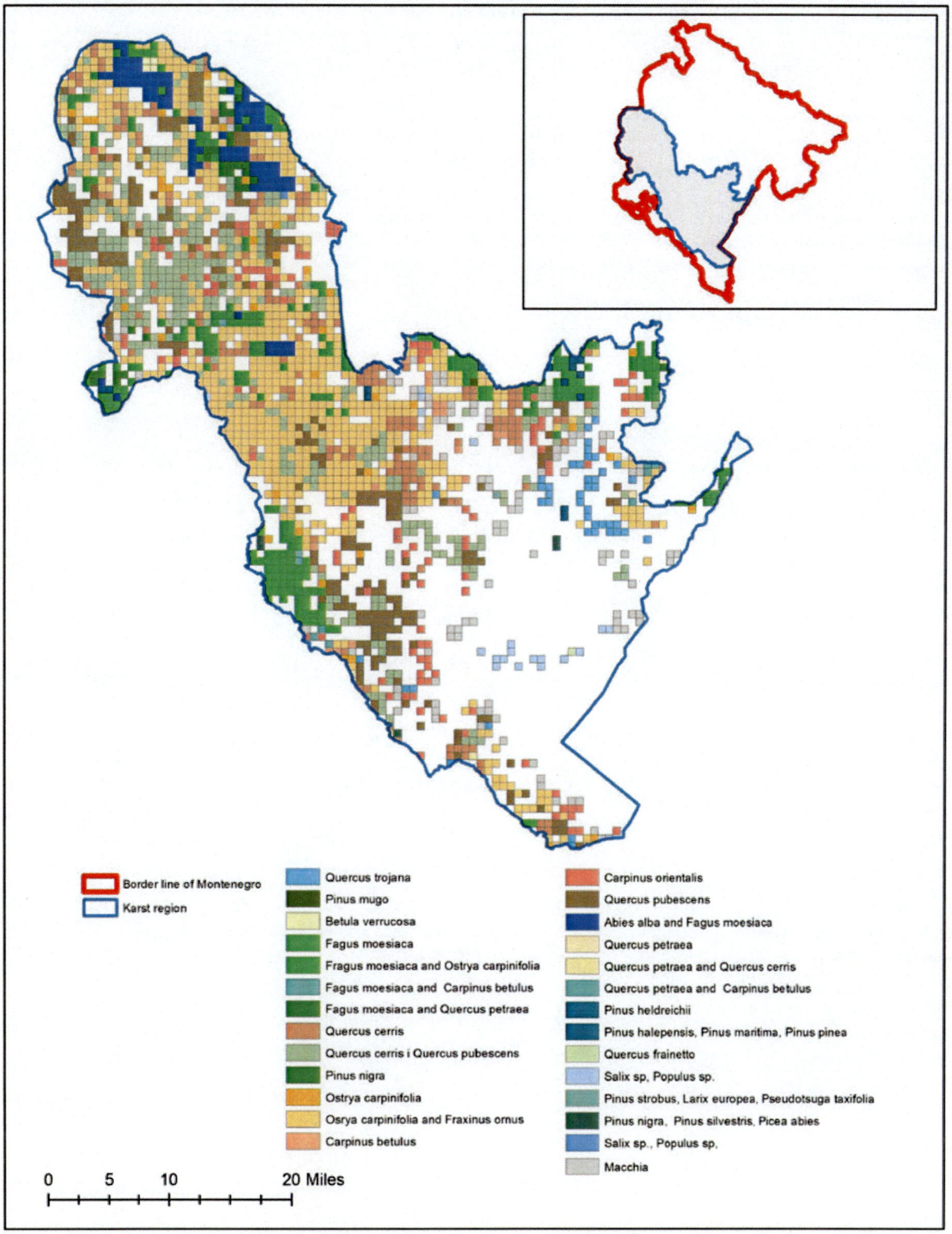

Fig. 1 Distribution of dominant tree species in the forests of the Montenegrin karst region

In recent decades, the frequency and severity of wildfires have increased throughout the Adriatic belt and represent the main factor threatening the stability of karst forests (Pavlek et al. 2017; Šturm and Podobnikar 2017). They most often occur in dry periods of the year. Forest fires are becoming more and more frequent, which, especially in the karst part of Montenegro, often take on large proportions and, in addition to forests, threaten other natural ecosystems, settlements and human lives. Montenegro still don't have an efficient system of combating wildfires and in the

Table 1 Growing stock by forest type of the Montenegrin karst region (Dees et al. 2013)

Forest type	P (ha)	P (%)	N/ha	V (m³/ha)	V (%)	Zv (m³/ha)	Zv%	Ds	Hs
Willow and poplar forests (*Salix sp, Populus sp*)	1722,8	0,8	279,0	36,8	0,7	1,2	0,9	23,3	13,9
Oriental hornbeam forests (*Carpinus orientalis*)	8954,9	4,2	1625,0	25,5	2,6	0,7	2,8	10,4	6,4
Macedonian Oak forests (*Quercus trojana*)	4218,2	2,0	641,0	16,5	0,8	0,5	0,9	12,9	6,7
Downy oak forests (*Quercus pubescens*)	19,711,3	9,2	706,0	14,9	3,3	0,5	4,1	12,3	7,0
Bitter oak and downy oak forests (*Qu. cerris and Qu. pubescens*)	30,701,9	14,4	903,0	25,5	8,8	0,8	10,4	14,0	8,2
Bitter oak forests (*Quercus cerris*)	18,650,3	8,8	887,0	42,0	8,8	1,1	8,6	16,1	9,8
Sessile oak forests (*Quercus petraea*)	199,4	0,1	958,0	50,0	0,1	1,2	0,1	14,3	7,1
Sessile oak and bitter oak forests (*Qu.petraea and Qu.cerris*)	296,1	0,1	560,0	38,4	0,1	1,3	0,2	19,7	9,4
Sessile oak and hornbeam forests (*Qu. petraea and Carpinus betulus*)	512,0	0,2	1960,0	35,1	0,2	1,4	0,3	8,3	7,2
Hornbeam forests (*Carpinus betulus*)	1484,4	0,7	1873,0	38,4	0,6	1,5	0,9	9,8	8,0
Hop-hornbeam forests (*Ostrya carpinifolia*)	3078,1	1,4	1241,0	46,7	1,6	1,4	1,8	15,4	9,5
Hop-hornbeam and flowering ash forests (*O. carpinifolia and Fraxinus ornus*)	51,593,8	24,2	1059,0	22,7	13,2	0,7	15,4	12,7	7,7
European beech and hop-hornbeam forests (*Fagus moesiaca and Ostrya carpinifolia*)	4361,5	2,0	860,0	88,4	4,3	1,8	3,4	25,4	13,0
European beech and sessile oak forests (*Fagus moesiaca and Quercus petraea*)	296,1	0,1	1017,0	203,4	0,7	6,0	0,8	20,8	15,2
European beech and hornbeam forests (*Fagus moesiaca and Carpinus* betulus)	398,8	0,2	1265,0	69,1	0,3	2,8	0,5	12,3	10,7
European beech forests (*Fagus moesiaca*)	17,301,2	8,1	875,0	112,3	21,9	2,7	19,9	19,3	12,4
Silver fir and European beech forests (*Abies alba and Fagus moesiaca*)	9054,7	4,2	537,0	175,4	17,9	3,9	15,0	26,3	14,8
Black pine forests (*Pinus nigra*)	2385,7	1,1	419,0	121,5	3,3	3,0	3,1	25,6	13,3
Bosnian pine forests (*Pinus heldreichii*)	896,2	0,4	235,0	113,5	1,1	1,9	0,7	33,8	13,4
Maritime pine forests (*Pinus halepensis, Pinus maritima, Pinus pinea*)	197,4	0,1	385,0	153,3	0,3	4,3	0,4	29,0	12,7
Silver birch forests (*Betula verrucosa*)	499,4	0,2	357,0	38,0	0,2	1,1	0,2	23,3	11,6
Forest cultures of autochtonous species (*Pinus nigra and/or Pinus sylvestris*)	400,3	0,2	687,0	310,2	1,4	6,3	1,1	27,0	16,5
Forest cultures of allochtonous species (*Pinus strobus and/or Larix europea*)	104,2	0,0	20,0	20,6	0,0	0,5	0,0	48,9	10,6
Forest cultures of allochtonous species, deciduous species	98,7	0,0	1140,0	126,0	0,1	3,9	0,2	14,1	12,0
Other forests with coniferous types of trees predominant	204,9	0,1	64,0	22,5	0,1	0,8	0,1	32,8	12,5
Other forests with deciduous types of trees predominant	35,776,8	16,8	580,0	18,1	7,3	0,5	8,3	14,9	8,7
Total	213,098,9	100,0	871,0	41,6	100,0	1,1	100,0	16,9	9,9

following period it will be necessary to pay special attention to early warning systems and efficient wildfire risk management (Bozovic et al. 2022). Mainly due to the negative influence of man (Ager et al. 2014), especially during long dry periods, there is an increased frequency of fires, causing incalculable damage to forest resources. Wildfires are causing significant changes in hydrological and erosion regimes (Curovic et al. 2021). Although the karst region is only one of five regions in Montenegro, as much as 47.4% of the burnt area in the period 2005–2010 was recorded in this region (Dees et al. 2013).

Apart from anthropogenic influences, numerous unfavorable environmental conditions affect the degradation processes in these forests (Parise et al. 2004). Unfavorable conditions should be mentioned here, such as undeveloped pedological cover, shallow and skeletal soils and often exposed karst rock. In addition, there is abundant precipitation during the period of rest of the vegetation when the washing of the pedological layer is most intense in deciduous forests, a marked lack of precipitation in the summer period, very steep slopes, a poorly developed hydrographic network, which also affect the difficult

Fig. 2 Forest of South European flowering ash (*Fraxinus ornus*) and European hop-hornbeam (*Ostrya carpinifolia*) in the Montenegrian karst area (photo by Čurović M.)

maintenance of forest vegetation and the improvement of its quality. Especially in recent decades, the negative impact of climate change on the water regime in karst areas (Sivelle et al. 2021), as well as on the stability of these forests (Dorado-Liñán et al. 2019), has been pronounced. Extended droughts are more frequent which will certainly in the future have increasing negative effects on the water regime, productivity and mortality of karst forests (Tague et al. 2019).

Expected climate changes will result in the displacement of certain vegetation zones (forest types) both in latitude and altitude. In some areas, increased drying of trees as a result of stress and attacks by pests and plant diseases, reduction of growth, difficult natural regeneration and increased damage caused by forest fires and atmospheric disasters can be expected. It is still not clear all the possible consequences for the biodiversity of the Mediterranean forests on the karst that will result from the climatic changes (Gil-Tena et al. 2009).

For these degraded forests, even the secondary vegetation with which they are overgrown is very important for the protection of those terrains from the continuation of the previously started erosion. In addition, these species are mostly honey-bearing, so they are also important for beekeeping, and due to the high content of essential oils, they can be used for the needs of the pharmaceutical industry. However, apart from the pioneer species characteristic of the karst area, invasive species are increasingly appearing (Pötzelsberger et al. 2020).

Intensive grazing has been a major driver of vegetation change of Eastern Mediterranean oak woodlands (Plieninger et al. 2011). Depopulation of the rural areas in the previous period significantly reduced the negative impacts on the karst forests, so the processes of forest degradation and devastation due to felling and keeping goats were stopped, which has led to a natural progressive succession of vegetation in recent decades (Nyssen et al. 2014; O'Hara et al. 2018). In recent decades, there has been an increase in the area of forests in karst area of Montenegro. The progressive succession of vegetation is clearly visible and takes place naturally. On the other hand, due to the lack of logging and maintenance of local roads, these forests are becoming increasingly difficult to access. Among other things, this makes it difficult to localize and extinguish fires, which, as already mentioned, currently present the greatest danger to forest ecosystems in karst area.

4 Conclusion

Sustainable management of forest resources implies the fulfillment of the social, ecological and economic needs of current and future generations for forest products and services, such as wood and wood products, water, food, energy, medicinal plants, habitat for game, sequestration of carbon dioxide, space for recreation and tourism, landscape, etc.

Stable and healthy forest ecosystems are one of the basic prerequisites for combat desertification, protection of soil and maintenance of surface and underground water regimes, and indirectly of processes and phenomena below the soil surface of the Montenegrin karst. This is particularly evident that highly vulnerable ecosystems, such as karst forests will require a very careful management approach in the future.

References

Ager AA, Preisler HK, Arca B, Spano D, Salis M (2014). Wildfire risk estimation in the Mediterranean area. Environmetrics 25, 384–396. https://doi.org/10.1002/env.2269

Blecic V., Lakusic R. (1976): Prodromus biljnih zajednica Crne Gore, Glasn.Republ. Zav.Zast. Prir.—Prirod. Muz. (Titograd) 9: 57-98

Bou Kheir, R., Abdallah, C. and Khawlie, M. (2008) Assessing Soil Erosion in Mediterranean Karst Landscapes of Lebanon Using Remote Sensing and GIS. Engineering Geology, 99, 239-254. https://doi.org/10.1016/j.enggeo.2007.11.012

Bozovic, D., Knezevic, M., Aleksic, M., Iker, O., Gostimirovic, L. (2022). Forest fire risk management information systems. Agriculture and Forestry, 68 (2): 65–81. https://doi.org/10.17707/AgricultForest.68.2.05

Carrière, S.D., Danquigny, C., Davi, H., Chalikakis, K., Ollivier, C., Martin-StPaul, N.K., Emblanch, C. (2017). Process-Based Vegetation Models Improve Karst Recharge Simulation Under Mediterranean Forest. In: Renard, P., Bertrand, C. (eds) EuroKarst 2016, Neuchâtel. Advances in Karst Science. Springer, Cham. https://doi.org/10.1007/978-3-319-45465-8_12

Chaideftou, E., Thanos, C.A., Bergmeier, E., Kallimanis, A., Dimopoulos, P. (2008). Seed bank composition and above-ground vegetation in response to grazing in sub-Mediterranean oak forests (NW Greece). In: Van der Valk, A.G. (eds) Forest Ecology. Springer, Dordrecht. https://doi.org/10.1007/978-90-481-2795-5_20

Curovic, M., Jovancevic, M., Balijagic, J. (2019). Wild Fruit Tree Species of Montenegrin Forests. in Monograph: Forests of Southeast Europe Under a Changing Climate, 21–28, Springer, ISBN: 978–3–319–95266–6, https://doi.org/10.1007/978-3-319-95267-3_2

Curovic, M., Spalevic, V., Stijovic, A., Curovic, Z., Buskovic, V. (2021). Forests of the Coast of Montenegro. In: Joksimović, D., Đurović, M., Zonn, I.S., Kostianoy, A.G., Semenov, A.V. (eds) The Montenegrin Adriatic Coast. The Handbook of Environmental Chemistry, vol.110. Springer, Cham. 51- 67. https://doi.org/10.1007/698_2020_711

De Waele, J. Evaluating disturbance on mediterranean karst areas: the example of Sardinia (Italy). Environ Geol 58, 239–255 (2009). https://doi.org/10.1007/s00254-008-1600-x

Dees, M., Andjelic, M., Fetic, A., Jokanovic, B., Tepavcevic ,V., Borota, D., Vasiljevic, A., Weinreih, A., Hahn, N., Markovic, D., Terzic, D. (2013). The First National Forest Inventory of Montenegro. Ministry of Agriculture and Rural Development, Government of Montenegro, Podgorica, 1–349

Dorado-Liñán I, Piovesan G, Martínez-Sancho E, Gea-Izquierdo G, Zang C, Cañellas I, Castagneri D, Di Filippo A, Gutiérrez E, Ewald J, Fernández-de-Uña L, Hornstein D, Jantsch MC, Levanič T, Mellert KH, Vacchiano G, Zlatanov T, Menzel A. (2019). Geographical adaptation prevails over species-specific determinism in trees' vulnerability to climate change at Mediterranean rear-edge forests. Glob Chang Biol. Vol.25 (4). 1296-1314. https://doi.org/10.1111/gcb.14544

Dorale, J.A. and Liu, Z. (2009). Limitations of hendy test criteria in judging the paleoclimatic suitability of speleothems and the need for replication. Journal of Cave and Karst Studies, v. 71, no. 1, p. 73–80.

Gams, I., Gabrovec, M. (1999). Land use and human impact in the Dinaric karst. International Journal of Speleology, 28: 55-70. https://doi.org/10.5038/1827-806X.28.1.4

Gil-Tena, A., Brotons, L., Saura, S. (2009). Mediterranean forest dynamics and forest bird distribution changes in the late 20th century. Global Change Biol. 15, 474–485. https://doi.org/10.1111/j.1365-2486.2008.01730.x

Kerckhof, A., Spalevic, V., Van Eetvelde, V., Nyssen, J. (2016). Factors of land abandonment in mountainous Mediterranean areas: the case of Montenegrin settlements. SpringerPlus 5, 485. https://doi.org/10.1186/s40064-016-2079-7

Kranjc, A. (2009). History of Deforestation and Reforestation in the Dinaric Karst. Geographical Research. 47 (1), 15-23. https://doi.org/10.1111/j.1745-5871.2008.00552.x

Lakusic R. (1987). Sumske zajednice Jugoslavije—SR Crna Gora in Sumarska enciklopedija, Zagreb,Vol III pp. 388-395

Lifei, Y., Shouqian, Z., Jingzhong, Y., Luming, W., Zhengren, C. (2002). Dynamics of a degraded karst forest in the process of natural restoration Scientia Silvae Sinicae. 38(1):1-7.

Nyssen, J., Van den Branden, J., Spalević, V., Frankl, A., Van de Velde, L., Čurović, M. and Billi, P. (2014). Twentieth century land resilience in Montenegro and consequent hydrological response; Land Degradation & Development, Vol.25, issue 4, pp 336-349. https://doi.org/10.1002/ldr.2143

O'Hara, K.L., Ina, A.B., Diaci, J., Anic, I., Boydak, M., Curovic, M., et al. (2018). Culture and silviculture: origins and evolution of silviculture in Southeast Europe. Int. Forest. Rev. 20(1): 130–143. https://doi.org/10.1505/146554818822824228.

Ozgul, M., Dindaroglu, T. (2021). Multi-criteria analysis for mapping of environmentally sensitive areas in a karst ecosystem. Environ Dev Sustain 23, 16529–16559. https://doi.org/10.1007/s10668-021-01363-7

Parise, M., Pascali, V. (2003). Surface and subsurface environmental degradation in the karst of Apulia (southern Italy). Env Geol 44, 247–256. https://doi.org/10.1007/s00254-003-0773-6

Parise, M., Qiriazi, P., and Sala, S. (2004). Natural and anthropogenic hazards in karst areas of Albania, Nat. Hazards Earth Syst. Sci., 4, 569–581. https://doi.org/10.5194/nhess-4-569-2004

Pavlek, K., Bišćević, F., Furčić, P., Grđan, A., Gugić, V., Malešić, N., Moharić, P., Vragović, V., Fuerst-Bjeliš, B. & Cvitanović M. (2017). Spatial patterns and drivers of fire occurrence in a Mediterranean environment: a case study of southern Croatia, Geografisk Tidsskrift-Danish Journal of Geography, 117:1, 22–35. https://doi.org/10.1080/00167223.2016.1266272

Plieninger, T., Schaich, H. & Kizos, T. (2011). Land-use legacies in the forest structure of silvopastoral oak woodlands in the Eastern Mediterranean. Reg Environ Change 11, 603–615. https://doi.org/10.1007/s10113-010-0192-7

Pötzelsberger, E., Lapin, K., Brundu, G., Adriaens, T., Andonovski, V. et al. (2020): Mapping the patchy legislative landscape of non-native tree species in Europe, Forestry: An International Journal of Forest Research, Volume 93, Issue 4, July 2020, Pages 567–586. https://doi.org/10.1093/forestry/cpaa009

Sivelle, V., Jourde, H., Bittner, D., Mazzilli, N., Tramblay, Y. (2021). Assessment of the relative impacts of climate changes and anthropogenic forcing on spring discharge of a Mediterranean karst system. Journal of Hydrology. Volume 598. https://doi.org/10.1016/j.jhydrol.2021.126396

Stefanidis, S., Alexandridis, V., Spalevic, V., Mincato, R.L. (2022). Wildfire Effects on Soil Erosion Dynamics: The Case of 2021 Megafires in Greece. Agriculture and Forestry, 68 (2): 49–63. https://doi.org/10.17707/AgricultForest.68.2.04

Šturm, T., Podobnikar, T. (2017). A probability model for long-term forest fire occurrence in the Karst Forest management area of Slovenia. International Journal of Wildland Fire 26(5) 399-412. https://doi.org/10.1071/WF15192

Tague, C.L., Moritz, M., Hanan, E. (2019). The changing water cycle: The eco-hydrologic impacts of forest density reduction in Mediterranean (seasonally dry) regions. Wires Water 6(4). e1350. https://doi.org/10.1002/wat2.1350

Vitali, A., Garbarino, M., Camarero, J., Malandra, F., Toromani, E., Spalevic, V., Curovic, M., Urbinati, C. (2019). Pine recolonization dynamics in Mediterranean human-disturbed treeline ecotones; Forest Ecology and Management; 435, 28-37. https://doi.org/10.1016/j.foreco.2018.12.039

Archaeological Studies of Caves in Montenegro

Dragica Mijanović, Branka Manojlović and Đurđica Perović

1 Introduction

A significant number of caves (notches and rock shelters) in Montenegro were the habitat of people in the prehistoric period, as evidenced by numerous remains of pottery, tools and drawings on the walls. In some of them, places of worship (churches) were also present. Their common characteristic is that they were all relatively spacious, deep enough and with larger openings, thus providing a safe shelter not only from weather disasters, but also from possible enemies. As a rule, in front of them is a relatively spacious plateau, which is sunlit most of the day and was most often used for making tools, sometimes as a sanctuary. Their openings are mostly facing south, southwest or southeast, only the opening of Mališina Stijena rock shelter is facing north with a small turn to the east" (Marković 2006; 46).

The first archeological excavations in Montenegro were started by Arthur Evans at the end of the nineteenth century (1885), after that, on the initiative of Prince Nikola, extensive research of ancient Duklja began, led by P. A. Rovinski. Modern archaeological research was initiated in the middle of the XX century, however, systematic excavations of some prehistoric caves in Montenegro commenced only in the 1970s (Spila near Perast, Odmut in the Piva canyon, Vranjaj on the Orjen, Koronina cave near Cetinje, Mališina Stijena in the Ćehotina valley, Bioče in the Morača canyon) (Marković 2006).

In this chapter, we will look at the archaeological research of caves inhabited by prehistoric man and the two famous monasteries that were built in the caves.

2 Caves as Human Habitats

Certainly the most significant, best studied from the archaeological aspect and with the richest collection of remains of material culture in Montenegro is the **Crvena stijena** rock shelter near Petrovići. Crvena stijena ("Red Rock") is one of the most important archeological sites in the southeastern part of Europe, named after the red color of limestone and dolomite, which originates from iron oxides. The cave is located at 700 m MSL on the left bank of Trebišnjica (today Lake Bileća), it is 26 m wide, 15 m high and 25 m deep. It contains over 20 m of archaeological deposits, starting from the Middle Paleolithic period (whose depth reaches 10 m) to the Bronze Age (Monnier et al. 2020).

The first excavations at this site began after 1954 and lasted until 1963, later continuing in the period 2004–2015. The Homeland Museum in Nikšić organized and financed the first excavations, which began in 1955 under the leadership of Alojz Benac,[1] when layers I-V were discovered and 3.5–4 m were excavated, and in 1956 a depth of 9 m was reached. Excavations were continued by M. Brodar from Ljubljana in 1958 and reached a depth of 12 m, on which occasion over 3,000 artifacts were found in layers XI-XVIII. Since the 1960s, the works have been taken over by Đ. Basler (from the Homeland Museum in Sarajevo) and reached a depth of over 12 m, reaching the XXIV layer. In 1961, he removed all the layers of the Upper Paleolithic in order to go deeper, and the following year he removed the layers of the Mesolithic and the Late Upper Paleolithic in order to reach the deepest layers. Although he got to a depth of 20.7 m and exposed the XXXI layer, he did not

D. Mijanović (✉)
Department of Geography, Faculty of Philosophy Nikšić, Niksic, Montenegro
e-mail: mdragica@t-com.me

B. Manojlović
Professor of Geography, Niksic, Montenegro

Đ. Perović
Faculty of Tourism Kotor, Niksic, Montenegro

[1] Homeland Museum in Sarajevo.

G. Barovic (ed.), *Speleology of Montenegro*, Cave and Karst Systems of the World,
https://doi.org/10.1007/978-3-031-49375-1_11

European culture of the Lower Paleolithic, although there is some resemblance to the Paleolithic culture known in southern France as the Tayacien. Proto-Mousterian, represented by layers XXV-XXVIII has scarce material culture findings and is difficult to identify culturally, while Mousterian, represented by layers XI-XXIV, whose layer thickness is about 8 m, is represented by numerous scrapers with a smaller core and a smaller number of flake tools and cutters. This period of development is connected with the sites on the Apennine Peninsula—the Pontinien. The final phase of the Mousterian is marked by finds from layers XII and XI, dominated by scrapers. After that, Crvena stijena is left by or at the end of the Würm II stadial. The transition from the Middle to the Upper Paleolithic is marked by layer XI consisting of eruptive volcanic material. The break until resettlement was long because the new (Upper Paleolithic) population was very different from the Middle Paleolithic population. This period is represented by layers IX and VIII. Some significant tools are bone awls from the shins of large mammals, there are various tools made of flint (scrapers, lamellae, cutters, pebbles, drills, pointy flakes). Scrapers, drills, awls were also found in layers VII-V. This period of development in Crvena stijena cannot be connected with any culture in the neighboring areas (Marković 2006.).

About 40,000 years ago, significant climate change took place on Earth, which caused the extinction of Neanderthals and the appearance of modern human Homo sapiens. A layer of volcanic ash was found in Crvena stijena, on the basis of which researchers conclude that the Middle Paleolithic culture ended in a cataclysm. It is assumed that Homo sapiens relocated to Crvena stijena from another area. Hence, man settled here occasionally and in stages, primarily due to the favorable position of the site. The main occupation of Paleolithic man in Crvena stijena was hunting and fishing, as evidenced by the remains of animal bones in some layers of the settlement (Radović http://www.montenegrina.net/).

In the Mesolithic period represented in layer IV (which is divided into several phases), in addition to hunting, people in Crvena stijena collected shells and snails, which is reflected in the tools made of stone and bone. Massive tools made out of horns also appeared in this period. Although there are no remains of tools that would indicate that the inhabitants of Crvena stijena were engaged in agriculture, Đ. Bazler claims that this is not proof that they were not engaged in it, but that there was probably no need to bring tools for agriculture to the cave, and that they probably left it in the fields ((Bazler Đ, 1975; 17–18). It should be emphasized that these Mesolithic tools are closely related in form and technique to occurrences in the Mediterranean and North Africa and to some extent in Central Asia, which indicates a cultural and ethnic connection "with areas of the Middle East, which is even more pronounced in the following, Neolithic development phase" (Garašanin 1967 41).

Pottery and numerous bones of wild animals from the Neolithic period were found in Crvena stijena (in layer III). However, no bones of domestic animals, which would indicate the development of cattle breeding, were found. There is also no evidence to suggest that the inhabitants were engaged in agriculture, although this period is mainly (and especially in the Middle East) related to the development of agriculture. The Neolithic is also associated with Cardium pottery that was widespread in the Mediterranean—"from Mesopotamia, through the Balkans, the northern and southern Mediterranean coast, to the Iberian Peninsula", which indicates the takeover of cultural influences caused by contact with other groups, but without changes in ethnic composition and without significant relocations. Ceramics of the so-called Danilo culture were found in layer II in Crvena stijena. They differ from ceramics from the earlier period; have richer and more diverse forms and a processing technique that is mainly based on incised lines, with spirals standing out (Garašanin 1967; 46–7). From the fragments of pottery found in this layer, it is not possible to say what shape it was because it was not possible to assemble any entire profile (Benac 1975).

According to A. Benac, at the end of the period to which layer II belongs, Crvena stijena is abandoned twice for a longer period (between the Early and Middle Neolithic and again between the Middle Neolithic and the youngest stratum) for two possible reasons: first is relocation in pursuit of better pastures and hunting sites, and second, which the author prefers, is that a certain group (from the Middle Neolithic) came here from the Adriatic and "brought with them a special cult with rhytons and settled in the previously abandoned Crvena stijena rock shelter" (Benac 1975; 145.). It was re-inhabited in the Late Bronze and Early Iron Ages—they belong to layer I with which life in Crvena stijena ends. It is assumed that in Crvena stijena, just as in other Balkan regions, the life of Neolithic groups ceased to exist at the end of the 3rd millennium BC following the great migration from the east (Indo-European migration). New groups of people who came to this area occasionally used the cave (as a cult or shelter), but did not stay permanently (Benac 1975) (Fig. 2).

The archeological findings in Crvena stijena consist mostly of stone tools of various shapes and types. In the layers belonging to the Mesolithic and Neolithic, tools made of bone were found, and in layers III, II and I, remains of pottery.

In addition, Crvena stijena is considered to be the site with the largest number of remains of bones of extinct animals from the Quaternary period in the southeastern part of Europe. There are remains of about 40 genera of animals, of which 13 species are extinct, 11 species nowadays

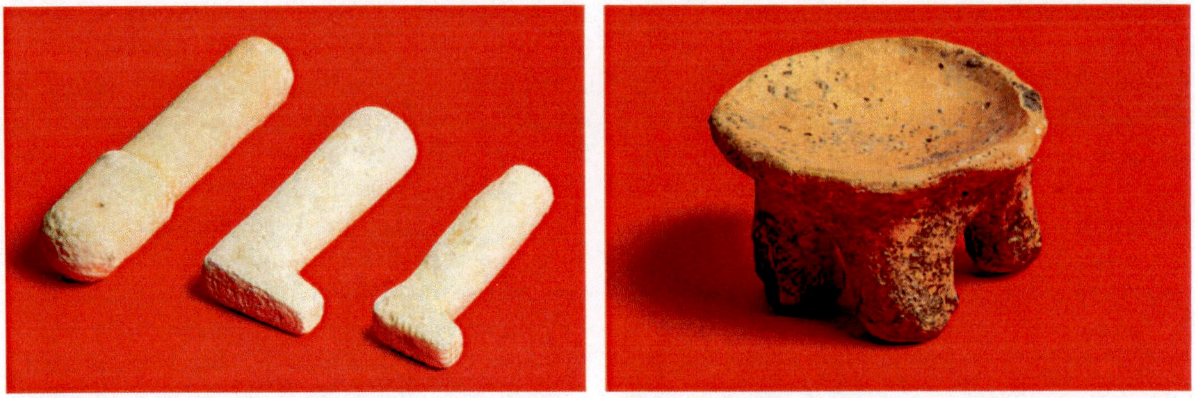

Fig. 2 Stone tools from Crvena stijena. *Source* https://www.b92.net/putovanja/zanimljivosti

inhabit some more remote areas, while the rest still live in the wider vicinity of the cave. Animals representative of a very warm climate were found in some strata of the cave, and those of a cold climate in others (Radović http://www.montenegrina.net/).

After Crvena stijena, **the Lipci rock shelter** is certainly the most important archeological site from the prehistoric period in Montenegro, although it is less studied and rather neglected. It is located in the Bay of Kotor, on the northern side of the Risan Bay not far from Bijela near Risan, about 500 m from the sea shore. Pavle Mijović discovered paintings in the cave in 1961 (probably prompted by a boy's accidental discovery in 1955) and that is when the first and most significant excavations of the cave by M. Garašanin and P. Mijović began. The rock shelter is located in a naturally formed amphitheater above the settlement of Lipci, where a rocky overhang rises abruptly from the surrounding ground some 7 m (Vučković A. https://www.ancient-origins.net/). It had an opening to the southwest, although it is visible that one part of its walls was torn down. To the west is a stream that interflows into the sea, representing good settling conditions, given the fact that the surrounding area is quite scarce in potable water. On the wall of the cave, there are drawings of animals shown walking—deer arranged in two rows. The upper row on the left shows a four-legged animal with short ears, which the researchers assume may be a dog or a deer cub, because behind it is a representation of another deer. Behind the second row of deer is a depiction of an old symbol—the swastika,[3] which was also found on one of the fallen blocks of rock. The dimensions of the swastika symbol are about 40 cm, and the deer up to 50 cm. The carcasses of all deer are hatched transversely (Garašanin 1967; 72.) (Fig. 3).

Fig. 3 Drawing of a deer in the Lipci rock shelter. *Source* http://montenegrina.net/)

According to P. Mijović, a hunting scene was painted in the cave in which there is a group of "seven deer running away from two hunters on horses. They each have a spear so schematic that it cannot be taken for any comparisons. A special decorative element consists of four swastikas with closed arms and a ship of the same size". He further states that the size of the figures is expressed according to symbolic belief: swastikas are largest, followed by deer, ship, and the smallest are hunters on horses (Mijović P., 1987; 11–12). All drawings are made with white paint based on kaolin bound with caseine,[4] which is actually a combination of cheese and lime (stone). This is an old technique and it is very resistant to all weather conditions—it is insoluble in water and unaffected by light (Mijović 1976). The convexity of these drawings is a consequence not only of the application of a layer of paint, but also of the effect of erosion on the limestone. This finding is a unique phenomenon in the Balkans. According to the depictions on the rock, it is the most similar to the site in Val Camonica near Brescia in northern Italy,[5] although the images at this site were carved.

[3] The swastika is thought to have been the primordial protohuman's symbol of the Sun.

[4] This technique was characteristic for prehistoric painting, because when this technique is applied to stone, it hardens and becomes resistant to erosion.

[5] Research has discovered over 20,000 rock paintings.

The site in Val Camonica had 4 development stages of rock paintings—the paintings in Lipci are linked to the third stage (Garašanin 1967).

About 50 m in front of the cave, unaltered stones of different sizes were found, arranged in a semicircle in the form of a wall up to 2 m wide. According to M. Garašanin, the construction technique is reminiscent of Illyrian fortifications—hillforts. Here, the locals were excavating in search of the treasure of Empress Teuta, which, according to legend, was buried in this area, and they found fragments of rougher ceramics belonging to the prehistoric period. It is assumed that there could have been a sanctuary here where magical rites[6] were performed, intended to achieve richer hunting and preserve the fertility of animals, thus it is supposed that these are the beginnings of the local cults of the Illyrians and Thracians, which will be differentiated much later (Garašanin 1967). P. Mijović supports the theory about the sanctuary as well, giving additional evidence that the swastikas, which are otherwise a cult symbol of the Illyrians, are facing "the first morning rays in the rock shelter in which the performance of sun rituals is expected—Lipci would, therefore, be one of the most monumental of our sanctuaries *sub divo*". He further states that deer are cult animals, so the scene with the deer would represent the sacrifice of the killed game, while the ship also has a symbolic meaning of transferring souls to the other world (Mijović 1987; 22).

Pavle Mijović dated the origin of the paintings in the cave to the eighth century BC (Mijović 1987). However, some authors assume, based on a comparison with the Val Camonica site, that the drawings were created in the tenth century BC (Vučković A. https://www.ancient-origins.net/), while G. Vukčević claims, based on the drawing of the ship, i.e. its appearance, that "ships with two masts were built first by the Phoenicians, and after them by the Greeks (Athenians) with whom they participated in the war against the Persians, in the battle of Salamis in 480 BC. This would mean that the ship could not be older than fifth century BC. The Romans had such ships in the 1st Punic War, in the middle of the third century BC. Hence, such a ship could have come to the Bay of Kotor after the victory over the Persians, to show its strength to the Illyrians who were allies of the Persians or during the Illyrian-Roman wars" (Vukčević 1992; 113).

Today, unfortunately, the site is neglected; it needs restoration and wider presentation because this is one of the rare sites of prehistoric art on the entire Adriatic coast.

The Vranjaj cave is located on Orjen at an altitude of 900 m; the opening of the cave is 6 m high, 12 m wide and faces southwest. Its depth is about 30 m and it is divided into several halls. It was found by archaeologist I. Pušić in 1980. In it, material remains were found that indicate population (permanent or occasional) in the period from the Middle Neolithic to the Late Bronze Age. In the Late Neolithic (from 4500 to 3000 BC), the population of the Bay of Kotor belonged to the Hvar-Lisičići group, as evidenced by the few material remains from this cave, as well as the Spila cave near Perast. The oldest remains found are parts of a larger dish (bowl) decorated with notches, and given that similar dishes were found in Crvena stijena, it is assumed that Vranjaj was inhabited either at the end of the 7th or the beginning of the 6th millennium BC (RTCG https://www.youtube.com/).

Three cultural horizons were identified in the cave: the first belongs to the Neolithic, which is divided into an older and a younger phase. Ceramics from the older phase belong to Cardium culture, while the younger phase belongs to the Middle Neolithic and to the Danilo-Kakanj culture complex (http://montenegrina.net/tag/nikola-borovinic/). The small number of findings from the Late Neolithic, as well as the fact that no traces of a hearth or any other findings of a longer stay in the cave were found, leads to the assumption that only small groups of hunters during hunting or herders during grazing occasionally settled there.

There is no evidence of the transition from the Stone Age to the Late Copper Age. At the beginning of the Bronze Age, with the emergence of the Ljubljana culture, the ethnic composition of the population on the eastern coast of the Adriatic and in the hinterland was significantly changed, due to the "breakthrough" of newcomers from the interior. The new inhabitants of the Bay of Kotor were engaged in cattle breeding, so the habitats had to be close to pastures and out of the reach of potential enemies, so Vranjaj then became one of the most important habitats of members of the so-called Dinaric culture (RTCG https://www.youtube.com/).

The Spila cave is located on the southwest side of the St. Ilija hill below the village of Glogovac, 1,5 km east of Perast, at an altitude of 320 m. In front of the entrance to the cave is a smaller leveled plateau. The opening is 6 m high and 9.5 m wide, facing southwest. From the entrance all the way to the end of the canal of the cave, the ground gradually descends, which affected the thickness of the layers—they are 0.6 m thick at the entrance, and up to 2 m toward the middle of the canal. The first explorations of the cave were carried out in 1968 by G. Brajković (who also found the cave), and continued in 1974. During the excavation, twelve fireplaces were found in the cave, which testifies to the life in the cave. The analysis of the material found in the cave revealed that it belongs to the Neolithic—stratum I (namely three phases older Ia, middle Ib and younger Ic) and Eneolithic stratum II (also three phases IIa, IIb and IIc) (Marković 2006).

[6] It should be noted that even today yule logs are lit here on the Christmas Eve, which can be linked to its ritual meaning.

In the layers that belong to the Early Neolithic (Ia phase), baked clay pottery both with coarse or fine texture were found. Based on the fragments, it was concluded that these were deeper oval pots, as well as large and small spherical bowls. These were decorated mainly by incising and embossing, most often using jagged or flat seashells or snail shells. Finer pottery is well-fired; it has thinner walls with polished surfaces. Spherical, non-foliated bowls with a flared rim and oval bowls with a short cylindrical neck are dominant here. There was no decoration on this pottery. Among the tools from this phase, several longer and larger knives made of lamellae and two awls made of mammal bone were found. These findings place Spila in stage II of the Early Neolithic of the Adriatic (http://montenegrina.net/tag/nikola-borovinic/).

From the Middle Neolithic period (Ib phase) there is a small number of fragments, mainly coarse, transitional and fine pottery. Coarse pottery is predominantly made of clay with a lot of sand, well-fired, gray-brown to brown in color. Transitional ceramics are thinner-walled, well-fired, also mixed with sand and dark gray, light brown to brown in color. Fine ceramics are scarce; they are characterized by good cleanliness with an insignificant amount of sand and mica, and their surfaces are polished and dark gray in color. The presence of pear-shaped dishes can be reliably established. Decoration in the form of rhombuses was found only on one piece. The Middle Neolithic lasted a short time and, according to the analyzed material, it belonged to the Danilo group. The remains of tools from this period were not found (http://montenegrina.net/tag/nikola-borovinic/).

The remains of pottery from the Late Neolithic (Ic phase) are numerous, one complete bowl was also found. It is made of insufficiently refined clay with a lot of sand and mica, well-fired and is mostly brown and dark gray in color. Bowls that are decorated in a number of ways are predominant: engraving, painting, fluting and profile ribbing. All found pieces belong to monochrome pottery with polished surfaces. This phase of development belongs to the Adriatic group Hvar-Lisičići. No flint tools were found in this phase either, and only one bone tool—an awl made of deer horn was found. Of animal bones found, most belong to domestic animals (sheep and goats). Hunting was important in this period as well, as evidenced by numerous remains of bones. Not a single fish bone was found, even though the cave is relatively close to the sea, but there are plenty of shell remains (http://montenegrina.net/tag/nikola-borovinic/).

About 500 m to the east of Spila is the **Tamnica cave**, with a low, barely visible entrance. It was very important for the inhabitants of Spila at the time. It consists of several canals with water-filled indentations at the end and one large hall. Based on ceramic fragments found along the access canal and on the floor of the hall, it is assumed that the inhabitants of Spila used it for water supply for most of the year (Marković 2006).

The Daletina cave is located above the settlement of Sveti Stasija in Dobrota near Kotor, at an altitude of about 400 m. The opening of the cave faces the sea, direction southwest. The opening is 4 m high and 8 m wide. Part of the entrance to the cave is covered by vegetation. The cave has an S-shape which "vanishes" into the chasm. It has four halls with a gallery. Entering it is difficult due to fallen rocks. According to the sources, there is water in the interior (Kovačević 1995–1998).

Between the first and second galleries, a cultural layer was registered where many ceramic fragments, smaller animal bones, as well as the remains of sea shells (mussels) and snails were found. Ceramic material mostly consists of parts of dishes made from unfiltered dirt with admixtures of sand. The bowls are coated on the outside and polished to make them smooth. Some of the dishes are decorated with engraving of simple lines (panicle ornament and net ornament), so based on this it can be concluded that this is the Adriatic type of the Ljubljana culture, which dates the habitation of the cave to the period of the early and developed Bronze Age (Kovačević 1995–1998).

The Dučići cave is located at an altitude of 700 m on a rocky cliff above the villages of Peuta and Gornje Vrbice on the northern edge of the Skadar valley. The opening faces southwest, is 0.8 m wide and 3 m long. A roughly knapped axe with a hole for the handle in the middle and a long (20 cm) curved flint knife—associated with the Eneolithic period, were found inside. It is assumed that there are more material remains in the cave, but research has never been carried out (http://montenegrina.net/tag/nikola-borovinic/).

The Bioče notch on the left bank of Morača at the foot of Gradina hill some 100 m upstream from the mouth of the Mala river into Morača is a site with Middle Paleolithic layers. The cave is located about 50 m above the river, the opening is of small dimensions facing the southwest, which allows this notch to sunbathe throughout the day. The cave was explored in 1986, then periodically from 1988 to 1996 and again in 2010. The cultural deposit discovered is almost 3 m thick with 8 cultural layers from the Middle Paleolithic, which is roughly the time period between 70,000 and 40,000 BC (Marković 2006).

It should be emphasized that the excavation of deposits in the cave did not reach the solid rock base, so it cannot be said with certainty whether there are more cultural layers under the clay base. Life in the cave took place mainly during the Mousterian period, that is, the Middle Paleolithic. Those layers are located immediately on the surface, which may indicate that life did not take place in the cave in other periods, although there are also assumptions that the upper layers in the cave, which were removed by human action, may have contained layers from the Middle Stone Age of the Mesolithic. It is likely that later investigations will provide more evidence about the periods of life in this notch.

In the material found, eight cultural layers were identified, which were divided into two series, different in texture and color. In the first series, which mainly consist of layers of reddish and grayish-colored clay with admixtures of humus, soot and ash, a lot of stone artifacts, mostly chert, and a significant number of animal bones were found. In the second series of yellowish color, there were much fewer animal bones. Stone artefacts were mostly made of pebbles and considerably less of chert. These findings show similarity to the Crvena stijena Mousterian (Marković 2006).

The Vruća cave is located on the left bank of the Mala Rijeka near the mouth of the Morača. The cave is in the form of a smaller tunnel, 11 m long and 3–4 m wide. The entrance is 3.5 m high and 4.5 m wide, facing south. There is a small terrace in front of the cave. The cave was explored twice in 1988–89 and 1996–97. It was established that the cultural layer is 1.5 m thick, and the material findings belong to the Mesolithic and Neolithic. In the Mesolithic layer, a hearth, pieces of pottery, a small number of animal bones and a lot of shells and snails were found. Among the tools, knapped stone tools, various scrapers and a harpoon were discovered. In the Neolithic layer, a dozen fragments of Cardium wares, animal bones, typically from deer and roe deer, harpoons, stone tools—mostly cutters and trowels with narrow blades—were found. It is assumed that the tools from this period are part of the same tradition as in the Odmut cave in the Piva canyon (Borić et al. 2019).

Odmut in the Piva canyon, next to Crvena stijena, is the most-researched cave in Montenegro. It is located at the foot of the limestone hill Kuline, by the mouth of the Vrbnica into Piva, at an altitude of 558 m, some 3 km from the old Plužine—nowadays it is submerged at the bottom of the Piva Lake. The width of the cave opening is 20 m, and the height is 14 m. Systematic archaeological excavations and research began in 1972 and lasted until 1974. The opening of the cave faces south, so the cave has very favorable climatic conditions, which caused it to be continuously inhabited for a long period of time. Based on the cultural material found, it was established that life in the cave lasted from 8300 to 1800 BC. Seven strata were discovered, the oldest of which belongs to the Mesolithic, the next two to the Neolithic, three to the Eneolithic, and the youngest layer that ended life in the cave belongs to the early Bronze Age (Marković 2006).

The Mesolithic layer, with a total thickness of 1.5 m, is divided into older Ia, 1.1 m thick, and younger Ib, 0.40 m thick layers. Layer Ia was formed in the Preboreal period between 8100 and 6700 BC, and layer Ib was formed in the Boreal and early Atlantic between 6700 and 5200 BC. A large number of stone tools were discovered in both layers: knives, scrapers and flakes made of gray-green sandstone and gray and reddish stone, few of which had regular shapes, as well as a significant number of tools made of bones, deer antlers and wild boar teeth. In this cave, the Mesolithic was distinct, because the aforementioned tools, especially harpoons,[7] have peculiarities that were found neither in Montenegro nor on the Balkan Peninsula. Flat and cylindrical harpoons are found in the late Magdalenian or Asselian phase of the southwestern and western European cultures. Also "a bone fragment with an engraved geometric ornament and a river pebble with an ocher color decoration can only be linked to the Final Paleolithic of the Franco-Cantabrian area on the Apennine Peninsula" (Marković 2006; 77).

The oldest layer (Odmut I), 1.50 m thick, lies on alluvial-eolian sediments and is clearly separated from the younger layers (Odmut II-VII) by its yellow color and loess-like structure, which consist of dark loose soil mixed with a larger amount of soot, ash and broken stone. Only the bones of wild animals (ibex, deer, bear, wild boar, roe deer, etc.) were found in the Odmut I layer, which indicates that the first generations of the inhabitants of Piva were engaged in a hunter-gatherer system. Among the stone tools of the Odmut I layer, made mainly of greenish chert or grayish stone, five basic types can be distinguished: knives, scrapers, flakes and the so-called geometric microliths (Marković 2006).

The Odmut II layers belong to the Early Neolithic, which had two development phases: the older Odmut IIa and the younger Odmut IIb. Three circular hearths were found there, one of which was surrounded by several upright stones. The Neolithic and Eneolithic layers in the cave confirm the close ties with the layers of cultures of the same period that developed along the Adriatic, on the islands and the wider Mediterranean, which is especially reflected in the pottery from the Early Neolithic, decorated impressing (Cardium pottery). Close links of the Danilo culture with the cultures of the Middle Neolithic, as well as with the cultures of the Late Neolithic, which are known as the Hvar-Lisičići culture, have also been established. Ceramics are characteristic of phase IIa, while in phase IIb they are very rare, and toward the end they completely disappear. The most common forms of pottery are spherical or pear-shaped pots, deep hemispherical bowls and larger conical bowls. The tools found from this period were mainly made using the knapping technique, while there are significantly fewer polished stone[8] tools. Findings

[7] "Harpoons made of deer horn pay a special homage to the oldest culture of the Odmut cave. In layer I, 56 complete and fragmented harpoons were found. Two basic forms are observed: flat harpoons (52 pieces) and cylindrical harpoons (four pieces)." Radović G. http://www.montenegrina.net/pages/pages1/arheologija/pecina_odmut.htm.

[8] Only one polished stone ax was found in the cave from this period.

from the Middle Neolithic and the beginning of the Late Neolithic are missing in Odmut, so it can be assumed that the cave was not inhabited in that period for some reason. Findings from the Late Neolithic are very poor as well, which indicates that this phase in the development of the Neolithic was short-lived. In addition to hunting, cattle breeding was a secondary economic branch of the inhabitants in this period. From the remains of the material culture, clay dishes, pots and bowls were found, and from the tools, which are much rarer than in earlier periods, stone arrows and axes (Marković 2006).

The rich Eneolithic layers, which are the first documented findings of the existence of the Eneolithic in Montenegro, made it possible to observe this epoch in three development phases. They are connected with the Eneolithic cultures of the Adriatic, as well as with cultures that came from other regions, mainly from the east—the area of the Black Sea steppes, who brought with them the remains of their culture, which had an impact both on the changes of the traditional Eneolithic cultures, and on the formation of a new ethnic composition. 3 strata belong to the Eneolithic period: Odmut IV, V and VI, but Odmut IV has characteristics of the final Neolithic and the beginning of the Eneolithic. In stratum V, one hearth was found with the remains of a stone wreath, which indicates a longer stay in the cave. In this period, a new type of pottery appears "which consists of more or less rounded dishes with uneven or smoothed surfaces with a specially separated and accentuated neck and a rolled-back rim over the outer edge of the dish" (Marković 2006 161). The inhabitants of the cave were primarily engaged in hunting, but, based on the remains of the fauna, it can be concluded that cattle breeding was also developed as a secondary activity.

In stratum VI, which belongs to the developed and late Eneolithic period, finer ceramics rarely appear; mainly those with a rough texture dominate. Very few finds from stratum VIb are proof that the cave was only occasionally used in the Early Bronze Age. At the end of the Early Bronze Age, stratum VII was inhabited again, only to be abandoned around 1800 BC (Marković 2006).

The Na Sastavcima cave is located at the mouth of Komarnica into Piva, that is, at the confluence (sastavci) of Komarnica and Sinjac on the rocky slope of Dečić rock. Several ceramic fragments were found in it, which, based on their texture and shape, belonged to one of the earlier phases of the Neolithic. The analysis of the archaeological finds revealed that the cultural material from the cave has the most similarities with one of the phases of the Starčevo culture (Marković 2006).

The Grad cave is located in the Radmanska ravine in the village of Radmanci near Berane. The entrance is about 10 m wide and ends in a 4.5 m deep corridor. A probing radar test revealed a layer 0.5 m thick, in which there are mainly ceramics from three periods. The lowest and oldest has characteristics of the Eneolithic culture, the middle of the Bronze Age, while the youngest has elements of Roman culture (Marković 2006).

The Minina cave is located in the middle of the hill of the Kaludarska river canyon near Berane. The cave was not examined, but only its plateau, where a cultural layer consisting mainly of ceramics that most likely belong to the Eneolithic period was found. It is assumed that this cave was inhabited by a new population that moved there from the plains (Marković 2006).

The Vezačka cave is a notch 40 m long, 35 m wide and 7.5 m deep. It is located on the slope of Vezac hill in the settlement of Rvaši on the shore of Lake Skadar, municipality of Cetinje. It stretches in a southwest-northeast direction. The first examinations were carried out in 2013 and 2016. In 2017, a more detailed research of the cave was conducted. Two geological layers with a thickness of 1.8 m were found which were divided into six phases. In the second geological layer, which namely has five phases, fragments of ceramics, bones of deer and roe deer, as well as of some domestic animals (goats, sheep, pigs, horses and dogs) were found (Gazivoda et al. 2014). The examination of the discovered layers in the cave revealed cultural layers from the Iron and Bronze Ages, the Roman period and the Middle Ages (Gazivoda 2018). An expert analysis carried out in Great Britain established that the Vezačka Cave was inhabited in the periods from 4746 to 4546 BC, from 2471 to 2289 BC, from 192 to 37 BC and around 1000 AD (Gazivoda et al. 2015).

The Seocka cave is located in the village of Prevlaka, municipality of Cetinje. It stretches in the northwest-southeast direction; is 41 m long, up to 23 m wide and 7.5 m high. The floor of the cave descends from the entrance toward the southeastern part, where a small lake forms in the cold half of the year (autumn–winter). Probes for the exploration of the cave were placed in 2013, and the research continued in 2014. In the investigated layer was found a large number of animal bones (deer, beaver, pig and lamb), flint artifacts and several fragments of ceramics from the Bronze Age (Gazivoda et al. 2014). Examination of samples at the Scottish Universities Environmental Research Centre in Great Britain and OSL analysis of microfauna, rock and chert, revealed that Seocka Cave was inhabited in the period 8000–7600 BC and around 1500 AD (Gazivoda et al. 2015).

The Koronina cave is located about 3.5 km northeast of Cetinje at the bottom of the Dobrštak slope at an altitude of 750 m. It is a rock shelter with a 6 m wide and 4.5 m high opening, facing southwest. There is a small flat plateau in front. The depth of the cave is 3.2 m, while the depth of its layers reaches up to 80 cm. In the eastern part of the cave was found a layer 35 cm thick, which contained archaeological finds that indicate that it was used for a relatively short time during the Early Neolithic period as a shelter for a small group of hunters. Analysis of the ceramic findings determined that they were larger oval dishes with a rounded and slightly disheveled rim. This pottery belongs to the Cardium culture. Three animal horns were also found, one of which had traces of processing (Marković 2006).

The Čaja cave is located on the slopes of mount Garač in the village of Do Pješivački, municipality of Danilovgrad. It was registered as an archaeological site in 1988. It stretches in the southeast-northwest direction, consisting of one room up to 29.5 m long and up to 12.5 m wide. The entrance is 5.85 m high. The floor of the cave descends toward the interior in the northwest direction. In the layer of dark gray soil, interlayered with small stones, located at a depth of 0.6 to 1.65 m, a hearth was found in a layer of gray soil, and some mobile material around the hearth, mostly fragments of ceramics, goat, sheep and deer bones and animal teeth. A second hearth was found in the clayey terra rossa layer at a depth of 1.68–1.8 m. The works were conducted up to the depth of 2 m until they were suspended. Examining the central part of the probe at a depth of 2.25 to 2.7 m in the brownish-beige sediment reveals ceramic fragments, animal teeth and bones (Gazivoda et al. 2015). The cave was definitely inhabited during the Neolithic period—a more detailed periodization will be determined when the results of samples sent to the UK for detailed analysis arrive.

The Vrbička cave near Nikšić was discovered in 2010 during the inspection of the territory of our country for the purpose of creating the Archaeological Map of Montenegro; it was recorded as a potentially interesting archaeological site (Pobjeda, 30.10.2015). The cave is located in Duga,[9] about 18 km northwest of Nikšić at an altitude of 948 m. The research was carried out in the period 2012–2013 and 2015–2017. Based on the analysis of the found material, it was established that they originate from different periods of human history: Upper Paleolithic, Mesolithic, Early Neolithic, Late Neolithic, Copper and Bronze Age (Borić D. et al., 2019). Remains from the "younger", that is, Upper Paleolithic in the Vrbička cave, belong to the period of the early or middle Epigravettian (about 20,000–14,000 years ago). Tools made from solid flint or from bones of various animal species were found, as well as a large number of rodent bones, most likely marmots,[10] which were hunted at that time. Flakes with steep retouch and smaller, retouched blades, thumbnail scrapers and right-angle blades were also found (Pobjeda, 19 August 2012). Based on the remains found (mostly findings with traces of cutting and burning) from the Paleolithic period, it was concluded that the cave was a transit stop for Paleolithic hunters who hunted marmots in that period. (https://www.rtcg.me/vijesti/).

Researches assume that there was a hiatus from the Upper Paleolithic to the Late Mesolithic (Pobjeda 30.10.2015). The layer that belongs to the Mesolithic is 10 cm thick, where scrapers, trowels and the remains deer, roe deer, chamois, wild boar and ibex bones were found. This period is dated to the end of the eighth millennium and the first three centuries of the seventh millennium. The discovery of a bead made from the pharyngeal tooth of a carp is interesting; similar ones were found in the vicinity of Đerdap, which indicates the connection between these cultures (Borić et al. 2019). It is assumed that one tooth stems from the Mesolithic period (about 7000 years old). One hearth from this period was also discovered (https://www.cdm.me/).

From the Early Neolithic period, which is related to the Cardium culture of the eastern Adriatic, pieces of ceramics were found (https://old.dan.co.me/), and there are also remains related to the Late Neolithic (about 6500 BC) which are linked to the Hvar culture of the eastern Adriatic region (https://www.vijesti.me/). The layer from the Late Neolithic is well represented in the cave with many fragments of pottery with ornaments typical of Hvar tradition. It should be emphasized that this cave is the richest in finds associated with the Hvar culture. In addition, large quantities of animal bones, flint arrowheads knapped on both sides, sandstone grinding wheels, discoid-shaped beads made of seashells or stone were found from this period (Pobjeda, 30.10.2015). There is also a tooth found in the cave from the Neolithic period. The aforementioned objects probably belonged to the first farmers and herders who came from the Middle East and Anatolia, spreading across the Balkans and further into Europe. It is assumed that there was also a coastal route along the Aegean, Ionian and Adriatic coasts (https://www.cdm.me/).

The cave was also inhabited in the Bronze Age, as evidenced by numerous pieces of ceramic fragments and 3600 years old human bones with traces of cutting. This was probably done for ritual purposes, as this was a part of the rites in the burial practice. It is assumed that they can be linked to the so-called Cetina culture (https://old.dan.co.me/). Bones of a child from this period were also found—it was probably buried in the cave (https://www.cdm.me/).

Cave under the "Gospića Vrh" peak is a notch located in the Ćehotina valley between Vrulja and Ljutić. Archaeological research conducted in 1982 established that the cave was inhabited during the Paleolithic—the Old Stone Age. Unfortunately, very little cultural material has been preserved—a flint tool, i.e. a stepped retouched side

[9] Settlement Presjeka.

[10] One marmot bone, was dated to 23,000–24,000 BC (Pobjeda 30. 10. 2015).

scraper, on the basis of which it can be determined that the cave was inhabited in the period from 50,000 to 40,000 BC, and, that the material has characteristics of the Mousterian culture, which indicates a connection with the localities Krapina in Slovenia and Veternica in Croatia, as well as localities in northeastern Bosnia and Šumadija (https://www.youtube.com/).

Mališina stijena in the gorge of the Ćehotina river on the left bank is a notch 15 m deep, 35 m wide and 10 m high. The opening faces north, in contrast to other caves whose opening faces south or southeast. It was investigated in the period 1980–1986. According to the results obtained from 16 cultural layers, it was concluded that the cave was continuously inhabited from the Middle throughout the entire Late Paleolithic. There is a similarity of the tools found with those from Crvena stijena and with caves from the Paleolithic period in Slovenia and Croatia (Marković 2006). According to the excavations and analysis, there are four archaeological stages in the cave; the lower three with stone artifacts from the Middle and Lounger Paleolithic, and the upper, fourth layer from the Final Paleolithic. Fauna remains were also found in the layers (https://www.youtube.com/).

Medena stijena is located at the foot of the hill on the right bank of the Ćehotina canyon, at 780 m MSL. The opening of the cave, 30 m wide, faces south; the interior is 10 m high and 8 m deep. Systematic investigations carried out between 1983 and 1991 revealed a stratum that could be divided into ten cultural layers, of which the five oldest belonged to the cultures of the Late Paleolithic, the fourth belonged to the Mesolithic, the third and the second to the Eneolithic and the Early Bronze Age, while the first belongs to the modern age. Stone tools from the cave belonging to Paleolithic layers share similarities with those from Crvena stijena and Franchthi cave in Greece (Marković 2006).

The tools from the Mesolithic layers are made of flint, mainly green and red jasper, gray and black flint of poor quality and brown stone of good quality. These are roughly jagged flakes, scrapers, short scrapers and short blades; they have similarities with those from Lepenski vir on Đerdap, Crvena stijena and Odmut. The youngest layer in Medena stijena was formed above the Mesolithic and belongs to the Copper Age, and not to the Neolithic as expected, which indicates that the cave was abandoned during the Neolithic, then briefly inhabited at the beginning of the Metal Age (Marković 2006).

The Trlica cave near Pljevlja is located 1.6 km northeast of the mouth of the Ćehotina canyon at an altitude of 925 m on the slope of a mountain depression of tectonic origin. The cave was opened and partially destroyed during the construction of the Pljevlja-Prijepolje road. In 2010 and 2011, probing examinations were carried out that proved the presence of bones of small and large mammals in the upper layers from the Middle Pleistocene, and in the lower layers from the Early Pleistocene period (Derevjanko et al. 2015). The remains in the cave are between 600,000 and 700,000 years old. In this fossil cave, sediments with bones and teeth of extinct vertebrates (ancestors of today's animals) are deposited. In addition to other findings, the following were discovered: an ancestral species of elephant, rhinoceros, giant Balkan sheep (previously unknown to science), several species of chamois, giant deer, several species of bison from the Pleistocene period, several species of wolf, bear, two species of horse, a cave panther, a sabre-toothed tiger, porcupine, African dog, beaver (https://www.portalanalitika.me/).

3 Monasteries Founded in Caves

The Ostrog Monastery was built in a cave above the village of Zagorak in Ostroške Grede at an altitude of 900 m. According to historical data, Ostrog-grad was mentioned for the first time "near Onogošt in Gornja Zeta" in the muniments of the King of Naples in 1444 and 1454, and in 1601 in Mavro Orbini's work "Kingdom of the Slavs". It is assumed that it was built in the early Middle Ages (in the time before Prince Vlastimir) and existed as a border fortification until the end of the fifteenth century when it was destroyed by the Ottomans. Today, of the former settlement remains only the name of the mountain and the monastery, which was built two centuries later in the ascetics' caves of the Venerable Isaiah of Onogošt and his fellow eremites at the beginning of the seventeenth century, with the blessing of the Orthodox archpastor of Zahumlje and Skenderija, commonly known as Saint Basil (Vasilije) of Ostrog (http://manastirostrog.com/). Based on the data on the stay of Saint Basil of Ostrog, who spent the last five years of his life there "in a prayerful feat" (http://manastirostrog.com/) and died on May 12, 1671, it may be concluded that the Ostrog Monastery, that is, the cave church of the Holy Cross, was built in 1665–6, and the iconostasis completed two years later. Also, there is no precise information about the origin of the Lower Monastery. According to popular belief, there used to be a village there, and on the property that Vasilije bought from two brothers, he built a billet for young monks, which is considered the beginning foundation of the Lower Monastery. The Church of the Holy Trinity—the Matica of today's Lower Monastery was built in 1824 (https://me.visit-montenegro.com/).

According to folklore, Vasilije dreamed that the Lord ordered him to start building a monastery in Ostroške Grede. He went there the following day and, since he liked the place, he built the Temple of the Presentation in the first cave, a storeroom and a guest room in the second, and a chapel where things for worship were kept in the third.

Fig. 4 Monastery of Ostrog. *Source* https://dobrocinstvo.rs/manastir/manastir-ostrog/

Below the monastery (about half an hour's walk) there was a small church with a chamber for young monks and a storehouse for grain (https://me.visit-montenegro.com/).

The monastery and the area of Ostrog have a rich history spanning all the way from the Roman times when roads crossed here, throughout the period of medieval development of the then state of Zeta, to the period of Ottoman reign. In 1850s, while defending themselves from the Ottomans, Mirko Petrović and 17 Montenegrin fighters locked themselves in the monastery. On that occasion, they had to take out the relics of St. Basil and hide them, so that the Ottomans would not burn them during the invasion of the monastery. At that time a significant part of the cultural treasure was destroyed by the invaders. Also, during the Second World War, in February 1942, an assembly of PLW (People's Liberation War) fighters was held in the monastery regarding the further course of the struggle for liberation (https://me.visit-montenegro.com/).

Important cultural treasure is kept in the monastery. The monks of Ostrog were gifted a fragment of the cross upon which Christ[11] was crucified, which is an invaluable relic of the Christian world. The frescoes in the cave church of the Holy Cross on the top of the Upper Monastery, which were painted in the seventeenth century, are also a very valuable cultural treasure. The relics of St. Basil are kept in the Church of the Presentation of the Virgin Mary, which is considered the most important shrine in Montenegro and Herzegovina. It was built by upgrading a vaulted stone overhang on the cliff. Until the Second World War, a copy of the Octoechos from 1494 was kept here (https://me.visit-montenegro.com/). Some believe that King Petar II Karađorđević left chests with gold in Ostrog at the beginning of the Second World War, during his stay in the monastery before going to Greece, as Ostrog was supposed to be the war seat of the Kingdom of Yugoslavia. However, these claims have never been proven. What is known is that, in addition to the fragment of the Holy Cross, two other relics of immeasurable value were found here: the hand of St. John the Baptist and the icon of Our Lady of Philermos (https://www.nezavisne.com) (Fig. 4).

The monastery is one of the most important sites of pilgrimage in Montenegro and beyond. During the summer months alone (July, August, September) it is visited by 250,000 to 300,000 believers and tourists. Up to 20,000 believers gather in the monastery on the day of St. Basil of Ostrog on May 12 every year (https://www.nezavisne.com).

The Dajbabe Monastery is located 3–4 km from Podgorica on the road to Petrovac, named after the village of the same name in which it is located; it is dedicated to the Dormition of the Mother of God. The monastery is buried in the ground; the space for the church is a natural cave. According to the words of Archimandrite S. Popović, who is the founder of the monastery, a boy from Zeta, shepherd Petko, came to him in Ostrog (where he served at the time), and told him that he had a vision in which a saint appeared

[11] It is assumed that this is the reason why the temple is dedicated to the Holy Cross.

Fig. 5 Dajbabe Monastery. *Source* https://waytomonte.com/rs/p-3178-monastery-dajbabe

to him and told him that he was a holy disciple of a Saint who lived before the Battle of Kosovo (Saint Sava), that he had been the bishop of this place and that he wanted a monastery built for him. Simeon Popović went to Dajbabe and started excavating with several villagers from Dajbabe. He found a cave in the shape of a cross in the rocks, which is believed to have been the catacombs from the third century AD where the first Christians may have taken refuge. He expanded the cave by blasting, and after 2 years of works, he consecrated it on December 22, 1897 (Popović S. 1928). He stayed there for the rest of his life.

The church in the cave, which is buried in the ground, is small—21.5 m long, with an average width of 2.5 m. The uneven parts of the walls are plastered, and in the central part next to the northern wall is the resting place of the monastery's founder, Simeon Dajbabski. From the main nave the church bifurcates into several chapels. On the walls are frescoes painted by the founder Simeon Dajbabski (https://www.panacomp.net/) (Fig. 5).

A Russian shroud from the eleventh century and an icon of Our Lady of Jerusalem, brought by hieromonk Sergej from Skopje in 1915 and left to Simeon Dajbabski for safekeeping are kept in the monastery's church (https://www.panacomp.net/).

4 Conclusion

Caves were the natural habitats of people especially in earlier historical periods. It has been reliably archaeologically established that the first human communities inhabited the territory of Montenegro before the beginning of the Middle Pleistocene (about 150,000 years ago). The deepest stratum in Crvena stijena with layers XXXI-XXIX belongs to the culture that precedes the Middle Paleolithic. Layers from the Paleolithic period were found not only in Crvena stijena, but also in the Bioče cave near Podgorica, the cave under the "Gospića Vrh" peak, the Mališina and Medena caves in the Ćehotina valley near Pljevlja, and the Vrbička cave in Duga near Nikšić. Traces of fauna between 600,000 and 700,000 years old were found in the Trlica cave near Pljevlja. The single occupation of the inhabitants back then was the hunting-gathering economy.

Cultural remains from the Mesolithic period were found in the caves of Crvena stijena near Petrovići, Spila near Risan, Vrbička cave, Odmut cave near Plužine, Bioče cave and Vruća cave in the Morača valley and Medena stijena. Compared to the previous period, there are no major changes regarding the occupations of the inhabitants; hunting is still the main activity, gathering is significantly represented, while fishing is also present in some places.

Cultural layers from the Neolithic period were found in Crvena stijena, Čaja cave near Danilovgrad, Vranjaj cave on the Orjen mount, Spila near Perast, Odmut cave, Koronina cave near Cetinje, Vruća and Vrbička caves. The "economic life" of the inhabitants of the Neolithic period in Montenegro is mostly similar to that in the Mesolithic period—except for the fact that in the Middle Neolithic there are traces of cattle breeding, and in the Late Neolithic there are also traces of agriculture, but only in localities that were not of the cave type.

Traces of material culture from the Eneolithic—Copper Age period can also be found in the following caves of Montenegro: Crvena stijena, Vrbička Cave, Medena stijena, Odmut cave, Spila cave, Grad cave and Minina cave near Berane. During the Bronze Age, four caves were mainly inhabited in Montenegro: Crvena stijena, Odmut, Vranjaj and Grad. The basis of the economy in this period still consists of cattle breeding, hunting and fishing. Although the gathering economy has persisted in some areas, agriculture is scarcely represented—traces of a millstone for grinding grain were found only in Crvena stijena.

In later periods of human development, traces of life in caves become increasingly rare. Thus, traces of material culture from the Iron Age are found only in the Vezačka cave near Lake Skadar (Cetinje municipality). Traces from the Roman period, in addition to the Vezačka cave, are also found in the Daletina cave near Kotor.

References

Basler Đ. (1975): Stariji litički periodi u Crvenoj stijeni, u Crvena stijena – zbornik radova, Nikšić.

Benac A. (1975): Mlađi praistorijski period u Crvenoj stijeni, u Crvena stijena – zbornik radova, Nikšić.

Borić D., Borovinić N., Đuričić LJ., Bulatović J., Gerometa K., Filipović D,. Allué E., Vušović-Lučić Z., Criatiani E. (2019): Spearheading into the Neolithic: Last Foragers and First Farmers in the Dinaric Alps of Montenegro, European Journal of Archaeology 22 (4) 2019, 470–498, https://doi.org/10.1017/eaa.2019.14. Accessed 31 May 2022.

Derevjanko A.P., Šunjkov M.V., Agadžanjan A.K., Bulatović L., Vislobokova I. A,, Uljanov V.A., Anojkin A.A, Medenica I. (2012): Proučavanje pleistocenih slojeva u pećini Trlica na sjeveru Crne Gore, Davnine-arheologija u Crnoj Gori, br. I, str. 45–61, Društvo arheologa Crne Gore, Cetinje.

Derevjanko A.P., Šunjkov M.V., Agadžanjan A.K., Vislobokova I. A,, Uljanov V.A., Anojkin A.A, Živanović M.M. (2015): Proučavanje pećine Trlica u blizini grada Pljevlja na sjeveru Crne Gore, Glasnik zavičajnog muzeja, knj. 10, str. 73–92, Pljevlja.

Garašanin M. (1967): Crna Gora u praistorijsko doba, u Istorija Crne Gore knjiga I, od najstarijih vremena do kraja XII vijeka, Titograd.

Garašanin M. (1976): Crna Gora do doseljavanja Slovena, u Monografija Crna Gora, „Književne novine", Beograd.

Gazivoda D., Linden M.V., Mihailović D. (2014): Sondažna arheološka istraživanja Vezačke pećine u rvašima i Seoske pećine na Prevlaci, opština Cetinje, Glasnik Narodnog muzeja Crne Gore, nova serija knj. X, str. 139–143, Cetinje.

Gazivoda D. (2018): Sondažna arheološka istraživanja Vezačke pećine u Rvašima, prijestonica Cetinje, u okviru međunarodnog projekta EUROFARM, JU Centar za konzervaciju i arheologiju Crne Gore, Godišnjak br. 1, str. 22–24, Cetinje.

Gazivoda D., Linden M.V., Mihailović D. (2015): Sondažna arheološka istraživanja Seocke pećine na Prevlaci i pećine Čaja u Dolu Pješivačkom, opštine Cetinje i Danilovgrad, Glasnik Narodnog muzeja Crne Gore, nova serija knj. XI, str. 117–125, Cetinje.

Kovačević V. (1995–1998): Rezultati obilaska Daletine pećine iznad Dobrote u Kotoru, Godišnjak Pomorskog muzeja u Kotoru, br. XLIII-XLVI, str.295–299, Kotor.

Marković Č. (2006): Arheologija Crne Gore, Podgorica.

Marijanović B; (2007): Neka pitanja ranog neolitika istočnog Jadrana, ARCHEOLOGIA ADRIATICA I, str. 7–54, PDF, ISSN (Online) 1848–9281 https://doi.org/10.15291/archeo. Accessed 15 June 2022.

Mijović P. (1976): Pregled umjetnosti Crne Gore, u Monografija Crna Gora, „Književne novine", Beograd.

Mijović P. (1987): Pradavne i davne kulture Crne Gore, Leksikografski zavod Crne Gore, Titograd.

Monnier G., Tostevin G., Pajović G., Borovinić M,. Baković M., (2020): Nova istraživanja paleolitskog nalazišta Crvena stijena, istorijski kontekst, Istorijski zapisi XCIII (1–2):71, str. 72–108, Podgorica. UDK 902/904(497.16).

Pobjeda, dnevne novine, 19. avgust 2012. godine.

Pobjeda, dnevne novine, 30. oktobar 2015. godine.

Popović S. (1928): Postanak sveto-uspenskog manastira, Dajbabe, Podgorica.

Vukčević G. (1992): O porijeklu Ilira–multidisciplinarna istraživanja porijekla stanovništva zapadnog Balkana iz praistorijskog doba, Unireks Nikšić", Podgorica.

Online Sources

http://montenegrina.net/tag/nikola-borovinic/ (Borovinić N. Neolitizacija Crne Gore PDF). Accessed 2 June 2022.

http://www.montenegrina.net/pages/pages1/arheologija/ (Radović G. Svjedočanstvo o postojanju praistorijskih kultura u Crnoj Gori). Accessed 2 June 2022.

http://www.montenegrina.net/pages/pages1/arheologija/ (Radović G. Pećina Odmut). Accessed 5 June 2022.

https://www.b92.net/putovanja/zanimljivosti (Crvena stijena u Crnoj Gori: Svedočanstvo istorije i kulture pračoveka 17.05.2021.). Accessed 5 June 2022.

https://www.youtube.com/ RTCG Tragovi - Pećina Vranjaj preistorijsko stanište iznad Herceg Novog (2. 1. 2018). Accessed 12 June 2022.

https://www.portalanalitika.me/clanak/281421 (Nastavljena istraživanja: Koje tajne krije Vezačka pećina u Rvašima 09. 2017). Accessed 7 June 2022.

https://www.portalanalitika.me/clanak/206176 (Šta krije pećina Čaja 23. 10. 2015). Accessed 7 June 2022.

https://www.rtcg.me/vijesti/drustvo/294752/ (Ostaci ljudi iz mezolita pronađeni u Nikšiću 03. 10. 2020). Accessed 8 June 2022.

https://www.cdm.me/drustvo/ (Senzacionalno otkriće kod Nikšića: Pronađeni najstariji ljudski ostaci iz praistorije u Crnoj Gori 30/09/2020). Accessed 11 June 2022.

https://www.vijesti.me/vijesti/drustvo/115140/kopaju-tajne-stare-do-24-milenijuma (Kopaju tajne stare do 24 milenijuma 06.09.2016). Accessed 8 June 2022.

https://old.dan.co.me/?nivo=3&rubrika=Kultura&clanak=563478&datum=2016-09-08 (Arheološka istraživanja na lokalitetu kod Nikšića dala zanimljive rezultate izdanje: 08–09–2016). Accessed 9 June 2022.

https://www.youtube.com/ (Paleolitske pećine: Vruljansko Maočki kraj - emisija RTV Pljevlja 27 decembra, 2004). Accessed 10 June 2022.

https://www.portalanalitika.me/clanak/98874 (Istraživanja Trlice: Otkriveni ostaci afričkog psa i vrste nepoznate nauci 26.04.2013). Accessed 5 June 2022.

https://me.visit-montenegro.com/ (Gradnja manastir Ostrog). Accessed 10 June 2022.

https://www.nezavisne.com/zivot-stil/putovanje/ (Manastir Ostrog: Mjesto na kojem se mole ljudi svih vjera 18.08.2015). Accessed 12 June 2022.

https://www.panacomp.net/ (Manastir Dajbabe). Accessed 3 June 2022.

https://waytomonte.com/rs/p-3178 (Manastir Dajbabe). Accessed 22 June 2022.

http://manastirostrog.com/ (Istorija manastira). Accessed 16 June 2022.

Underwater Cave Systems in the Montenegrin Littoral

Vladan Dubljević

Abstract

The diversity and specificity of underwater cave systems in the Montenegrin Littoral is conditioned by the complex geological structure, tectonic structure and geomorphology of the terrain, as well as the climatic characteristics of the area. There are several gaping cave channels on the Montenegrin coast from which water occasionally emerges. The most known are Sopot, Spila Risanska, Gurdić and Orahovačka Ljuta. Insufficient volumes of drinking water during summer months and the importance of the problem from the aspect of hydrogeology and related scientific disciplines conditioned the organizing of exploration of these structures on several occasions. Numerous and diverse factors that determine the hydrogeological characteristics of an area required the use of numerous scientific methods and methodological research procedures during three years of work by the Geological Survey of Podgorica and other collaborators from the country and abroad (from 1996 to 1999). Only some of them are listed, including hydrogeological mapping, various geophysical surveys, hydrochemical, speleological, cave diving and other research, whose characteristics classify them as complex geological research aimed at defining specific characteristics of this area. The increasingly complex endeavours required increasingly modern technology, so the most modern achievements used in the world at the time were applied in the fields of cave diving and geophysical surveys.

Keywords

Underwater cave systems · Speleology · Karst · Montenegrin littoral

V. Dubljević (✉)
Geological Survey of Montenegro, Podgorica, Montenegro
e-mail: vladan@t-com.me

1 Instead of an Introduction

The reaction to the huge volume of precipitation saturating limestone surfaces formed and rising from the sea about 60 million years ago is the creation of incredible shapes and forms. Those are primarily karrens. The simplest and the smallest karst forms. Their presence often makes the terrain hard to pass. It is easiest to describe them as shallow grooves with sharp capes. Karrens are surface forms, and their existence is linked exclusively with the chemical effect of water. Running over a healthy limestone surface, waters react with it and hollow tight channels by chemical corrosion and then widen the tight channels (Mirković et al. 1985).

By definition, caves are larger or smaller karst underground continuously connected cavities hollowed in the slopes of limestone and dolomite hills, or in the sides of river valleys. The solvent effect of water with the interaction of mechanical erosion has an important role in the development of cavities at the very beginning of formation. After this first phase, the mechanical erosion that is widening channels by caving of walls and arches of existing cavities has a dominant role in the development of cave forms as part of the further evolution of karst cavities (Ford and Cullingford 1976).

Caves are one or more horizontal channels of different lengths, cross-section and spread. The caves with one karst channel are very rare. Their general characteristic is that they are mostly very short. In branched channels, whose origin is predisposed by a complex tectonic composition which is manifested through a system of composite cracks, we distinguish the main channel and several lateral channels which are locally intertwined (Ford and Cullingford 1976).

When the lateral channels are branched in two and more horizons, we talk about cave systems. Individual cave horizons are created as a consequence of the deeper lowering of the water level and, by default, such horizons become dry,

G. Barovic (ed.), *Speleology of Montenegro*, Cave and Karst Systems of the World,
https://doi.org/10.1007/978-3-031-49375-1_12

and we can expect the occurrence of an underground watercourse, sometimes permanent, sometimes intermittent, only in the lowest section of such cave. Strong vortex movement occurs in some horizons in the period of high waters, and it carries large volumes of bed load forming various erosion forms along the bottom, on the walls and arches. Calcium carbonate deposits where water courses are slower, and where water is saturated with carbonates. Where water courses are formed, in dry and abandoned parts of cave systems, cave formations are deposited as a consequence of leaking and draining of water from narrow cracks in their arches. The cave formations are highly important for the aesthetic valuing of caves and their inclusion in the tourist offer (Gunn 2004).

Unlike caves, pits on karst terrains develop vertically and represent steep channels which may be very deep, from around ten meters to several tens and even hundreds of metres. Their formation is caused mainly by the abyss function; however, there are also pits whose genesis is linked with zones of karst water discharge. Pit openings may have various forms and sizes: round, oval, rectangular and elongated. It often happens that an entrance opening means nothing when it comes to the dimensions and characteristics of the structure. Sometimes it happens that a very small opening, which is discovered accidentally and which a person can barely pass through, hides large cave channels, as the case is with several pits in Kameno. The deepest pits usually do not have a simple form, but are cascading in profile, since they developed along variously intersected cracks even several hundreds of metres deep (Ford and Cullingford 1976; Gunn 2004).

Due to high and rapid technological development and advancing of exploration equipment and resources, a new discipline within standard speleology has occurred in recent years, which is cave diving or exploring of cave systems which are filled with water. Cave diving is only one of a series of exploration methods applied during the execution of works within the project of the Republic Geological Survey.

This idea was translated into an official scientific research project back in 1989 and it involved researchers from several countries with a view to conducting a scientific comparative analysis of our karst terrain with theirs. The project was generally accepted by scientists from France, Romania, Italy, etc., but its implementation was prevented by the introduction of the United Nations sanctions against the state of Yugoslavia at the time, which also applied to science, so the planned research had to wait for better times.

Only in 1995, the University of Montpellier requested a continuation of the research and the Republic Geological Survey from Podgorica accepted cooperation, and implementation of the project named Hydrogeological Research and Protection of the Basin of the Bay of Kotor started.

The numerous and various factors which define hydrogeological characteristics of an area required the use of numerous scientific methods and methodological procedures of the research during three years of work (from 1996 to 1999). We indicate only some of them, including hydrogeological mapping, various geophysical surveys, hydrochemical, speleological, cave diving and other research, whose characteristics classify them as complex geological research aimed at defining specific characteristics of this area. The increasingly complex endeavours required increasingly modern technology, so the most modern achievements used in the world at the time were applied in the fields of cave diving and geophysical surveys.

2 Orahovačka Ljuta

The Orahovačka Ljuta (Fig. 1) is certainly breathtaking and poses a real challenge to both scientists and cave divers. Ljuta is located right next to the main road that leads from Kotor to Risan, near the settlement of Orahovac. Travellers often cross the small bridge overarching this unusual river without being aware of it. Immediate proximity of steep cliffs which rise almost vertically to around one thousand metres, does not allow the human mind to link a river with the existence of wide valleys to think that a phenomenon such as Orahovačka Ljuta is hidden here. This seemingly small river with the total length of around 100 m offers a unique answer to the question of why karst and karst phenomena are the main cause of the problems of a lack of drinking water in summer months in the territory of the Montenegrin coast. The yield of Ljuta in the period of maximum equals 183 m^3/s, while some free estimates increase this value even to 300 m^3/s. When this is compared to the minimum in summer months when the yield drops by more than three times, and when discharge takes place at the level of sea surface, it is easy to conclude that the entire rainfall in its hinterland disappears very quickly through branched systems of cracks. Longer stay underground is prevented by the steep gradient of the level of groundwaters and high throughput of formed channels and caverns in the basin. The regime of aquifer waters is directly dependent on rainfall, so it happens that its yield increases from 100 1/c to around 15 m^3/s after only one rainy night in the hinterland (Marić 1996; Dubljević 2001).

However, not only the surface flow of Orahovačka Ljuta is a specific phenomenon. Undoubtedly, this is also a cave channel which descends almost vertically below this small groundwater lake to a depth of more than 120 m. This is the depth which was explored by the Yugoslavian-French diving team of the time (Dubljević 2001). During these explorations, karst channels representing typical karrens were observed in the zone which is now below the sea level

Fig. 1 Orahovačka Ljuta in minimumu and maximum

(they are presented by an arrow in the figure, recorded at the depth of 50 m), and they indicate that the cave channel of Orahovačka Ljuta used to be in a dry zone. This leads to the conclusion that the erosion base, i.e., the real discharge zone of Orahovačka Ljuta is considerably lower than the current one. Nowadays probably deep below the sea surface. These phenomena are often associated with solid earthquakes that must have been cataclysmic in nature (Glavatović 2019).

3 Spila Risanska

The gaping cave opening from which large volumes of water emerge in the period of intense precipitation, called Spila Risanska by people, is located in the northwest part of the town of Risan. Spila or Spilja Risanska (Fig. 2) creates periodical water flow led to the sea by the water that occasionally emerges in this spot (Barović et al. 2021).

The existence of an underground cave channel from which Spila emerges occasionally is a testimony that the existing karst process has advanced a lot. In hydrological minimum, discharge of Spilja is estimated at around 30 m^3/s, while in the period of the minimum, after lasting and enduring explorations, we can now claim that this spring dries out and gets exposed to the intense influence of the sea.

This spring is very important from the aspect of supplying water to Risan with the surrounding, irrespective of the salinization identified in the period of low waters. Regardless of a relatively large catchment area covering around 120 m^2, this spring cannot be considered as

Fig. 2 Spila Risanska in maximum and minimum

potentially good for addressing the water supply problem in this area.

The regime of aquifer water which is directly dependent on the rainfall regime indicates their highly variable distribution and highly variable summary volumes of water in the hinterland of the Risan Bay where this spring recharges from. A direct connection between Spila Risanska and abysses in the Grahovsko Field was established by the method of dying (Dubljević 2001).

4 Sopot Cave

Opening of the Sopot cave (Fig. 3) is located on the main road from Risan to Herceg Novi, around two kilometres away from Risan. The road slowly rises in this section and passes right by the cave opening from which large volumes of water occasionally emerge and plunge into the sea creating huge noise (Dubljević 2001; Parise et al. 2018).

Limestone sediment of the zone of Visoki krš constitutes a highly favourable environment for the development of karst relief forms. Their development is conditioned by a favourable composition of sediments and a prominent system of neo-tectonic faults. Rainwater that falls in the territory of Orjen sinks very quickly from the surface and their emergence concentrates also in the territory of Sopot. The hydrogeological research so far has assumed that the Sopot basin covers around 90 km^2, however, the marking of groundwaters has not yet accurately defined the catchment area of this vast karst aquifer.

The two largest known submarine springs are located in a small bay next to Sopot, at the depth ranging from 28 and 23 m.

These phenomena, known as spring holes, can often be seen from the said main road. Not the spring holes, since they are located at great depth, but their surface effects are represented in the form of concentric circles denoting the discharge zone.

The opening of the Sopot cave is 4 × 10 m and the channel keeps these dimensions in the first 100 m, where a large gallery whose dimensions exceed 30 m is reached. After descending for about 20 m between large limestone blocks, one reaches the siphon lake whose level is at the height of several metres relative to the sea level. This was at the same time the beginning of exploration by applying the cave diving method.

Chemical analyses of water were performed during exploration, and it was established that intense penetration of sea water and salinization according to the Ghyben–Herzberg principle takes place even in the most distant point, at about 500 m from the sea. That prevents any drinking water intake from this spring in the conditions of hydrological minimum.

Due to high oscillations of the sea level in the last 30–40 thousand years, many of the springs of that time are not deep below the sea surface. Testimony to that are the positions of many submarines springs both within the Mediterranean basin and worldwide. Near the border between Italy and France, several springs emerge below the sea level from the depth of around 39 m, at the distance

Fig. 3 Sopot spring in maximum and minimum

of about 700 m from the coast. In the Atlantic Ocean, near Florida, there is a known submarine spring in the waters of Crescent Beach, 4.5 km from the coast, at the depth of 38 m. In Greece, Anavalos Aktros is located at the distance of 400 m from the coast, at the level of 39 m, and many others. The spring holes in the Risan Bay undoubtedly belong to this group (Dubljević 2001).

Exploring the said spring holes established specific mechanisms of movement of these waters, which essentially obey hydraulic laws of motion of two-phase mixing liquid in pressurized pipes and open channels. While exploring the terrain, it was established that two separate water discharge spots were located at a mutual distance of about ten metres.

Exploration of these specific hydrogeological phenomena is possible only in the conditions of hydrological minimum when they do not discharge water. During the dive, it was possible to enter the opening, located at a depth of 36 m and has an irregular shape of 1.5 × 2 m. The channel is passable for about ten metres, and the further passage ends at around 44 m by stone blocks, preventing different courses. That was the end of the exploration by cave diving method.

5 Gurdić Cave System

The submarine estavelle Gurdić is located along the walls of the Old Kotor, below the eastern gate of the town (Fig. 4). A small water surface which reminds passers-by of a little lake shelters the entrance to the Gurdić cave system, at the depth of 12 m. Depending on the precipitation regime, this spring acts as a strong karst aquifer in periods of high precipitation. In dry periods, sea water easily penetrates into the system giving this aquifer characteristics of an abyss. This means that the entire system acts as an estavelle. The Greeks established a term which relates to the sinking of sea water, which sometimes can be found in our literature as well: it is the katavotra (Đurović and Đurović 2021; Dubljević 2001).

Exploration of Gurdić has been the subject of work of many scientists. This phenomenon was explored also by geologists, geographers, divers and many others, and every one of them contributed in their own way to a better definition of this phenomenon (Dubljević 2001; Barović et al. 2021). It was known before our research that this spring together with the spring of the Škurda River, which encompasses the western walls of the Old Town Kotor, constitutes a single hydraulically connected system. This broken spring system drains wasters whose basin area covers around 90 km^2 and includes the slopes of Lovćen, Njeguško Field and its surrounding. This assumption has been confirmed by dying of the abyss in Erakovići, Koritnik and in the pit Duboki Do in Njeguši. Also, the connection with the abyss in Ivanova Korita has been established by dying.

As in other coastal springs, this one is also under a strong influence of the sea during the summer period. Sea water sinks in Gurdić due to the tendency to achieve verified and established principles of balance with standard water. The penetration of water through lower underground karst channels is a consequence of achieving balanced conditions which are subject to the laws of difference between the specific gravity of sea water and fresh water. Basically, this is how the salinization of coastal karst springs Škurda and Gurdić occurs.

Fig. 4 Grudić

However, due to this specific work regime, the Gurdić spring has numerous limiting factors when it comes to access to its interior. As noted earlier, breaking through to the inside is impossible due to strong currents which drain water out in the rainy period of the year, in a situation when this spring acts as a very karstic spring. In the period when Gurdić acts as an abyss and swallows sea water, strong currents are directed towards the inside, and high caution is required in that situation. It could easily happen that you get practically sucked into the system without the ability to return since swimming against the current at the speed of more than 5 km/h is practically impossible. Exploration of Gurdića requires the right moment. That moment is at the time of transitional regime of operation, when the spring slowly retreats, and has not yet taken the function of the abyss.

References

Barovic, G., Vujacic, D., Spalevic, V. (2021). Rivers of the Coast of Montenegro. In: Joksimović, D., Đurović, M., Zonn, I.S., Kostianoy, A.G., Semenov, A.V. (eds) *The Montenegrin Adriatic Coast*. The Handbook of Environmental Chemistry, vol 110. Springer, Cham. https://doi.org/10.1007/698_2020_707

Bognar, A. (1987). Geomorfologija-položaj, razvoj i problemi. *Hrvatski geografski glasnik*, 49(1.), 78–79.

Dubljević, V. (2001) *Hidrogeološke karakteristike sliva Boke Kotorske (magistarski rad)*. Beograd: Univerzitet u Beogradu - Rudarsko-geološki fakultet.

Đurović, M., & Đurović, P. (2021). Review of the most significant caves in Montenegro. *Acta Carsologica*, 50(1), 49–65.

Ford, T. D., & Cullingford, C. H. D. (1976). *The science of speleology*. London-New York-San Francisco: Academic Press.

Glavatović, B. (2019). 40 godina od katastrofalnog zemljotresa u Crnoj Gori: osvrt na seizmičnost i nacionalnu seizmološku djelatnost. *Geološki glasnik*, 17, 83–99.

Gunn, J. (2004). *Encyclopedia of caves and karst science*. Taylor & Francis.

Marić, M. (1996). *Karstna izdan Ljute, Spile, Sopota i Gurdića: fondovski material*. Podgorica: Republički zavod za geološka istraživanja.

Mirković, M., Živaljevic M., Djukić, V., Petrović, Z., Kalezić, M., Pajović, M. (1985). *Tumač Geološke Karte SR Crne Gore 1:200.000*. Titograd: Zavod za geološka istraživanja.

Parise, M., Gabrovsek, F., Kaufmann, G., Ravbar, N. (2018). Recent advances in karst research: from theory to fieldwork and applications. *Geological Society, London, Special Publications*, 466(1), 1–24.

Marine Caves—Biodiversity and Threats

Mačić Vesna, Petović Slavica, Đorđević Nikola, Varda Dušan and Panou Aliki

Abstract

Marine caves are considered an important and endangered habitat type and they are listed in Annex I of the EU Habitat Directive as one of the priority habitats. Here, a synthetized knowledge on this type of habitat in Montenegro is presented. The biodiversity data include 23 marine caves hosting more than 250 species. The main threats are physical damage (caused by divers and new infrastructure building), disturbance of the species (by marine litter and noise), non-indigenous species and climate change. Further research is needed in order to better document the biodiversity of marine caves, fill the gaps in knowledge of completely submersed caves and establish appropriate monitoring and management of this specific habitat.

Keywords

Marine caves · Biodiversity · Threats · Adriatic Sea

M. Vesna (✉) · P. Slavica · Đ. Nikola
Institute of Marine Biology, University of Montenegro, Podgorica, Montenegro
e-mail: macic.v@ucg.ac.me

P. Slavica
e-mail: kascelanslavica@gmail.com

Đ. Nikola
e-mail: nikoladj@ucg.ac.me

V. Dušan
NGO Mediterranean Center for Environmental Monitoring, I. Milutinovića 7, Sutomore, Montenegro
e-mail: vardad@yahoo.com

P. Aliki
Archipelagos—Environment and Development (NGO), 28100 Lourdata, Kefalonia, Greece
e-mail: aliki.panou@yahoo.gr

1 Introduction

The relief of Montenegro belongs to the southeastern part of Dinarides and is particularly complex. In a relatively small area (13,812 km^2), height varies remarkably, so that altitudes up to 200 m above sea level make up only 10% of the territory, between 200 and 1000 m above sea level cover 35% of the territory, between 1000 and 1500 m cover 40% and altitudes higher than 1500 m above sea level make up 15% of the national territory (Barovicetal. 2021). Various geological, lithological and erosion processes resulted in the creation of very diverse geological structures and more than 1500 caves in Montenegro, of which only several hundred are described (Barović 2011; Presetnik et al. 2014). Unfortunately, although the need for creating speleological maps and the inclusion of caves in the touristic offer are stated (Barović 2011), so far only Lipska cave on land is open for tourists to visit, while the marine caves Plava špilja (Blue cave) and Sumporna cave (Sulfur cave) are also visited by tourists, however not in an organized manner.

Surveys of marine caves in general have begun in parallel with the development of SCUBA diving in the second half of the last century (Gerovasileiou and NikeBianchi 2021). In Montenegro, the first surveys of semi-submersed marine caves were performed during research on bat species close to Ulcinj (Mirić 1973; Paunović and Stamenković 1998) and later for the needs of water supply for the city of Kotor along with a regional hydrogeological characterization (Petrović and Rudaković 1980; Milanović 2005; Milanović et al. 2014). However, the biodiversity of marine caves started being surveyed only in the last decade (Mačić et al. 2013, 2014, 2015, 2019; Petović et al. 2019; Petović and Mačić 2021).

As there are various definitions for land caves, so it is with sea caves, although a lot has been done in recent years to standardize some basic concepts (Gerovasileiou et al. 2016a, b; SPA/RAC-UN Environment/MAP, OCEANA 2017; Gerovasileiou et al. 2020). The term "cave" is

G. Barovic (ed.), *Speleology of Montenegro*, Cave and Karst Systems of the World,
https://doi.org/10.1007/978-3-031-49375-1_13

referring to an opening into a natural underground or underwater hollow, which is large enough for a human to enter, and its longest dimension (i.e. length or depth) is longer than the cross-sectional dimensions at the entrance (Gunn 2004; Gerovasileiou et al. 2016a, b). The main classification of marine caves is according to their submersion level: semi-submerged caves which are partially below the sea level and submerged caves which are completely below the sea level. Marine caves are also known as littoral or sea caves. Waves and sea surface dynamics are just some of the possible mechanical actions responsible for their formation. They may be formed by different processes and their morphology may also vary significantly (blind-ended caves, tunnels, pits, complex combination of different forms) (Gerovasileiou and Nike Bianchi 2021).

Not only the cave morphology but also marine cave biota can be very specific. They are divided into four main ecological categories:

1. Stygobionts (troglobionts)—obligatory cavernicolous organisms, adapted to subterranean life (e.g. loss of pigmentation and vision).
2. Stygophiles (troglophiles)—organisms that can live and complete their life cycle inside caves, but can also be found in suitable habitats outside caves (e.g. undersides of rocks, deep waters, etc.).
3. Stygoxenes (trogloxenes)—organisms that occur in caves but do not complete their life cycle inside caves as they periodically move out (e.g. finding shelter in caves but foraging in open waters).
4. Accidental—organisms which may enter caves by chance and are able to survive in this environment just for a short period of time.

In general, the prefix "stygo" should be preferred for aquatic cave biota while "troglo" should be used for the terrestrial species (Culver and White 2012; Gerovasileiou et al. 2016a, b; Gerovasileiou and Nike Bianchi 2021). Furthermore, due to the continuity of the aquatic environment, marine caves are less isolated from the surrounding environment than terrestrial caves. That is why the majority of species recorded in the marine caves are stygophiles (e.g. species that also occur in deep waters, i.e. dim-light environments) and stygoxenes (species that come into and out of the caves) (Gerovasileiou and Nike Bianchi 2021). Due to the species that are commonly found in marine caves but originate from external marine environments such as deep waters and coralligenous crevices, these species are characterized as "secondary stygobionts" (Iliffe 1990). We should also keep in mind that many species now characterized as exclusive to caves might be found in other, deep-sea habitats that are still not explored enough. But it is also possible that some of those species are relicts and results of ecological specialization due to this very specific and fragmented habitat (Gerovasileiou and Nike Bianchi 2021).

Another characteristic of marine caves is that environmental conditions such as light intensity and water movement may dramatically vary within a few meters, compared to the external environment where such changes are happening along tens or even hundreds of meters (Bianchi 2003; Gerovasileiou and Nike Bianchi 2021). Due to such environmental differences, species are not inhabiting caves homogeneously, but prefer distinct cave sections (Bianchi and Morri 1994). Due to the great influence of the French school on Mediterranean marine ecologists, the cave zone classification proposed by Peres and Picard (1964) has been the most widely accepted to date (Gerovasileiou and Nike Bianchi 2021). They divided cave biocenosis into the semi-dark biocenosis (or GSO, from French "grotte semi-obscure") and the dark cave biocenosis (or GO, from French "grotte obscure"). The semi-dark cave biocenosis is typically found at the entrance of a cave, but also in the holes and under overhangs outside caves where environmental conditions are similar to those in the cave interior. This biocenosis is dominated by sessile animals, i.e. sponges, anthozoans and bryozoans but some sciaphilic macroalgae may also be present. The dark cave biocenosis, present in the inner parts of the caves, is characterized by sponges, serpulid polychaetes, bryozoans and brachiopods (Peres 1967). The transition from the semi-dark to the dark cave biocenosis can be observed through the sharp decrease in biotic cover and abrupt changes in species richness, biomass and three-dimensional complexity. Due to the specific morphology of a cave and its environmental conditions (mainly light and currents), different variations may occur (Peres 1967; Harmelin 1985).

The more detailed classification was provided by Riedl (1966) who describes six biotic zones based on species replacement across the outside–inside gradient of blind-ended caves (Fig. 1). This classification scheme begins with the phytal zone at/around the entrance; with the decrease of light toward the interior, the abundance of algae is decreasing while sessile fauna becomes more abundant. The last biotic class is the "empty quarter" that is characterized by usually calm water, complete darkness, the almost complete absence of flora and fauna and scarcity of food supply. According to Riedl (1966), the most important factor for the distribution of the biota is the shape of the caves, but depth and size are very important factors as well.

Similar to Riedl (1966), there are six ecological zones described by Bianchi and Morri (1994) and Morri (2003). Instead of species replacement they considered a change in growing forms, trophic guilds, three-dimensional structure and biotic cover (Fig. 2).

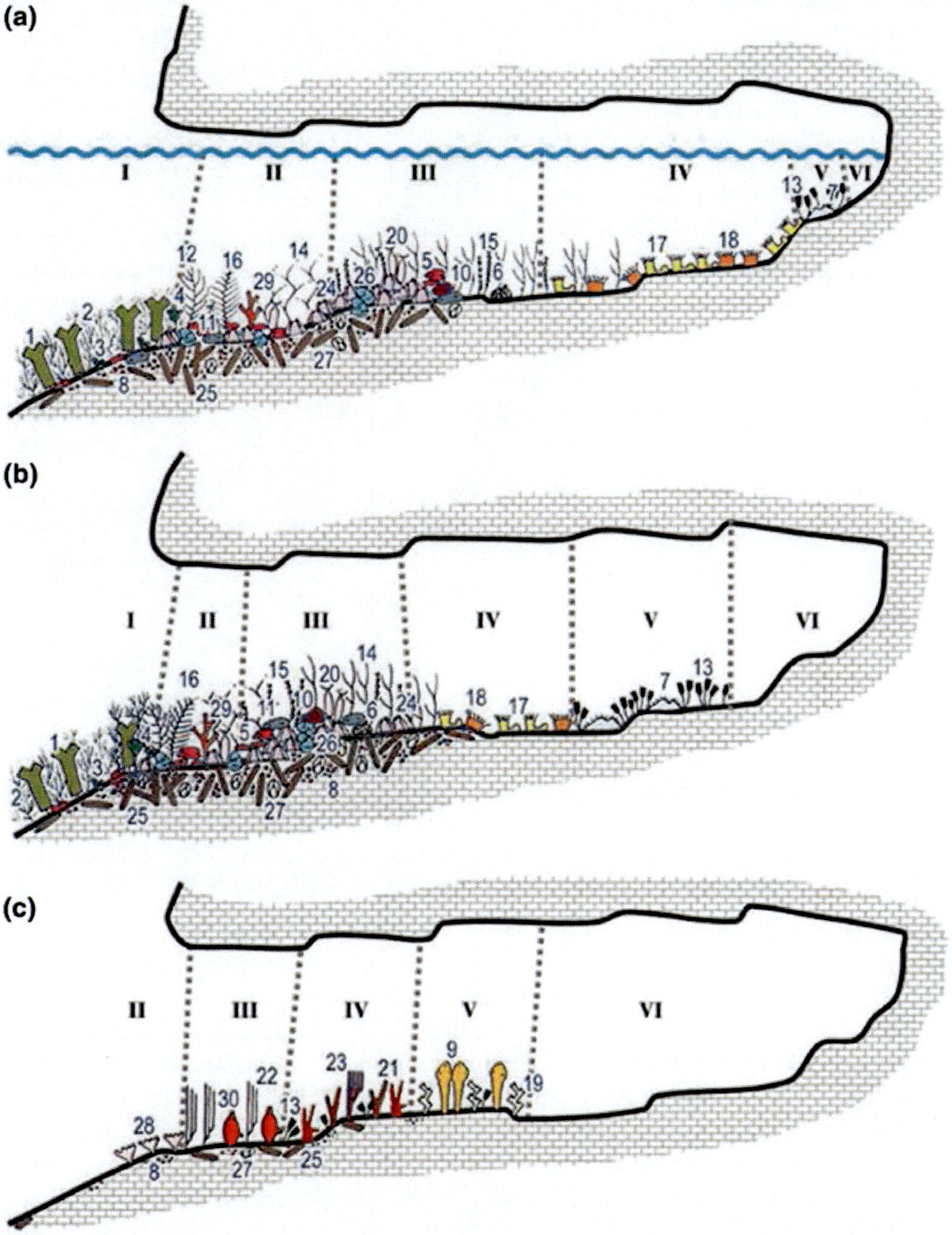

Fig. 1 Biotic zones in semi-submerged caves (**a**), in infralittoral (**b**) and circalittoral (**c**) submerged caves according to Riedl (1966) and modified by Gerovasileiou and Nike Bianchi (2021). The inner zones can occur close to the cave entrance as the depth increases. I: algae-dominated zone, II: entrance zone, III: anterior zone, IV: central zone, V: posterior zone and VI: "empty quarter." For simplicity and graphical clarity, organisms have been schematically illustrated only on the cave floor while walls and ceilings are also colonized: 1. Dictyopteris; 2. Cystoseira; 3. Halimeda; 4. Flabellia; 5. encrusting Rhodophyta; 6. Clathrina; 7. Petrobiona; 8. Cliona; 9. Axinella; 10.Petrosia; 11. Ircinia; 12.Pennaria; 13. Campanularia; 14. Eudendrium; 15. Dynamena; 16. Aglaophenia; 17. Parazoanthus; 18. Dendrophylliidae; 19. other Scleractinia; 20. Cornularia; 21. Corallium; 22. Eunicella; 23. Paramuricea; 24. Balanidae; 25. Lithophaga; 26. Ostrea; 27. Rocellaria; 28. Reteporella; 29. Myriapora; and 30. Halocynthia. Redrawn and modified from Riedl (1966)

Finally, characteristics like the absence of light and primary production in dark caves, qualify them as an easily accessible model for the study of deep marine ecosystems.

Harmelin and Vacelet (1997), which occupy over the half of the surface of our planet and still represent an area less known than the surface of the Moon.

The Mediterranean Sea is one of the best-surveyed areas on our planet and its rocky coasts are hosting more than 3000 marine caves mostly in the northern part of the basin. According to Gerovasileiou and Bianchi (2021), the overview of 360 literature sources revealed 2369 taxa belonging to 58 major taxonomic groups reported from 404 marine caves in 15 Mediterranean countries and confirmed marine caves as biodiversity reservoirs and refuge habitats for several endemic species and living fossils.

In the Adriatic Sea, the number of marine caves is probably exceeding 800 and the great majority is located along the eastern coast (Giacomo et al. 2013; Mačić et al. 2019). On the Croatian coast, marine caves are very well studied in terms of their geology and biodiversity, but also as a valuable archive of the evidence of sea level changes (Gardasil 1991; Baric et al. 2008; Sure et al. 2010; Bakran-Petricioli et al. 2012; Petricioli and Bakran-Petricioli 2019). The Montenegrin coast is mostly built of carbonate rocks, limestone and dolomite. This narrow area is separated from the rest of the country by the mountains Orjen, Lovćen,

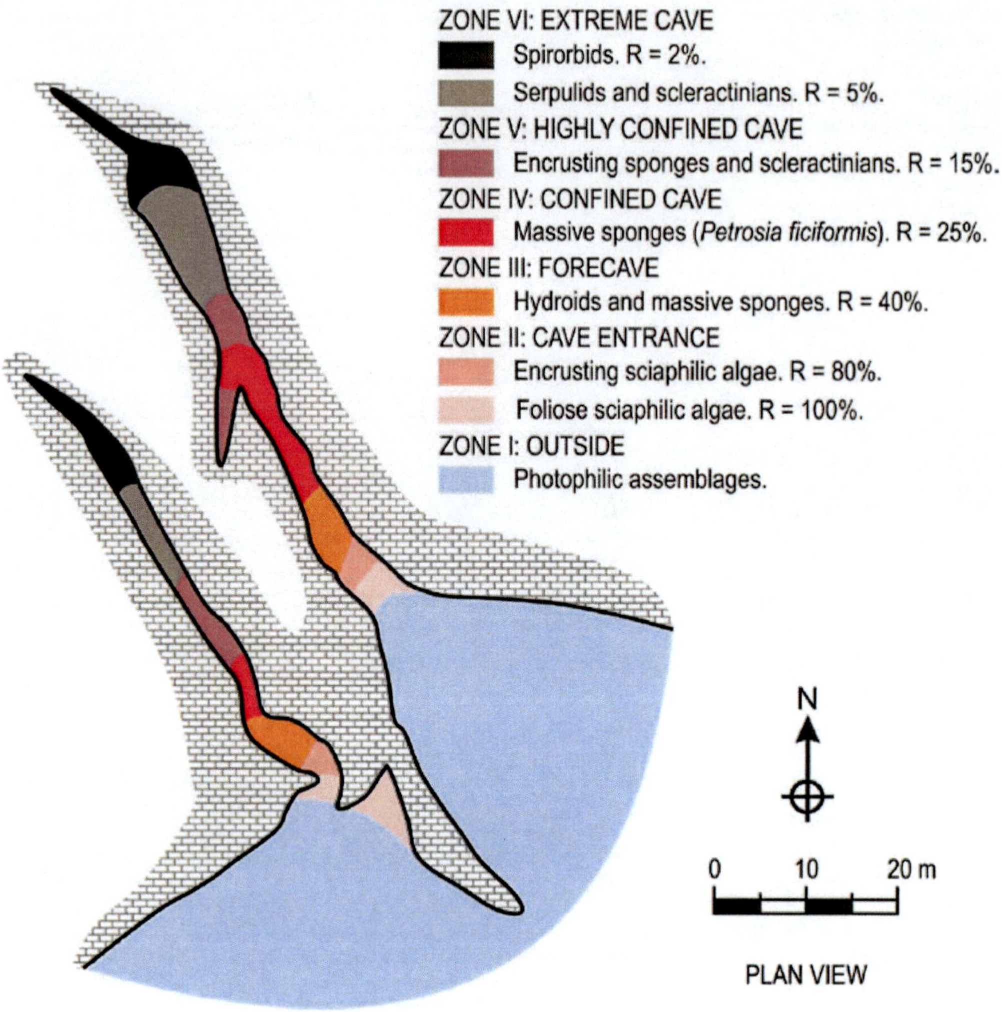

Fig. 2 Zonation of the biotic assemblages of the twin caves named "Grotte Del Bue Marino," in the Gorgona Island (Tyrrhenian Sea, Italy) according to the confinement gradient. R = percent cover. Redrawn from Bianchi and Morri (1994) and Morri (2003) and modified by Gerovasileiou and Nike Bianchi (2021)

Sutorman and Rumija (Barovic et al. 2021). Unfortunately, research on marine caves has been scarce up to the present time and biological surveys of marine caves started practically in the last decade (Mačić et al. 2013, 2014, 2015; Petović et al. 2019). Having in mind that marine caves are classified as a biodiversity hotspot and priority habitat according to the EU Habitat Directive (1992) and caves, in general, are protected by the Montenegrin legislation as well (Official Gazette RCG 2016), the great scientific interest for their further research and protection should be stressed here. Furthermore, The European Red List of Habitats (Rodwell et al. 2017) provided an assessment of 257 marine benthic habitat types and communities of Mediterranean medio-littoral caves and overhangs (EUNIS code A1.44) along with other 53% of marine habitat types classified as Data Deficient. Thus, the urgent need for further research is evident. The aim of the present work is to present the existing knowledge on biodiversity in marine caves in Montenegro and to present the main threats to this endangered and protected habitat.

2 Biodiversity

Based on the overview of available literature we may state that terrestrial caves in Montenegro are much better explored than marine ones. So far, 250 taxa belonging to 14 taxonomic groups have been recorded in 23 Montenegrin marine caves (Annex 1). The small research capacities of Montenegro are obvious and only with much more intensive surveys including more taxonomic experts we will be able to present a more complete list of species and to create a baseline for appropriate monitoring, management and protection plans.

The microbial diversity of marine caves is in general poorly studied (Gerovasileiou and Nike Bianchi 2021). In three marine caves (Sumporna cave, Plava špilja and Niska cave) analysis of intestinal enterococcus, fecal coliforms and *Escherichia coli* were performed several times during 2018 and 2019 (Mačić et al. 2018a, b; Petović et al. 2019). Results of very high concentrations of *E. coli* (7400/100 mL) and intestinal enterococcus (5400/100 mL) in Sumporna cave and additional analysis in the surroundings indicated that most probably their origin is from the bat populations resting in that cave (Mačić et al. 2018a, b; Djurović 2019). Only a small number of studies were dealing with pathogen bacteria living in the intestine of bats up to the present time (Muhldorfer 2013). Having in mind the possibility of bacterial transmission between bats, humans and other animals and the fact that the Sumporna cave and even more so Plava špilja are regularly visited by tourists, further surveys and the monitoring of the sanitary quality of the seawater in these caves are needed.

Phytoplankton species were studied in 5 semi-submersed caves: Sumporna, Nudistička, Jošova, Niska cave and Plava špilja. A total of 51 taxa were recorded, and the most numerous were Diatoms (28 taxa), followed by Dinoflagellates (4) and Coccolithophorids (4) (Mačić et al. 2018a, b; Petović et al. 2019) (Annex 1). Having in mind that these species are not capable of active movements, it is clear that theywere brought into the caves by waves and sea currents. Furthermore, it is questionable for how long they could survive in caves where light intensity is considerably decreased. Further studies over a whole year and in different parts of the caves are necessary for a better understanding of the functional role of these species in cave biocenosis and also to evaluate mixotrophy of microalgae and cyanobacteria inhabiting caves (Abdullin and Bagmet 2016).

Zooplankton is another group of organisms usually neglected in cave studies. In five semi-submersed marine caves surveyed in Montenegro (Sumporna, Nudistička, Jošova, Niska and Plava špilja) 34 zooplankton taxa were registered (Petović et al. 2019; Mačić et al. 2018a, b) (Annex 1). The most numerous class was Copepoda, numbering 18 taxa, while only two taxa were registered for each of the classes Pteropoda, Appendicularia, Chaetognata, Cladocera, as well as for Hydromedusae and Siphonophorae. In addition to those species, several larval forms of mollusks, crustaceans and polychaeta were registered. It is obvious that further surveys of cave zooplankton are necessary. An additional difficulty here poses the sampling by filtration; this is certainly one of the reasons for the relatively small number of zooplankton cave studies in general (Moscatello and Belmonte 2007).

Although algae are not typical organisms in marine caves, macroalgae are usually present at the entrance of the cave and in the semi-dark parts where there is enough light for photosynthesis. In total, only 25 species were registered (Annex 1) and the most numerous were Rhodophyta (15), followed byChlorophyta (5) and Ochrophyta (5) (UNEP/MAP-RAC/SPA 2016; Mačić et al. 2019; Petović et al. 2019). This small number of registered taxa is indicative of the fact that they were not in the focus of the performed surveys.

Porifera is the group with the highest number of species in the Montenegrin marine caves. So far, 39 species were recorded (Annex 1) representing 65% of the total number of sponge species reported in Montenegro (Petović et al. 2021). Bakran-Petricioli et al. (2012) reported 77 species in marine caves from a total of 283 species (Demospongie and Homoscleromorpha) reported from the Adriatic Sea. Having in mind that Gerovasileiou and NikeBianchi (2021) reported 329 sponge species from at least 185 Mediterranean marine caves and that marine caves in the Mediterranean are defined as "sponge biodiversity reservoirs" containing a high proportion (41%) of Mediterranean Porifera endemics, further surveys should focus on sponges and their protection (Gerovasileiou and Voultsiadou 2012).

Cnidarians, and especially Anthozoans, represent one of the most abundant taxa in marine caves. A total of 165 species have been recorded in Mediterranean marine caves (Gerovasileiou and Nike Bianchi 2021), while in Montenegro, only 17 taxa were recorded up to the present time (Annex 1). One of the main reasons for such a small number is the fact that the great majority of the surveyed marine caves in Montenegro are semi-submersed and shallow, thus not providing the typical habitat. The most frequently reported species was the red beadlet anemone (*Actinia equina*), typical for the mediolitoral zone of rocky coasts. It is important to underline the presence of the Mediterranean endemic species *Actinia cari* in the shallow caves close to Ulcinjand the protected *Cladocora caespitosa* in the Krekavicacavesituated in the Marine Protected Area (MPA) Platamuni (UNEP/MAP-RAC/SPA 2016; Mačić et al. 2019). The Krekavica cave is representing the biggest marine cave registered in Montenegro and the richest in biodiversity. Furthermore, 7 Madreporaria species registered in this cave are listed in Appendix II of the CITES convention (1990), while *Phyllangia americana mouchezii* and *Polycyathus muellerae* are forming particularly large colonies, thus creating an extraordinary bioconstruction together with some other Madreporaria, Porifera and Bryozoa (UNEP/MAP-RAC/SPA 2016).

High numbers of annelids have been registered in Mediterranean caves and, while oligochaetes were rarely found, 262 species of Polychaeta were recorded (Gerovasileiou and Nike Bianchi 2021). In Montenegro, only 6 species were registered so far in marine caves (Annex 1) and the most frequent species were *Protula* sp. and *Serpula vermicularis* which thrive on hard substrata. The recorded taxa were not cave-exclusive species and belong to some of the well-known representatives of the phylum; this may be the reason for a higher number of records. It is interesting that Sanfilippo et al. (2017) showed a clear trend of variability of Serpulidae abundance increasing inwards, specifically that of sciaphilic and deep-sea species. In dark caves, serpulids might be one of the most abundant species; thus, it is expected that this phylum will receive more scientific attention in further cave surveys. However, the involvement of taxonomic experts is needed here.

Mollusks are a very diverse and numerous phylum represented by 243 species in Mediterranean caves (Gerovasileiou and NikeBianchi 2021). In Montenegrin marine caves, only 21 gastropod and bivalve taxa were registered up to the present time (UNEP/MAP-RAC/SPA 2016; Mačić et al. 2019; Petović et al. 2019) (Annex 1). Almost all surveyed marine caves in Montenegro are semi-submersed and shallow—additionally, the scientific effort has been generally low: these could be the main reasons for such small numbers of reported mollusk species. In relation to that it is not strange that the most frequent species were *Patella rustica*, *Mytillus galoprovincialis* and the protected *Lithophaga lithophaga*. Another protected species, *Tonna galea,* was recorded only in the Krekavica cave which is situated in the newly established MPA Platamuni (UNEP/MAP-RAC/SPA 2016).

Arthropods are a very diverse phylum with a large number of species; this is one of the reasons why they are represented by the highest number of species in marine caves, along with Porifera, Polychaeta and Mollusca (Gerovasileiou and Nike Bianchi 2021). The groups with the highest number of species in the Mediterranean marine caves are Copepoda (113), Amphipoda (83), Decapoda (75), Isopoda (26), Mysida (21) and Pycnogonida (15), with all other groups comprising <10 taxa and making a total of 375 arthropod species (Gerovasileiou and Nike Bianchi 2021). We present planktonic arthropod species in Annex 1, separately from other arthropods, while from the other taxa of this phylum we registered only 10 Decapod species and 1 Cirripedia taxa (Annex 1). The most frequent species in our surveys was *Cirripedia sp.,* followed by *Palaemon serratus* and *Pachygrapsus marmoratus* (Mačić et al. 2018a, b, 2019; Petović et al. 2019). Additionally, one invasive species, the blue crab (*Callinectes sapidus*), was recorded (Mačić et al. 2018a, b).

Bryozoans are one of the dominant phyla in marine caves (228 species recorded in Mediterranean caves), especially in the transitional zone between the semi-dark and dark cave biocenosis (Harmelin, 1985; Harmelin 1997; Rosso and Di Martino 2016; Gerovasileiou and Nike Bianchi 2021). However, in Montenegrin marine caves, only 13 species were registered up to the present time: the ones more conspicuous and easier to determine on pictures (Annex 1). Obviously, more detailed studies including some taxonomic specialists for this faunal group would provide more comprehensive information on these organisms.

The Echinoderm fauna is not particularly diverse nor abundant in marine caves except for some Ophiurids and, to a lesser extent, Asteroids (Tortonese 1978; Gerovasileiou and Nike Bianchi 2021). On the Montenegrin coast, 15 species were recorded so far (Annex 1), representing 42% of the Echinoderm species recorded in Mediterranean marine caves (Gerovasileiou and Nike Bianchi 2021). The most frequently reported Echinoderm species in Montenegrin caves was *Arbacia lixula* (Mačić et al. 2019; Petović et al. 2019). The protected species *Centrostephanus longispinus* was recorded in the Krekavica cave (within the MPA Platamuni area) while the protected *Ophidiaster ophidianus* and *Holothuria tubulosa*, *H. sanctori* and *H. forskali* were recorded in several caves (Official Gazette RCG 2006; UNEP/MAP-RAC/SPA 2016; Mačić et al. 2019; Petović et al. 2019).

The most common tunicate species was *Halocynthia papillosa* (Mačić et al. 2019). Its red color and characteristic shape are probably the reasons why this species was easily observed and registered not only in Montenegro but also in many other Mediterranean marine caves (Gerovasileiou and Nike Bianchi 2021). Besides this species, 7 more taxa were recorded (UNEP/MAP-RAC/SPA 2016; Mačić et al. 2019) (Annex 1) out of 29 so far recorded for the Montenegrin coast (Petović et al. 2021).

The most common fish species in our case as well as throughout the Mediterranean Sea in general is the cardinal fish (*Apogon imberbis*) (Bussotti et al. 2015; Bussotti et al. 2017; Mačić et al. 2019). This species is typical for the rocky bottom with crevices and fissures. In marine caves, it seems to have a very important role in the food chain due to the daily inside–outside migration (Bussotti et al. 2017). According to Gerovasileiou and NikeBianchi (2021), 112 fish species have been recorded in Mediterranean marine caves while for Montenegro we registered only10 species up to the present time (Annex 1). There are several reasons for the small number of registered fish species. First of all, the generally low scientific effort into the research of marine caves in Montenegro; additionally, during the surveys performed, more attention was given to the sessile fauna. Furthermore, divers were concentrated primarily on the exploration of the morphology of the caves per se and, consequently, vagile species such as fish were disturbed, so a visual census could not be considered as fully implemented.

Two different mammal orders are important for marine caves, bats and the Mediterranean monk seal. Surveys of the semi-submersed caves in Montenegro revealed the presence of at least six bat species: *Myotis myotis/oxygnatus*, *Myotis emarginatus*, *Myotis sp.* (possibly *M. capaccinii*), *Miniopterus schreibersii*, *Rhinolophus ferrumequinum*, *R. euryale/hipposideros* (Radonjić 2019). Presetnik et al. (2014) confirmed some previous reports by Mirić (1973) and Paunović and Stamenković (1998) for Sumporna and Jošova (El Kaminova) caves as important maternity roosts for some bat species, but also as transient or hibernacula roosts for some others. Đurović (2019) registered the presence of important bat roosts, however, less numerous than in previous periods, most probably as a consequence of habitat disturbance. Nevertheless, the maternity roost for *M. emarginatus* is only the second one confirmed in Montenegro, thus the need for its protection and further monitoring is even more urgent (Presetnik et al. 2014). Furthermore, recent surveys of Plava špilja and Niska caves provided data for at least four bat species, while a small group of *Tadarida teniotis* was registered between these two caves (Mačić et al., 2019; Petović et al. 2019). Presetnik et al. (2017) registered 26 bat species in Montenegro, thus, the records of at least 7 species in marine caves are confirming the importance of this habitat type and the need for its better protection.

Another mammal species characteristic of the marine caves is the Mediterranean monk seal (*Monachus monachus*). The presence of monk seals in this region is documented until the 70 s. Additionally, one marine cave close to Petrovac (nowadays within the MPA Katič) is known as the "Tuljanovapecina" ("Seal Cave") indicating the former presence of seals in the greater area (Panou et al. 2017). Panou et al. (2014) reported at least 11 caves along the Montenegrin coast suitable for the species' resting and reproduction. Furthermore, Panou et al. (2017) recorded a total of 14 monk seal sightings throughout the coastline of Montenegro covering the period 1985–2010. As this is the most endangered marine mammal in Europe, it is suggested that protection measures should include all marine caves suitable for monk seals in order to allow the re-establishment of the species in Montenegro as also the reconnection of distant populations or individual animals (Bundone et al. 2019). Luckily, many of the recorded suitable habitats in Montenegro are situated in the newly established 3 MPA zones and appropriate management plans will provide ecological corridors for this endangered and protected species.

3 Threats

Marine caves are not easily accessible and long-time series data are mostly missing; therefore, an evaluation of their ecological status is difficult. The most important direct pressures impacting marine cave biodiversity are: mechanical damage of fragile species caused by unregulated diving activities; physical damage and siltation due to marine and coastal activities and building of new coastal infrastructure; and collection and disturbance of some species inside the caves. Indirect pressures are climate change effects, invasion of non-indigenous species, oil spills, eutrophication, etc. (Parravicini et al. 2010; Gerovasileiou et al. 2016a, b).

Cave biocenosis is considered as poorly resilient because the development of the sessile communities in the caves is a slow process and the distribution of those communities is causing a long-term effect (Guarnieri et al. 2012; Ouerghi et al. 2019; Gerovasileiou and NikeBianchi 2021). Furthermore, most of the species are not only growing extremely slowly but also particularly fragile, especially those with an erect calcareous morphology. As a consequence, these species are very vulnerable to mechanical damage such as breaking and abrading caused by divers (usually unintentionally). Furthermore, a high number of not well-trained divers inside the caves cause an increase in the resuspension of the sediment, and the accumulation of exhaled air bubbles at the ceiling may also cause problems to filter feeding organisms (Milazzo et al. 2002, Guarnieri et al. 2012). Diving tourism is an exponentially growing activity worldwide and, in Montenegro, it is expected to be a growing activity, particularly in the recently established MPAs. It is already documented that diving activities may have cumulative negative effects on sensitive cave biocenosis causing a shifting from three-dimensional to flattened patterns and also decreasing biodiversity. Thus, appropriate management measures and continuous monitoring should be well-planned and implemented (Milazzo et al. 2002). Presently, there are almost no regulations concerning diving activities in Montenegro. Therefore, it is rather difficult to estimate how intensive this pressure is at present and how it may develop in the future. Illegal spearfishing and collecting of some species do exist everywhere in the Mediterranean but it seems that this type of pressure is not significant for the caves in Montenegro. Such activities will hopefully be directed into sustainable patterns, at least in the three newly established MPAs through the issue of adequate and effective legislation and its enforcement as also the appropriate management of these pressures.

In relation to human frequentation of marine caves we should havein mind that even the presence of humans alone could also have negative impacts. That is the case for the monk seals that use marine caves for resting and pupping: the presence of humans not only inside the caves but also in the vicinity may trigger the abandoning of such locations (Panou et al. 1993). In Montenegro, there are no recent records of monk seal presence except for one sighting in 2017 (D. Varda, unpublished data), but this might be due to low reporting effort. Based on the occasional presence in Albania and Croatia and the previous sightings in Montenegro, Panou et al. (2017) believe that monk seals might regularly use the coastline of Montenegro, at least in transition. Thus, it is to be expected that MPAs will play an important role in providing appropriate, undisturbed habitat and ecological corridors for these endangered and protected species.

Another group of species under negative impact from human frequentation of marine caves are bats. The noise caused by boats and humans may have a negative effect on bats and it is already documented that the bat population in Plava špilja (Blue cave) was more abundant before the touristic season in 2014 (Mačić et al. 2019).

Unfortunately, there is no monitoring of this location neither of other caves important for bats such as Niska cave, Ženska cave and Jošova (El Kaminova) cave where huge bat populations were recorded (Mačić et al. 2019; Petović et al. 2019; Djurović et al. 2019). Plava špilja is the most popular marine cave in Montenegro, visited by an extremely high number of tourists. Assessments of more than 200 boats per day during the high season in 2018 indicate the urgent need for management measures, not only because of the protection of the cave environment per se, but also because of the safety of tourists and boats entering the cave (Petović et al. 2019). Furthermore, we have recently witnessed the still ongoing pandemic of SARS-CoV-2 and it is suspected that the virus originated from horseshoe bats, but does not have a certain origin and source of viral transmission to humans (Zhou et al. 2020). A large reservoir of SARS-CoV-like viruses has been reported in horseshoe bats, therefore the possibility of a re-emergence of SARS and other novel viruses from animals should not be ignored (Cheng et al. 2007). Thus, increased human activity in these caves along with the presence of bats could theoretically cause a possible outbreak of new viruses.

Unfortunately, important quantities of marine litter are entering marine caves, mostly by currents. At particular risk here are semi-submersed caves exposed to the main wind and wave directions. Huge quantities of marine litter in the caves could provoke abrasion of the living organisms from the cave walls, but also could hinder the use of this habitat by some species like monk seals who needs cave beaches for resting and pupping (Panou et al. 1993). The example of cleaning the marine cave "Papuča" ("Slippers") in Montenegro showed, as expected, the huge majority of plastics in the total amount of litter removed (95.2%) while 5 other categories were present in <5% (rubber 2.8%, glass 1.2%, cloths 0.6%, wood 0.2%, metal 0.1%) (Mačić et al. 2018a, b). In general, cleaning activities of this type of habitat are difficult and expensive, therefore, the prevention and appropriate disposal of the waste on land should be prioritized. The huge number of slippers (76) collected from this cave gave it its name "Papuča" and we believe that this specific name and the many pictures of marine litter collected from its interior will greatly help to better promote cave cleaning activities and underline the problems caused by marine litter. Obviously, much more effort is needed for awareness-rising campaigns and the appropriate management of marine litter.

Coastal infrastructure such as marinas and jetties are changing currents and sediment transport, thereby possibly increasing sediment deposition in the innermost parts of the caves with negative consequences for the sessile, filter feeding animals (Nepote et al. 2017; Gerovasileiou and NikeBianchi 2021). The Montenegrin coast is facing rapid urbanization and various activities are being developed in several coastal areas. For instance, one cave in Trašte bay has been destroyed due to the construction of a new marina. The inventory of marine caves carried out in 2013–2015 has already provided the necessary baseline for appropriate conservation measures that will hopefully be taken in the near future (Mačić et al. 2013; Mačić et al. 2014; Mačić et al. 2015).

An indirect threat to marine cave biodiversity is non-indigenous species (NIS). Gerovasileiou et al. (2016a, b) reported 56 NIS from 50 marine caves and tunnels in the Mediterranean, while we registered only one invasive species (*Callinectes sapidus*) (Mačić et al. 2018a, b). The impacts of these taxa on native cave biocenosis are still not very well known and should be further studied. Furthermore, seawater warming has already been documented as a factor causing changes in cave communities (Costa et al. 2018; Sempere-Valverde et al. 2019). The Mediterranean sponges are among the best-studied animal groups in marine caves and they seem to be a good indicator of such changes. Thanks to some old surveys and long-term series of observations at some locations in the Ligurian Sea, it was possible to conclude that some sponge species changed their structure because of heat waves: 3-dimensional growth forms were replaced by 2-dimensional ones causing homogenization of the cave biocenosis (Parravicini et al. 2010; Costa et al. 2018). Of course, other species should be monitored as well. Here, the lack of historical data is the main obstacle toward the evaluation of the ecological status of the cave biogenesis. Further surveys should focus on filling the gaps in taxonomic checklists of the marine cave biodiversity. Special attention should be given to the infralittoral caves of Montenegro because they were not surveyed during the detailed three-year project which focused on semi-submersed marine caves as a possible habitat for monk seals (Mačić et al. 2013, 2014, 2015). Finally, monitoring and management plans of several already known and valuable marine caves should be elaborated in order to assess potential changes and provide protection for this priority habitat type.

Acknowledgements Our special thanks go to the company Jugopetrol AD that financed the cave surveys in 2013–2015, by the decision of its director at that time, Mr. Jannis Geroulanos, followed by Mr. Andreas Triantopoulos. Furthermore, Public Enterprise Morsko dobro, Budva, Montenegro, financially supported projects

"Istraživanje i čišćenje morskih pećina u zoni Borove šume u opštini Ulcinj" contract no. 0204-362/1, and "Istraživanje morskih pećina na potezu od rta Arza do uvale Trašte, Luštica, Herceg Novi," contract no. 0204-3322/2.

Annex 1 List of Species Recorded in Marine Caves on the Montenegrin Coast

Bacteria
Rivularia sp. (Agardh 1886)
Foraminifera
Miniacina miniacea (Pallas 1766)
Phytoplankton
Achnanthes brevipes (Agardh 1824)
Bacteriastrum hyalinum (Lauder 1864)
Chaetoceros affinis (Lauder 1864)
Chaetoceros spp. (Ehrenberg 1844)
Cocconeis scutellum (Ehrenberg 1838)
Cyclotella striata (Kutzing) (Grunow 1880)
Cylindrotheca closterium (Ehrenberg) Reimann and Lewin (1964)
Entomoneis pulchra (Bailey) Reimer (1975)
Guinardia flaccida (Castracane) (Peragallo 1892)
Guinardia striata (Stolterfoth) (Hasle 1996)
Hemiaulus hauckii (Grunow ex Van Heurck 1882)
Leptocylindrus danicus (Cleve 1889)
Leptocylindrus minimus (Gran 1915)
Exilaria flabellata (Ehrenberg 1832)
Licmophora paradoxa (Lyngbye) (Agardh 1828)
Navicula spp. (Bory de Saint-Vincent 1822)
Nitzschia longissima (Brébisson) (Ralfs 1861)
Pleurosigma angulatum (Queckett) (Smith 1852)
Pleurosigma elongatum (Smith 1852)
Pleurosigma formosum (Smith 1852)
Proboscia alata (Brightwell) (Sundström 1986)
Pseudo-nitzschia sp H. Peragallo in Peragallo and Peragallo (1900)
Striatella unipunctata (Lyngbye) Agardh (1832)
Synedra ulna (Nitzsch) (Ehrenberg 1832)
Synedra sp. (Ehrenberg 1830)
Thalassionema frauenfeldii (Grunow) (Tempère and Peragallo 1910)
Thalassionema nitzschioides (Grunow) (Mereschkowsky 1902)
Toxarium undulatum (Bailey 1854)
Diplopsalis lenticula (Bergh 1881)
Gonyaulax sp. (Diesing 1866)
Gymnodinium sp (Stein 1878)
Gyrodinium fusiforme (Kofoid and Swezy 1921)
Lingulodinium polyedra (F. Stein) (Dodge 1989)
Oxytoxum sphaeroideum (Stein 1883)
Phalacroma rotundatum (Claparéde & Lachmann) (Kofoid and Michener 1911)
Prorocentrum cordatum (Ostenfeld) (Dodge 1976)
Prorocentrum scutellum (Schröder 1900)
Prorocentrum triestinum (Schiller 1918)
Protoperidinium pellucidum (Bergh 1881)
Protoperidinium steinii (Jørgensen 1899; Balech 1974)
Protoperidinium tuba (Schiller) (Balech 1974)
Protoperidinium conicum (Gran) (Balech 1974)
Scrippsiella sp. (Balech ex A. R. Loeblich III 1965)
Tripos horridus (Cleve) (Gómez 2013)
Tripos furca (Ehrenberg) (Gómez 2013)
Tripos fusus (Ehrenberg) (Gómez 2013)
Tripos kofoidii (E. G. Jørgensen) (Gómez 2013)
Calyptrosphaera oblonga (Lohmann 1902)
Syracosphaera pulchra (Lohmann 1902)
Helicosphaera wallichii (Lohmann) (Okada and McIntyre 1977)
Rhabdosphaera tignifer Schiller
Zooplankton
Obelia sp. (Péron and Lesueur 1810)
Sphaeronectes koellikeri (Huxley 1859)
Eudoxia sp. (Eschscholtz 1825)
Penilia avirostris (Dana 1849)
Evadne spinifera (Müller 1867)
Calanus helgolandicus (Claus 1863)
Calanus sp. (Leach 1816)
Paracalanus parvus parvus (Claus 1863)
Calocalanus sp. (Giesbrecht 1888)
Clausocalanus jobei (Frost and Fleminger 1968)
Mecynocera clausi (Thompson 1888)
Clausocalanus sp. (Giesbrecht 1888)
Centropages kroyeri (Giesbrecht 1893)
Isias clavipes (Boeck 1865)
Paraeuchaeta hebes (Giesbrecht 1888)
Acartia (Acartiura) clausi (Giesbrecht 1889)
Oncaeidae sp. (Giesbrecht 1893)
Sapphirina sp. (Thompson 1829)
Oithona sp. (Baird 1843)
Oithona nana (Giesbrecht 1893)
Oithona plumifera (Baird 1843)
Oithona similis (Claus 1866)
Monothula subtilis (Giesbrecht 1893)
Euterpina acutifrons (Dana 1847)
Microsetella sp. (Brady and Robertson 1873)
Corycaeus sp. (Dana 1845)
Creseis acicula (Rang 1828)
Creseis virgula (Rang 1828)
Oikopleura (Coecaria) longicauda (Vogt 1854)
Parasagitta setosa (Müller 1847)
Sagitta sp. (Quoy and Gaimard 1827)
Pelagia noctiluca (Forsskål 1775)

Macroalgae

Anadyomene stellata (Wulfen) (Agardh 1823)
Flabellia petiolata (Turra) (Nizamuddin 1987)
Palmophyllum crassum (Naccari) (Rabenhorst 1868)
Ulva sp. (Linnaeus 1753)
Halimeda tuna (J. Ellis & Solander) (Lamouroux 1816)
Cystoseira compressa (Esper) (Gerloff and Nizamuddin 1975)
Cladostephus spongiosus (Hudson) (Agardh 1817)
Dictyota dichotoma (Hudson) (Lamouroux 1809)
Lobophora variegata (J. V. Lamouroux) (Womersley ex Oliveira 1977)
Padina pavonica (Linnaeus) Thivy (1960)
Callithamnion granulatum (Ducluzeau) (Agardh 1828)
Corallina elongata (Ellis and Solander 1786)
Hildenbrandia rubra (Sommerfelt) (Meneghini 1841)
Lithophyllum sp. (Philippi 1837
Lithophyllum byssoides (Lamarck) (Foslie 1900)
Lithophyllum stictiforme (J. E. Areschoug) (Hauck 1877)
Mesophyllum alternans (Foslie) (Cabioch and Mendoza 1998)
Mesophyllum expansum (Foslie) (Cabioch and Mendoza 1998)
Osmundaria volubilis (Linnaeus) (Norris 1991)
Peyssonnelia bornetii (Boudouresque and Denizot 1973)
Peyssonnelia rosa-marina (Boudouresque and Denizot 1973)
Peyssonnelia rubra (Greville) (Agardh 1851)
Peyssonnelia squamaria (S. G. Gmelin) (Decaisne ex Agardh 1842)
Rhodymenia ardissonei (Kuntze) (Feldmann 1937)
Sphaerococcus coronopifolius (Stackhouse 1797)
Jania rubens (Linnaeus) (Lamouroux 1816)

Porifera

Acanthella acuta (Schmidt 1862)
Agelas oroides (Schmidt 1864)
Antho (Antho) inconstans (Topsent 1925)
Axinella cannabina (Esper 1794)[7]
Axinella damicornis (Esper 1794)[7]
Axinella vaceleti (Pansini 1984)
Axinella verrucosa (Esper 1794)[7]
Cacospongia mollior (Schmidt 1862)
Chondrosia reniformis (Nardo 1847)
Clathrina clathrus (Schmidt 1864)
Cliona celata (Grant 1826)
Cliona rhodensis (Rützler and Bromley 1981
Cliona schmidtii (Ridley 1881)
Cliona viridis (Schmidt 1862)
Crambe crambe (Schmidt 1862)
Crella (Crella) elegans (Schmidt 1862)
Crella (Grayella) pulvinar (Schmidt 1868)
Dendroxea sp. (Griessinger 1971)
Dysidea avara (Schmidt 1862)
Dysidea fragilis (Montagu 1814)
Dysidea sp. (Johnston 1842)
Fasciospongia cavernosa (Schmidt 1862)
Haliclona (Halichoclona) fulva (Topsent 1893)
Haliclona (Soestella) mucosa (Griessinger 1971)
Hemimycale columella (Bowerbank 1874)
Ircinia oros (Schmidt 1864)
Ircinia variabilis (Schmidt 1862)
Penares helleri (Schmidt 1864)
Petrosia (Petrosia) ficiformis (Poiret 1789)
Phorbas tenacior (Topsent 1925)
Pleraplysilla spinifera (Schulze 1879) *Sarcotragus foetidus* (Schmidt 1862)[2]
Sarcotragus spinosulus (Schmidt 1862)
Scalarispongia scalaris (Schmidt 1862)
Spongia (Spongia) officinalis (Linnaeus 1759)[3,5,7]
Terpios fugax (Duchassaing and Michelotti 1864)
Terpios gelatinosus (Bowerbank 1866)
Tethya aurantium (Pallas 1766)[7]

Cnidaria

Actinia cari (Delle Chiaje 1822)
Actinia equina (Linnaeus 1758)
Caryophyllia (Caryophyllia) inornata (Duncan 1878)[6]
Caryophyllia (Caryophyllia) smithii (Stokes and Broderip 1828)[6]
Cerianthus membranaceus (Gmelin 1791)
Cladocora caespitosa (Linnaeus 1767)
Clavularia sp. (Blainville 1830)
Eudendrium sp. (Ehrenberg 1834)
Hoplangia durotrix (Gosse 1860)
Leptopsammia pruvoti (Lacaze-Duthiers 1897)
Madracis pharensis (Heller 1868)
Nausithoe punctata (Kölliker 1853)
Paracyathus pulchellus (Philippi 1842)
Parazoanthus axinellae (Schmidt 1862)
Phyllangia americana mouchezii (Lacaze-Duthiers 1897)
Polycyathus muellerae (Abel 1959)
Sarcodictyon catenatum Forbes in (Johnston 1847)

Polychaeta

Bonellia viridis (Rolando 1822)
Filograna sp. (Berkeley 1835)
Hermodice carunculata (Pallas 1766)
Protula sp. (Risso 1826)
Serpula vermicularis (Linnaeus 1767)
Thylacodes arenarius (Linnaeus 1758)

Bryozoa

Adeonella calveti (Canu and Bassler 1930)
Beania magellanica (Busk 1852)
Bonellia viridis (Rolando 1822)
Caberea boryi (Audouin 1826)
Disporella hispida (Fleming 1828)
Frondipora verrucosa (Lamouroux 1821)
Membranipora membranacea (Linnaeus 1767)

Myriapora truncata (Pallas 1766)
Patinella radiata (Audouin 1826)
Reteporella sp. (Busk 1884)
Schizobrachiella sanguinea (Norman 1868)
Schizomavella (Schizomavella) mamillata (Hincks 1880)
Smittina cervicornis (Pallas 1766)
Mollusca
Calliostoma conulus (Linnaeus 1758) *Calmella cavolini* (Vérany 1846) *Clavagella* sp. (Blainville 1817)
Diaphorodoris papillata (Portmann and Sandmeier 1960)
Felimare orsinii (Vérany 1846)
Felimare picta (Philippi 1836)
Gibbula sp. (Risso 1826)
Glycymeris nummaria (Linnaeus 1758)
Lithophaga lithophaga (Linnaeus 1758)
Mytilus galloprovincialis (Lamarck 1819)
Neopycnodonte cochlear (Poli 1795)
Patella caerulea (Linnaeus 1758)
Patella rustica (Linnaeus 1758)
Peltodoris atromaculata (Bergh 1880)
Phyllidia flava (Aradas 1847)
Rocellaria dubia (Pennant 1777)
Tarantinaea lignaria (Linnaeus 1758)
Thylacodes arenarius (Linnaeus 1758)
Tonna galea (Linnaeus 1758)
Tritia incrassata (Strøm 1768)
Vermetus granulatus (Gravenhorst 1831)
Crustacea
Amphibalanus amphitrite (Darwin 1854)
Eriphia verrucosa (Forskål 1775)
Chthamalus montagui (Southward 1976)
Dromia personata (Linnaeus 1758)
Palaemon serratus (Pennant 1777)
Pachygrapsus marmoratus (Fabricius 1787)
Stenopus spinosus (Risso 1827) in Risso (1826–1827)
Callinectes sapidus (Rathbun 1896)
Dardanus calidus (Risso 1827) in Risso (1826–1827)
Pagurus anachoretus (Risso 1827) in Risso (1826–1827)
Perforatus perforatus (Bruguière 1789)
Echinodermata
Antedon mediterranea (Lamarck 1816)
Arbacia lixula (Linnaeus 1758)
Centrostephanus longispinus (Philippi 1845)
Coscinasterias tenuispina (Lamarck 1816)
Echinaster (Echinaster) sepositus (Retzius 1783) *Hacelia attenuata* (Gray 1840)
Holothuria (Holothuria) tubulosa (Gmelin 1791)
Holothuria (Panningothuria) forskali (Delle Chiaje 1823)
Holothuria (Platyperona) sanctori (Delle Chiaje 1823)
Marthasterias glacialis (Linnaeus 1758)
Ophidiaster ophidianus (Lamarck 1816)
Ophioderma longicaudum (Bruzelius 1805)
Ophiothrix fragilis (Abildgaard in Müller 1789)
Paracentrotus lividus (Lamarck 1816)
Sphaerechinus granularis (Lamarck 1816)
Tunicata
Clavelina dellavallei (Zirpolo 1925)
Diplosoma spongiforme (Giard 1872)
Diplosoma sp. (Macdonald 1859)
Lissoclinum weigelei (Lafargue 1968)
Halocynthia papillosa (Linnaeus 1767)
Microcosmus sp. (Heller 1877)
Pycnoclavella communis (Pérez-Portela et al. 2007)
Pisces
Apogon imberbis (Linnaeus 1758)
Coris julis (Linnaeus 1758)
Microlipophrys nigriceps (Vinciguerra 1883)
Mugil cephalus (Linnaeus 1758)
Mullus barbatus (Linnaeus 1758)
Muraena helena (Linnaeus 1758)
Scorpaena notata (Rafinesque 1810)
Serranus cabrilla (Linnaeus 1758)
Serranus scriba (Linnaeus 1758)
Thorogobius ephippiatus (Lowe 1839)
[1]Bern Convention 1979 Annex I; [2]SPA Protocol Annex II 2018; [3]SPA Protocol Annex III 2012; [4]Habitat Directive 1992; [5]Bern Convention 1979 Annex II–III; [6]CITES 1990 Annex II; [7]Official Gazette RCG 76/06

References

Abdullin, S.R., Bagmet, V.B. (2016). Mixotrophy of cyanobacteriae and algae inhabiting caves. *Zhurnal Obshcheĭ Biologii* 77, 54–62.

Bakran-Petricioli, T., Radolović, M., Petricioli,D. (2012). How diverse is sponge fauna in the Adriatic Sea? *Zootaxa*, 3172(1), 20-38.

Baric, A., Grbec, B., Bogner, D. (2008). Potential implications of sea-level rise for Croatia. *Journal of Coastal Research*, 24(2), 299–305.

Barović, G. (2011). Intenziviranje speleoloških istraživanja na prostoru Crne Gore u funkciji unapredjenja geoprostora, Zbornik radova, *7 Simpozijum o zaštiti karsta*, Bela Palanka,11–18.

Barovic, G., Vujacic, D., Radusinovic, S.: Physical-Geographical Characteristics of the Coast of Montenegro. In A. Joksimović, M. Đurović, I.S. Zonn, A.G. Kostianoy, A.V. Semenov (eds.), The Montenegrin Adriatic Coast: Marine Biology, pp.15–28. Springer Nature Switzerland AG, Cham (2021)

Bern Convention 1979. Convention on the Conservation of European Wildlife and Natural Habitats (ETS No. 104), Annex I, https://www.coe.int/en/web/conventions/full-list?module=treaty-detail&treatynum=104 Accessed 20 June 2020

Bianchi, C.N.: Flora e fauna: Lineamenti generali e prospettive. In F. Cicogna et al. (eds.), Grotte Marine: Cinquant'Anni di Ricerca in Italia, pp. 137–146. Ministero dell'Ambiente e della Tutela del Territorio, Rome (2003)

Bianchi, C.N., Morri, C. (1994). Studio bionomico comparativo di alcune grotte marine sommerse; definizione di una scala di confinamento. *Memorie dell'Istituto Italiano di Speleologia* serie II 6, 107–123.

Bundone, L., Panou, A., Molinaroli, E. (2019). On sightings of (vagrant?) monk seals, Monachus monachus, in the Mediterranean Basin and their importance for the conservation of the species. *Aquatic Conserv: Mar Freshw Ecosyst*. doi: https://doi.org/10.1002/aqc.3005.

Bussotti, S., Di Franco, A., Francour, P., Guidetti, P. (2015). Fish Assemblages of Mediterranean Marine Caves. *PloS ONE*. doi:https://doi.org/10.1371/journal.pone.0122632.

Bussotti, S., Di Franco, A., Pey, A., Vieux-Ingrassia, J.V., Planes, S., Guidetti, P. (2017). Distribution patterns of marine cave fishes and the potential role of the cardinal fish Apogon imberbis (Linnaeus, 1758) for cave ecosystem functioning in the western Mediterranean. *Aquatic Living Resources* 30, 1–9.

Cheng, V.C., Lau, S.K., Woo, P.C., Yuen, K. Y. (2007). Severe acute respiratory syndrome coronavirus as an agent of emerging and reemerging infection. *Clinical microbiology reviews*, 20(4), 660-694.

CITES Convention (1990). Madracis pharensis (Heller, 1868). UNEP World Conservation Monitoring Centre. http://checklist.cites.org/#/en 1990 CITES (1990). Accessed 20 June 2020

Costa, G., Betti, F., Nepote, E., Cattaneo-Vietti, R., Pansini, M., Bavestrello, G., Bertolino, M. (2018). Sponge community variations within two semi-submerged caves of the Ligurian Sea (Mediterranean Sea) over a half-century time span. *The European Zoological Journal* 85, 382–392.

Culver, D.C., White, W. (2012). Encyclopedia of Caves. Oxford: Elsevier.

Djurović, M. (2019). Ponovni pronalazak Jošove pećine i rezultati novijih pregleda Sumporne pećine - dva bitna ljetnja skloništa slijepih miševa u Crnoj Gori. *Hypsugo* IV, 1, 34-39.

Garašić, M. (1991). Morphological and hydrogeological classification of speleological structures (caves and pits) in the Croatian karst area. *Geološki Vjesnik*, 44, 289–300.

Gerovasileiou, V., Nike Bianchi, C. (2021). Mediterranean marine caves: A synthesis of current knowledge. *Oceanography and Marine Biology: An Annual Review*, 59, 1-88.

Gerovasileiou, V., Voultsiadou, E. (2012). Marine caves of the Mediterranean Sea: A sponge biodiversity reservoir within a biodiversity hotspot. *PloS One* 7, e39873

Gerovasileiou, V., Martínez, A., Álvarez, F., Boxshall, G., Humphreys, W., Jaume, D., Becking, L., Muricy, G., van Hengstum, P., Dekeyzer, S., Decock, W., Vanhoorne, B., Vandepitte, L., Bailly, N., Iliffe, T. (2016a). World Register of marine Cave Species (WoRCS): A new thematic species database for marine and anchialine cave biodiversity. *Research Ideas and Outcomes* 2, e10451.

Gerovasileiou, V., Voultsiadou, E., Issaris, Y., Zenetos, A. (2016b). Alien biodiversity in Mediterranean marine caves. *Marine Ecology* 37, 239–256.

Gerovasileiou, V., Martínez García, A., Álvarez Noguera, F., Boxshall, G., Humphreys, WF., Jaume, D., Becking, LE., Muricy, G., van Hengstum, PJ., Yamasaki,H., Bailly,N., Iliffe,T.M.(2020).World Register of MarineCaveSpecies (WoRCS). Oostend, Belgium: Vlaams Instituut voor de Zee. doi:https://doi.org/10.14284/351.

Giakoumi, S., Sini, M., Gerovasileiou, V., Mazor, T., Beher, J., Possingham, H.P., Abdulla, A., Cinar, M.E., Dendrinos, P., Gucu, A.C., Karamanlidis, A.A., Rodic, P., Panayotidis, P., Taskin, E., Jaklin, A., Voultsiadou, E., Webster, Ch., Zenetos, A., Katsanevakis, S. (2013). Ecoregion-based conservation planning in the Mediterranean: Dealing with large-scale heterogeneity. *PloS One* 8, e76449.

Guarnieri, G., Terlizzi, A., Bevilacqua, S., Fraschetti, S. (2012). Increasing heterogeneity of sensitive assemblages as a consequence of human impact in submarine caves. *Marine Biology* 159, 1155–1164.

Gunn, J. (2004). Encyclopedia of Caves and Karst Science. New York: Fitzroy Dearborn.

Habitat Directive (1992) Council Directive 92/43/EEC of 21 May 1992 on the conservation of natural habitats and of wild fauna and flora

Harmelin, J.G.: Bryozoan dominated assemblages in Mediterranean cryptic environments. In C. Nielsen, G.P. Larwood (eds.). Bryozoa: Ordovician to recent, pp. 135–143, Olsen & Olsen, Fredensborg (1985)

Harmelin,J.G.(1997).Diversityof bryozoans in aMediterranean sublittoral cave with bathyal like conditions: Role of dispersal processes and local factors. *Marine Ecology Progress Series* 153, 139–152.

Harmelin, J.G., Vacelet, J. (1997). Clues to deep-sea biodiversity in a nearshore cave. *Vie et Milieu* 47, 351–354.

Iliffe, T.M. (1990). Crevicular dispersal of marine cave faunas. *Mémoires de Biospéologie* 17, 93–96.

Mačić, V., Panou, A., Bundone, L., Varda, D. (2013). *Istraživanje budućeg zaštićenog područja u moru "Platamuni" i susjednog poluostrva Luštica sa špsebnom pažnjom na morske pećine kao potencijalno stanište za ugroženog sredozemnog tuljana*. Kotor, Jugopetrol AD.

Mačić, V., Panou, A., Varda, D., Lazarević, B. (2014). *Nastavak istraživanja morskih pećina i staništa za morske medvedice (rt Platamuni - rt Volujica)*. Kotor, Jugopetrol AD.

Mačić, V., Panou, A., Bundone, L., Varda, D., Pavićević, M. (2015). *Poslednje istraživanje morskih pećina uključujući staništa za morske medvedice u Crnoj Gori (od rta Volujica do rta Đeran)*. Kotor, Jugopetrol AD.

Mačić, V., Drakulović, D., Đurović, M., Huter, A., Joksimović, D., Lazarević, B., Pestorić, B., Petović, S. (2018a). *Istraživanje i čišćenje morskih pećina u zoni Borove šume u opštini Ulcinj. II dio Analiza sumporovite morske vode i biološki parametri u 3 pećine*. Kotor: Institut za biologiju mora.

Mačić, V., Đorđević, N., Petović, N., Malovrazić, N., Bajković, M. (2018b). Typology Of Marine Litter In „Papuča" (Slipper) Cave (Montenegro, South Adriatic Sea). *Studia Marina*, 31(2), 38-43.

Mačić, V., Panou, A., Bundone, L., Varda, D., Pavićević, M. (2019). First inventory of the semi-submerged marine caves in South Dinarides karst (Adriatic Coast) and preliminary list of species. *Turkish Journal of Fisheries and Aquatic Sciences* 19, 765–774.

Milanović, S. (2005). Hydrogeological characteristics of some deep siphonal springs in Serbia and Montenegro karst. *Water Resources and Environmental Problems in Karst - CVIJIĆ* 2005, 451–458.

Milanović, P., Stevanović, Z., Čokorilo Ilić, M.: Field Trip Guide: Trebinje, June 2014. Beograd: DIKTAS Project: Faculty of Mining and Geology, Department for Hydrogeology, (Bosna i Hercegovina), http://dr.rgf.bg.ac.rs/s/repo/item/755 (2014). Accessed 18 June 2020

Milazzo, M., Chemello, R., Badalamenti, F., Camarda, R., Riggio, S. (2002). The impact of human recreational activities in Marine Protected Areas: What lessons should be learnt in the Mediterranean Sea? *Marine Ecology* 23, 280–290.

Mirić, Đ., (1973). Prilog rasprostranjenju i sistematskom položaju dugonogog šišmiša (*Myotis capaccinii* Bonaparte, 1837) iz istočnih delova Jugoslavije. *Glasnik Prirodnjačkog muzeja u Beogradu*, B, 28, 179–205.

Morri, C. Confinamento idrologico. In F. Cicogna et al. (eds.) Grotte Marine: Cinquant'Anni di Ricerca in Italia, pp. 291–296, Ministero dell'Ambiente e della Tutela del Territorio, Rome (2003)

Moscatello, S., Belmonte, G. (2007). The plankton of a shallow submarine cave ('Grotta di Ciolo', Salento Peninsula, SE Italy). *Marine Ecology* 28, 47–59.

Muhldorfer, K. (2013). Bats and bacterial pathogens: A review. *Zoonoses Public Health* 60, 93–103.

Nepote, E., Bianchi, C. N., Morri, C., Ferrari, M., Montefalcone, M. (2017). Impact of a harbour construction on the benthic community of two shallow marine caves. *Marine Pollution Bulletin*, 114(1), 35-45.

Official Gazette RCG (2006). Rješenje o stavljanju pod zaštitupojedinih biljnih i životinjskih vrsta. Official Gazette RCG, no.76/2006 on 12.12.2006. year.

Official Gazette RCG No 54/16 (2016). *Zakon o zaštiti prirode* "Službeni list Crne Gore", Podgorica, br. 054/16 od 15.08.2016.

Ouerghi, A., Gerovasileiou, V., Bianchi, C.N.: Mediterranean marine caves: A synthesis of current knowledge and the Mediterranean Action Plan for the conservation of 'dark habitats'. In Öztürk B (eds.) Marine Caves of the Eastern Mediterranean Sea. Biodiversity, Threats and Conservation, pp. 1–13. Turkish Marine Research Foundation (TUDAV), Istanbul (2019)

Panou, A., Jacobs, J., Panos, D. (1993). The endangered Mediterranean monk seal Monachus monachus in the Ionian Sea, Greece. *Biological Conservation*, doi: https://doi.org/10.1016/0006-3207(93)90649-L

Panou, A. Bundone, L., Varda, D., Mačić, V. (2014). Habitat availability for the Mediterranean monk seal Monachus monachus in Montenegro. Abstract Book. *28th Annual Conference of the European Cetacean Society*, Liege (Belgium), 5–9 April 2014, p 205.

Panou, A., Varda, D., Bundone, L. (2017). The Mediterranean Monk Seal, Monachus monachus, in Montenegro. *Proceedings ISEM7*, October, 4–7, 2017. (pp: 94–101). Podgorica.

Parravicini, V., Guidetti, P., Morri, C., Montefalcone, M., Donato, M., Bianchi, C.N. (2010). Consequences of sea water temperature anomalies on a Mediterranean submarine cave ecosystem. *Estuarine and Coastal Marine Science* 86, 276–282.

Paunović, M., Stamenković, S. (1998). A revision of the distribution and status of *Rhinolophus euryale* Blasius, 1853 and *Rhinolophus blasii* Peters, 1866 (Rhinolophidae) in Yugoslavia, based on the discrimination properties of distinctive morphological characters. *Myotis*, 36, 7–23.

Peres, J.M. (1967). Mediterranean benthos. *Oceanography and Marine Biology: An Annual Review* 5, 449–533.

Peres, J.M., Picard, J. (1964). Nouveau manuel de bionomie benthique de la Mer Mediterranee. *Recueil des Travaux de la Station Marine d'Endoume* 47, 3–137.

Petović, S., Mačić, V.: Marine habitats of special importance along the Montenegrin coast. In A. Joksimović, M. Đurović, I.S. Zonn, A.G. Kostianoy, A.V. Semenov (eds.). The Montenegrin Adriatic Coast: Marine Biology, pp. 233–247. Springer Nature Switzerland AG, Cham (2021).

Petović, S., Mačić, V., Drakulović, D., Joksimović, D., Pestorić, B., Jokanović, S., Đurović, M., Đorđević, N., Lazarević, B. (2019). *Istraživanje morskih pećina na potezu od rta Arza do uvale Trašte, Luštica, Herceg Novi*. Kotor, Institut za biologiju mora za potrebe JP Morsko Dobro.

Petović, S., Marković, O., Đorđević, N.: Macrozoobenthic Species as a Part of the Benthic Communities Along the Montenegrin Adriatic Coast. In A. Joksimović, M. Đurović, I.S. Zonn, A.G. Kostianoy, A.V. Semenov (eds.). The Montenegrin Adriatic Coast: Marine Biology, pp. 153–192. Springer Nature, Cham (2021)

Petricioli, D., Bakran-Petricioli, T. (2019). Status of knowledge on marine cave environments of the Croatian Adriatic coast. In H. Langar, A. Ouerghi (eds.) *Proceedings of the 2nd Mediterranean Symposium on the Conservation of Dark Habitats*, pp. 19–24. Tunis: Regional Activity Centre for Specially Protected Areas.

Petrović, J., Rudaković, M. (1980). Značaj speleoloških istraživanja za vodosnadbjevanje u Bokokotorskom zalivu, *Sedmi Jugoslovenski speleološki kongres*, Titograd, 353–360.

Presetnik,P., Paunović,M., Karapandža,M., Đurović,M., Ivanović,Č., Ždralević, M., Benda, P., Budinski, I. (2014). Distribution of bats (Chiroptera) in Montenegro. *Vespertilio*, 17, 129–156.

Presetnik, P., Radonjić, M., Pavlovič, E., Gojznikar, J., Jovanović, M. (2017). Rezultati istraživanja faune slijepih miševa tokom istraživačkog kampa studenata biologije „Ekosistemi Balkana – Skadarsko jezero 2017" (Crna gora). *Hypsugo* 2(2), 41–52.

Radonjic, M. (2019). Rediscovery of Jašova cave and the latest bat survey of Sumporna cave- two important bat summer roosts in Montenetgro. *Hypsugo* 4(1), 1–7.

Riedl, R. (1966). *Biologie der Meereshöhlen*. Hamburg: Paul Parey.

Rodwell, J., García Criado, M., Gubbay, S., et al.: European red list of habitats. Part 1, Marinehabitats. PublicationsOffice. https://data.europa.eu/doi/https://doi.org/10.2779/032638 (2017). Accessed 15 June 2020

Rosso, A., Di Martino, E. (2016). Bryozoan diversity in the Mediterranean Sea: An up-date. *Mediterranean Marine Science* 17, 567–607.

Sanfilippo, R., Rosso, A., Guido, A., Gerovasileiou, V. (2017). Serpulid communities from two marine caves in the Aegean Sea, eastern Mediterranean. *Journal of the Marine Biological Association of the United Kingdom* 97, 1059–1068.

Sempere-Valverde, J., Lorenzo, Á.S., Espinosa, F., Gerovasileiou, V., Sánchez-Tocino, L., Navarro-Barranco, C. (2019). Taxonomic and morphological descriptors reveal high benthic temporal variability in a Mediterranean marine submerged cave over a decade. *Hydrobiologia*, *839*(1), 177–194.

SPA/RAC–UN Environment/MAP, OCEANA: Guidelines for inventorying and monitoring of dark habitats in the Mediterranean Sea. SPA/RAC Tunis. https://www.researchgate.net/publication/331788348_Guidelines_for_inventoring_and_monitoring_of_dark_habitats_in_the_Mediterranean_Sea (2017). Accessed 16. June 2020

SPA Protocol—Annex II 2018. List of endangered or threatened species, UNEP/MAP-SPA/RAC, 10p.

SPA Protocol—Annex III 2012. List of species whose exploitation is regulated, UNEP/MAP-SPA/RAC, 4p.

Surić, M., Lončarić, R., Lončar, N. (2010). Submerged caves of Croatia: distribution, classification and origin. *Environmental Earth Sciences*, 61, 1473–1480.

Tortonese, E. (1978). On two seastars living in the Mediterranean submarine caves. *Pubblicazioni della Stazione Zoologica di Napoli* 40, 506–515.

UNEP/MAP-RAC/SPA: Montenegro: Platamuni and Ratac areas. Mapping of marine key habitats and initiation of monitoring network. RAC/SPA –MedKeyHabitats Project, Tunis. https://www.rac-spa.org/sites/default/files/doc_mkh/montenegro/platamuni_ratac_areas_mapping.pdf (2016). Accessed 25 June 2020

Zhou, P., Yang, X.L., Wang, X.G., Hu, B., Zhang, L., Zhang, W., Si1, H.R., Zhu, Y., Li1, B., Huang, C.L., Chen, H.D., Chen, J., Luo, Y., Guo, H., Jiang, R.D., Liu, M.Q., Chen, Y., Shen, X.R., Wang, X., Zheng, X.S., Zhao, K., Chen, Q.J., Deng, F., Liu, L.L., Yan, B., Zhan, F.X., Wang, Y.Y., Xiao, G.F., Shi, Z.L. (2020). A pneumonia outbreak associated with a new coronavirus of probable bat origin. *Nature*, *579*(7798), 270-273.

Caves and Karst of the Prokletije Mountains

Ditta Kicińska, Krzysztof Najdek and Adam Łada

Abstract

The Prokletije Mountains are the highest range in the Dinaric Alps. Karst phenomena are developed in the carbonates of the Mesozoic age and belong to the High Karst and Durmitor units. Most of the caves are known from the Kolata and Belić massifs, where over 80 caves have been discovered, with a total length of over 36 km. In these massifs there are the longest and deepest caves of the entire Prokletije Mts. The Złota/Gold Cave System is the longest (with length of 8,202 m and vertical extent of 609 m) one and Górnicza/Mining Cave is the deepest (with vertical extent of 715 m and length of 2,633 m) one. A smaller number of caves occur in the vicinity of the Grbaja Valley. The oldest cave systems probably formed in the Pliocene, may be indicated by the morphology of the passages and the age of the highest level speleothem. The vast majority of caves are vertically developed and were formed during deglaciations. These are large pits, but also smaller, tighter passages, especially meanders. The Prokletije Mts with numerous karst and glacial forms are a mix of high-mountain glaciokarst and high-mountain karsts (alpine karsts).

1 Introduction

The Prokletije Mountains are one of the least known mountain areas in Europe. One of the first researchers was Cvijić who is the author of the first work on the glaciation of the Prokletije Mts (1903). The research was carried out by various authors in the 20-30 s, and then in the 70-80 s. In one hundred years, the Prokletije Mts were inaccessible to scientists and cavers due to the political situation (including two world wars and the Cold War). Only 20 last years have been a favorable period for both scientific research and cave exploration.

Many cave entrances of the Prokletije Mts, especially the lower ones, have long been known for local humans. These are very short caves or shelters. In the higher parts of the mountains, there are shallower and deeper pits.

The first information about the cave in this area comes from 1926 and deals with faunal research in Kriva Jama located above the village of Vusanje (Pretner 2011).

The largest known and horizontally developed cave in the vicinity of Gusinje is the Čardak Cave (in the Greben Mount). The cave was measured to the so-called I waterfall by members of the local tourist club. In the summer of 1973, an expedition of the Polski Klub Górski (Polish Mountain Club) explored the area near Gusinje. This expedition discovered further parts of the Čardak Cave and mapped the whole cave. At that time, 20 caves in the Grbaja Valley were also found (Kardaś 1978).

Ledena pečina/Jaskinia Lodowa/Ice Cave located near the hiking path to the highest peak of Montenegro—Zla Kolata (2534 m a.s.l.) was also known to the local people before. However, there is no information about its previous exploration. The cave is developed vertically and is inaccessible to tourists; they can visit only the first 10 m; further there is a deep pit, which can only be visited using specialized cave equipment.

The caves of this area were mapped and explored in the 70 s and 80 s (Karaman 2018).

Then the mountains were again inaccessible for political reasons. In 2017 a guide to the Prokletije Mts by Mulič (2017) was published. In "Plavsko-Gusinjske Prokletije—četrdeset planinarskih staza" the author mentions 10 caves known for the local people. These are Bezimena pećina na Mokroj planini, Pećina u izvorištu Veličke rijeke, Mirakova

D. Kicińska (✉)
Institute of Geology, Adam Mickiewicz University, Poznań, Poland
e-mail: ditta.kicinska@amu.edu.pl

K. Najdek · A. Łada
Wielkopolski Caving Club, Poznań, Poland

G. Barovic (ed.), *Speleology of Montenegro*, Cave and Karst Systems of the World,
https://doi.org/10.1007/978-3-031-49375-1_14

pećina na stjenovitim padinama Goleša u Murinskoj rijeci, Pećina u Malom vrhu na Čardaku iznad Grope, Pećina u Zabelju na Čardaku, Pećina u Refu na Čardaku, Ledena pećina u Ljug Preslopu na Bjeliču, Mirakova pećina na južnim padinama Lipovice, Đevojačka pećina na Sjekirici, Dženetska vrata kod Dvojnica na Bogićevici and Šaćirova pećina na Velikom Hridu.

Since 2006, systematic cave exploration has been conducted by the Wielkopolski Klub Taternictwa Jaskiniowego (WKTJ, Wielkopolski Caving Club, Poland) in co-operation the Speleoklub Świętokrzyski (SŚ, Świętokrzyski Speleoclub, Poland) and the Akademski speleološko-alpinistički klub (ASAK, Belgrad, Serbia). Cavers from Serbia, Croatia, Macedonia and France have also participated in expeditions organized by Poles. In 2010, Italian cavers in the middle of the Belić Massif were also explored in co-operation with WKTJ (Najdek and Kicińska 2017).

The cave expeditions were initiated and led by Krzysztof Najdek in 2006–2019 (except for 2013: the expedition was led by Aleksandra Maciejewska). In 2020, the expedition did not take place due to the coronavirus epidemic. Adam Łada has been the organizer of the expedition since 2021. From the beginning of exploration, geological research into the genesis and evolution of the discovered cave systems has been carried out.

The exploration and scientific activities have been conducted in co-operation with the National Park of Prokletije and the Agencija za zaštitu životne sredine/Environment Protection Agency of Montenegro.

2 Geographical and Geological Settings

The Prokletije Mts are located at the boundary between Montenegro (eastern part of the country), Albania and Kosovo (Fig. 1). The highest peaks are: Jezerski Vrh/Maja Jezerce (2694 m) in Albania, Djeravica (2656 m) in Kosovo and Zla Kolata (2534 m) in Montenegro. In the Montenegrin part the height difference between the highest peaks and the valley bottoms exceeds 1000 m. These mountains extend almost perpendicular to the Dinaric Alps. The Prokletije Mts are commonly considered the most south-eastern part of the Dinaric Alps, although many researchers emphasize their individuality (Cvijić 1899; Djurović and Djurović 2016) and do not include them in that mountain system or pay attention to their distinctiveness. The geographical division of both the Prokletije Mts belonging to the Dinaric Alps and the territory of Montenegro itself are unclear. According to some authors, the Prokletije Mts also include Visitor, Komovi, Žijevo and other massifs (Petrovič 2014; Djurović and Djurović 2016). These massifs will not be discussed in this paper.

Geologically, the Prokletije Mts belong to the High Karst and Durmitor units. The main karst area is the southernmost part of the Prokletije Mts. The karst areas of the High Karst Unit are mainly composed of limestone, dolomitic limestone, dolomite and limestone breccia (Lower Triassic to Upper Cretaceous age; Djokič et al. 1976). In the northern and northeastern parts of the mountains, the High Karst Unit is overthrusted by the Durmitor Unit. The Durmitor Unit comprises Paleozoic (Devonian and Carboniferous) phyllites and schists, the Permian–Triassic conglomerates and the Jurassic carbonate rocks with sandstones. The thickness of the karst rocks exceeds 1500 m (Djokič et al. 1976).

The Prokletije Mts elevation (above 2600 m above sea level) and their relief with reference to the surrounding areas is related to the Shkoder-Peja Fault and related discontinuities (Gemignani et al. 2022) to the south. According to Frasheri et al. (2009) the Shkoder-Peja transverse fault is a transition zone between the clockwise rotation of the Albanides and Hellenides and the counterclockwise one of the Dinarides. Hence the almost transverse extent of the Prokletije Mts in relation to the Dinaric Alps. During Oligocene–Miocene regional collisional deformation the Prokletije Mts were deformed into large high-angle reverse faults and folds (Roberson and Shallo, 2000).

Glacial and karst forms predominate in the morphology of the Prokletije Mts. There are such forms as U-shaped valleys (Fig. 2), glacier circuses, hanging valleys and remains of moraines, erratic blocks and fluvioglacial sediments. Glacial lakes in the Prokletije Mts include Ridsko, Ropojansko and Rikavičko. The largest glacial lake in Montenegro—Plavskie/Plav Lake is 2160 m long, 1490 m wide and with average depth of 3.86 m and maximum depth of 9.15 m. Pleistocene glaciations in the Prokletije Mts and in other areas of the Balkans were first observed and studied by Jovan Cvijić (1903, 1913). Milivojevič et al. (2008) distinguished three series of moraines and reconstructed the positions of former glaciers. According to the authors "during the most extensive glacial event, the average ELA was at 1750 m, in the second event at 1942 m, and during the youngest glacial event it was positioned at 2123 m altitude". Milivojevič et al. (2008) also defined the length of the glacier of the Ropojana Valley on 12.5 km and the length of the Grbaja Valley on 5 km.

Karst phenomena occur most frequently in the surroundings of the Ropojana and Grbaja valleys. These valleys are separated by the Karanfili-Brada range, a vast calcareous region drained by two vaucluse karst springs of great discharges: Alipašni Izvori/Ali Pasha and Oko Skakavica/Savino Oko. The Alipašni Izvori/Ali Pasha Spring, a 300-m-long zone of springs, is located in the Vrujskie Polje at 925 m a.s.l. It has discharge from 1–1,5 m^3/s in the

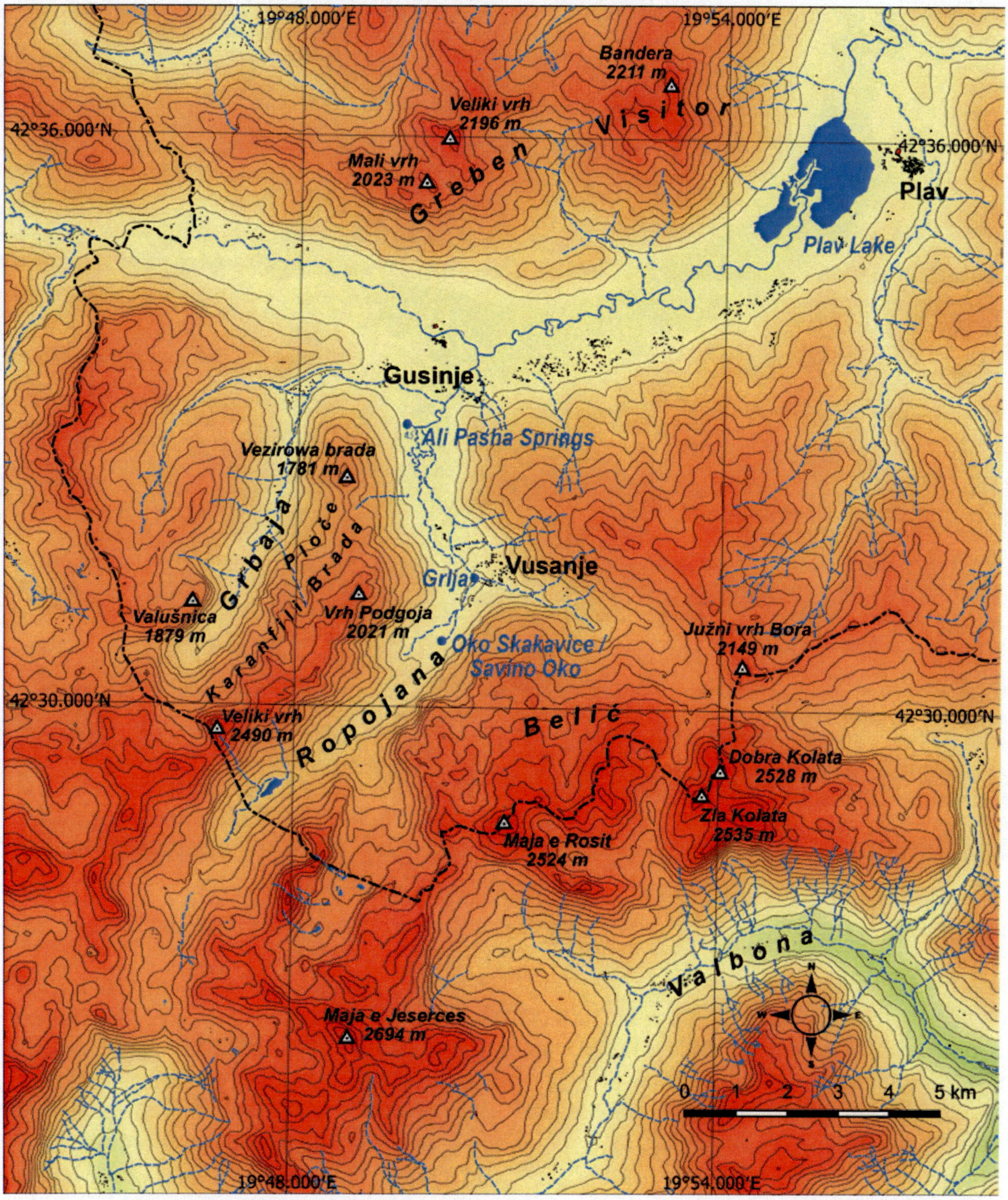

Fig. 1 Location of the studied area (author Krzysztof Najdek)

summertime up to 3 m^3/s in spring. From this point water flows as the Vruja stream toward the Ljuča river.

The estimated minimum efficiency of Oko Skakavica/Savino Oko is about 50 l/s (Fig. 3). The spring probably drains a large part of the Belić Massif. Dye-tracer experiments of the water in caves have not been done so far to establish how large the area is and define flow directions. The waters that flow from the Oko Skakavica/Savino Oko spring go into the Grlja Canyon (Fig. 4) after two kilometers. There are three more waterfalls in the canyon, with the highest one of about 25 m high. The Vruja River flows out of the canyon and together with Grnčar near Gusinje forms the Ljuča River. In the Grnčar valley, near Gusinje, there are the Bajrović Springs with a capacity of 49–73 l/s, according to measurements made in the 1970s (Radojević, 2012).

The Prokletije Mts are characterized by continental, mountainous and sub-alpine climate. The mean annual air temperature (MAAT) is calculated on the basis of the measurements in Vermosh, Albania (6.7 °C at 1152 m a.s.l.) and Kolašin, Montenegro (7.2 °C at 945 m a.s.l.) by Palmentola et al. (1995) and Milivojevič et al. (2008). The annual

Fig. 2 Ropojana Valley—example of U-shaped valley (photograph by Adam Łada)

rainfall varies, depending on the height, from 1000 to 2900 mm. It is 1,241 mm in Plav and 1,461 mm in Gusinje. Snowfall is quite heavy (Vladan et al. 2012). Despite its highest altitude, the Prokletije Mts has a lower precipitation rate than the regions of the Adriatic Sea (Djurović and Lješević 1994 fide Djurović and Djurović 2016).

3 Cave Exploration in 2006–2021

In 2006, the expedition explored the eastern part of the Grbaja Valley, in the Vezirova Brada-Karanfili ridge. Several dozen entrances were checked on the Ploče plateau and in the Mečin Do and Ljubokuč valleys. The expedition included cavers from WKTJ, Akademicki Klub Grołazów (AKG, Poland), ŚS, ASAK and Ponir Banja Luka (PBL) from Bosnia and Herzegovina (Kicińska and Najdek, 2007; Biegała et al., 2013; Kicińska and Najdek, 2013a). In 2007 the expedition took place in 3 different areas: Zastan Grbajski and Volušnica Mount in the Grbaja Valley and Belić Massif in the Ropojana Valley. Twenty pits have been surveyed, out of dozens of pits checked. The Belić Massif appeared to be the most prospective, with Setka/A Hundred and Lodowy Smok/Ice Dragon caves. The expedition was organized by WKTJ, AKG, SŚ and ASAK clubs (Najdek 2007; Biegała et al., 2013, Kicińska and Najdek, 2013a).

In 2008, two spring and summer expeditions were organized. The expedition made a documentation of 26 caves. The deepest ones are: Do Savino Oko/Cave to Savino Oko, Lodowy Smok/Ice Dragon Cave and Wrota Piekieł/Infernal Gate. The expedition was organized by WKTJ with the cooperation of SŚ. The expedition was also attended by cavers from other Polish clubs (AKG, Speleoklub Łódzki, Speleoklub Częstochowa, Klub Sportów Ekstremalnych—Jura, Speleoklub Gawra Gorzów) and Serbian clubs (Akademski Speleolosko-Alpinisticki Klub from Belgrad and Speleoklub Kraljevo). In 2009, two expeditions were organized by WKTJ and SŚ. The one in the spring period had the purpose of checking the entrance in the eastern part of the Belić Massif. During the summer, the expedition continued its exploration, both in the eastern and western parts of Belić (in the vicinity of Kuj Brešnik and Maja Rosit). In the eastern part of the massif, exploration was carried out in Lodowa/Ice and Kolektor/Collector caves (Kasza et al. 2010).

Fig. 3 Oko Skakavica/Savino Oko Spring (photograph by Ditta Kicińska)

In 2010–2011, the expedition explored the eastern part of the Belić Massif and the Kolata Massif. The exploration of Ledena pečina/Lodowa/Ice Cave continued, reaching 445 m of vertical extent and 1391 m in length, becoming the deepest cave in the massif. Among others, Gigant/Giant, Nibyczarna/Pseudoblack, Ziemia Obiecana/Promised Land, Kolektor/Collector and Górnicza/Mining caves have been discovered and explored. In 2010, more than 5,200 m of new cave passages were mapped. The expeditions were organized by WKTJ, SŚ, ASAK and Speleoklub Valjevo (SV). In 2010 cavers from Speleoklub Bobry and Speleoklub Tatrzański (Poland) also participated (Kicińska et al. 2011).

In 2012–2016, the exploration of Górnicza/Mining, Lodowa/Ice, Nibyczarna/Pseudoblack and Gigant/Giant caves continued. New caves such as W Trzech Kopcach/In Three Mounds, Entuzjastyczna/Enthusiastic, 03 143, Lincura, Grom Iż Vedra Neba/A bolt from the blue and Aladinova/Aladdin were discovered in 2012. In 2013, Obiboki/Slackers, Smocza Jama/Dragon's Den and Jastrzębia/Hawk caves were found. This season's activity was carried out in 14 caves (Filipiak and Filipiak 2013; Najdek and Tabaczyński 2013; Kicińska and Najdek, 2013a, 2013b).

In 2014, the main exploration focused on Entuzjastyczna/Enthusiastic Cave, which was connected to Nadzwyczajna/Extraordinary Cave and the 03 062 Cave, resulting in a system with 5 entrances. The exploration in Górnicza/Mining Cave was continued. The new discovered caves were: Gajgur, 03 061 and T5 (Kicińska and Najdek 2014; Najdek and Kicińska 2015).

In 2015, Smocza Jama/Dragon's Den was connected with the Entuzjastyczna/Enthusiast Cave System (the system reached 3,096 m in length and 225 m in depth). As a result of exploration, caves 03 061 and T5 were connected. The exploration was continued in Gajgur and Ledena Pečina/Lodowa/Ice caves (Najdek and Kicińska 2016; Najdek et al. 2016).

In 2012–2015 the organizers of the expedition were WKTJ, SŚ, ASAK and SV clubs. In these years, members of the following clubs also participated in the expeditions: V.M. Manda Valjevo and PK AS Belgrade (Serbia), PD Volujak Gacko and Ponir Banja Luka (PBL, Bosnia and Herzegovina), STJ KW Kraków (Poland) and one person from Bulgaria.

In 2016, two bivouacs were organized in Górnicza/Mining Cave. As a result of activities in the deeper parts, the cave was deepened to 585 m. The exploration was also carried out in the Entuzjastyczna/Enthusiastic Cave System. The expedition was organized by WKTJ and cavers from ASAK and PK AS Belgrade (Serbia), PBL (Bosnia and Herzegovina),

Fig. 4 The Skakavica Waterfall on the entrance of the Grlja Canyon (photograph by Adam Łada)

Speleoklub Warszawski (SW) and STJ KW Kraków also took part in it (Najdek et al. 2016; Najdek and Kicińska 2017).

In 2017–2021, the expeditions were organized by WKTJ, with the participation of ŚS (cavers from SW from Poland and one person from Macedonia participated in the expeditions). Exploration was conducted in the largest caves discovered in previous years. Górnicza/Mining Cave reached a depth of 696 m in 2017 and 715 m in 2018 (based on bivouac and diving in the end siphon). New passages were discovered in at the Red Path Cave, which became a large system with 9 entrances. Several new caves have been discovered, the largest of which are: Krucza/Raven, Złota/Gold, Kryształowa/Crystal, Ślubna/Wedding, Zielona/Green, Kraina Deszczowców/Land of the Rain, Oczy Sowy/Owl's Eyes, Niebieska/Blue and Kozi Hockey/Goat Hockey. In 2019, Złota/Gold Cave was connected to the previously known Frizider and Cedevita caves. In 2021–2022, after the discovery of a great chamber called the Skalne Miasto/Rock City, the Złota/Gold Cave System reached 7,515 m and became the longest in the Prokletije Mts (Najdek et al. 2017; Łada et al. 2019, 2021; Najdek and Łada 2019). In 2023, the lenght of the system reached 8,202 m.

Regardless of the expeditions organized above, the Oko Skakavica/Savino Oko spring was explored by various teams of divers. In the years 2007/2008 the Serbian-Polish cave diving expedition reached a depth of -64 m in the Oko Skakavica/Savino Oko spring. It was the first dive in this spring (Szerszeń 2008). In 2008, the Oko Skakavica/Savino Oko spring was explored by Jarosław Kur, which reached a depth of -90 m (Kur 2009).

In 2015, as part of an expedition organized by WKTJ, SŚ and ASAK, Krzysztof Starnawski dived in the Oko Skakavica/Savino Oko spring and reached the depth of -96 m (Najdek and Kicińska 2016).

4 Caves of the Prokletije Mts

Belić and Kolata Massifs

The caves of the Prokletije Mts are most numerous in the Kolata and Belić massifs (Figs. 5 and 6). In 2009–2023, more than 80 caves with a total length of over 36 km were discovered in this area, including 5 large cave systems connected during exploration, such as: the Złota/Gold, the Entuzjastyczna/Enthusiastic, the 03 061-T5, the Kolektor/Collector-Łezka/Tear and the Nibyczarna/Pseudoblack/03 131 (Table 1). These are objects with both pits and developed horizontal passages. The morphology of the passages indicates their multi-stage development in phreatic, epiphreatic and vadose zones. There are passages with circular or ellipsoidal cross sections, keyhole-type passages and meanders. The caves of the Belić and Kolata massifs are poor in speleothems, although there are accumulations of them in some caves such as Gigant/Giant, Kolektor/Collector-Łezka/Tear and the Złota/Gold cave systems. Selected caves are shown below.

The Złota/Gold Cave System is the longest cave in the entire Prokletije Mts (length of 8,202 m, vertical extent of 609 m). The system consists of 3 caves: Złota/Gold, Frizider and Cedevita which were connected during the expeditions

Fig. 5 Map of the Kolata and Belić massifs showing discovered caves (author Krzysztof Najdek)

Fig. 6 Kolata and Belić massifs (photograph by Adam Łada)

Table 1 The longest and the deepest caves of the Prokletije Mts

No	Cave	Lenght [m]	Vertical extent [m]
1	Złota/Gold system	8202	609
2	Entuzjastyczna/ Enthusiastic system	3680	269
3	03 061-T5 Cave system	3154	356
4	Górnicza/Mining	2633	715
5	Ice Cave/ Ledena Pečina/ Jaskinia Lodowa	2129	451
6	Przy Czerwonym/At the Red Path	1867	210
7	Gigant/Giant	1635	296
8	Nibyczarna/ Pseudoblack-03 131 system	1611	236
9	Abisso degli Uomini Liberi	849	402
10	Kolektor/Collector-Łezka/ Tear system	776	235
11	Wielka Mała Burza/Great Little Storm	813	154
12	W Trzech Kopcach/In Three Mounds	725	141
13	Do Savino Oko/To Savino Oko	588	256
14	Wrota Budvy/Budva Gate	491	130
15	Kraina Deszczowców/The Land of Rainforests	485	201
16	Gajgur	436	141
17	Pod Płatem/Under the Lobe	402	45
18	Lodowy Smok/Ice Dragon	396	183
19	Kolacka Niżnia/the Kolata Lower	344	40
20	Guzikowców/Button Makers	319	113

(Fig. 7). The system has 10 known entrances. It is the first such large horizontal cave system in the Kolata Massif (Fig. 8). The upper level of the cave is developed at an altitude between 1995 and 1987 m a.s.l. and the lower one at 1868–1888 m a.s.l. This system consists of both horizontal and vertical passages. Some of the horizontal parts have ellipsoidal cross sections, indicating that they were formed in the phreatic zone (Fig. 9). Clastic sediments are found in small amounts.

In 2021, a huge Rock City/Skalne Miasto Chamber was discovered with a length of approx. 250 m and width of approx. 100 m (at its widest point) and several meters high (Fig. 10). The chamber is located at an altitude of about 2000 m a.s.l. It is probably one of the largest chambers located at this height in alpine/high-mountain karst in Europe. The bottom of the chamber is sloped and covered with large blocks (several meters high and wide, and some of the size of single-family houses). In some places there are also located ice covers and one icefall (Fig. 11). According to preliminary observations, the chamber was built along the lithological border between limestone and clay shales (Norbert Skowroński, personal information).

Górnicza/Mining Cave is the deepest cave in the Prokletije Mts (Fig. 12). The cave entrance is located at an altitude of 2019 m a.s.l. in the western part of the Belić Massif. Górnicza/Mining Cave was discovered in 2010. In later years, exploration reached the lowest levels which are developed horizontally at depths over 690. These parts end with the WKTJ Siphon (Fig. 13). In 2018, in order to check the exploration possibilities in the siphon, a bivouac was organized in the cave and a telephone line was led to connect the base with the bivouac. This required the use of more than two kilometers of telephone cable. After diving in the WKTJ Siphon (Witold Hoffmann) the cave was deepened to 715 m. Further, the passages in the siphon continue but have not been checked.

Górnicza/Mining Cave is a typical vertical cave (Fig. 14). Huge pits, including the deepest Rosi Pit (80 m deep), are developed to a depth of 685 m. There is little clastic sediment in the cave. Larger blocks are piled up on the ledges between the successive pits. At an altitude of approx. 1500 m above sea level there are siphons filled with clastic sediments. In one of the side chambers, sandy deposits formed into columns with a height of about 20 cm are noteworthy.

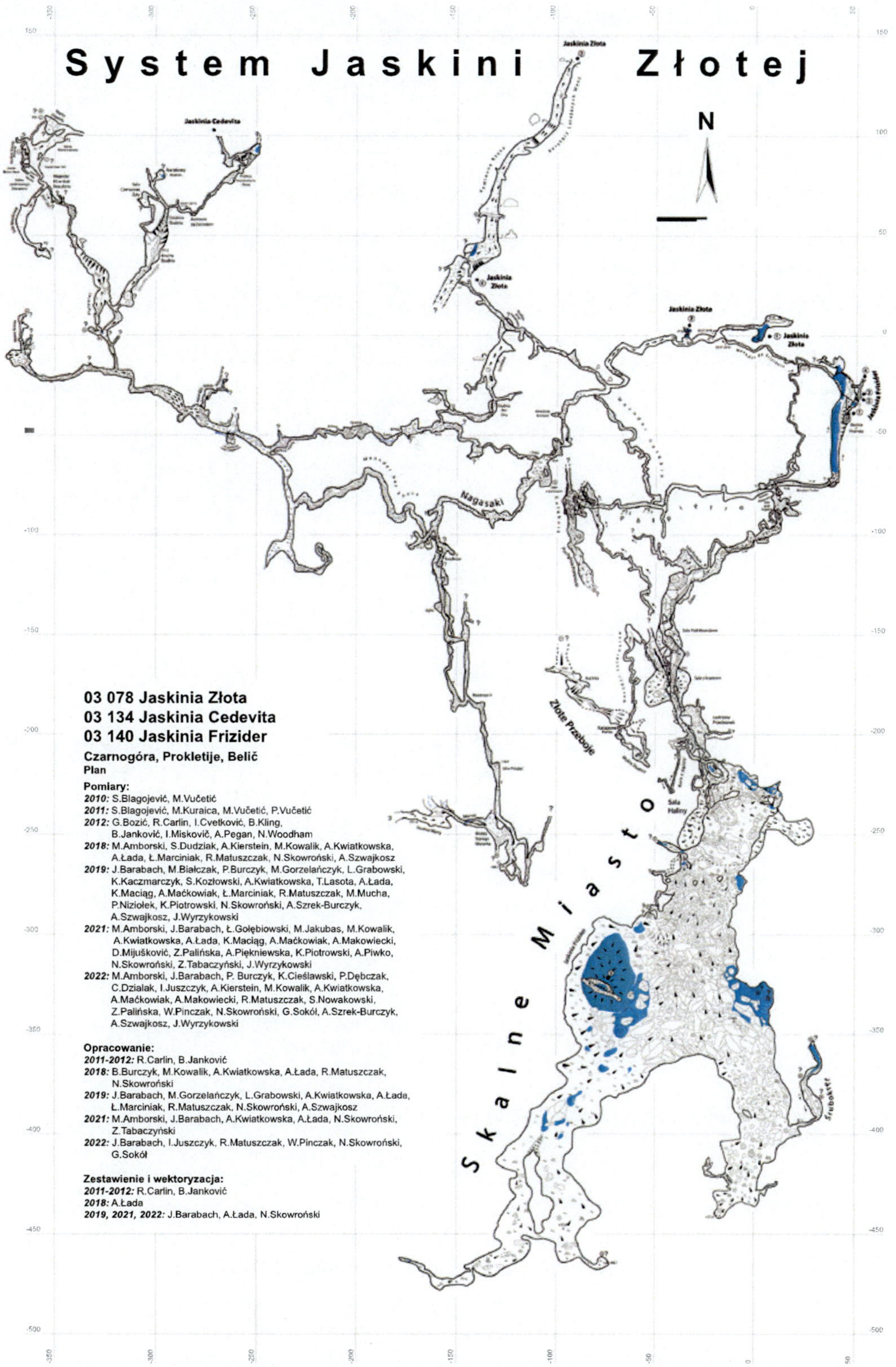

Fig. 7 The plan of the Złota/Gold Cave system (WKTJ Archive/Prokletije)

Ledena pečina/Jaskinia Lodowa/Ice Cave is the third deepest cave in the Montenegrin part of the Prokletije Mts (Fig. 15). Two entrances are located on the upper-western side of the Skripa Valley (the eastern part of the Belić Massif). The main entrance is located at an altitude of 1960 m a.s.l. (the second entrance is 8 m higher), on the tourist path to the Maja Kolata Peak (Fig. 16). In summer, strong and cold air blows from this entrance, which is most likely the reason why the cave was named "Ledena pečina" by the local people. It is a typical vertical cave with the deepest pit in the Prokletije Mts—the Evlijan Pit with a depth of 135 m. At a depth of −40 m, there is a horizontal

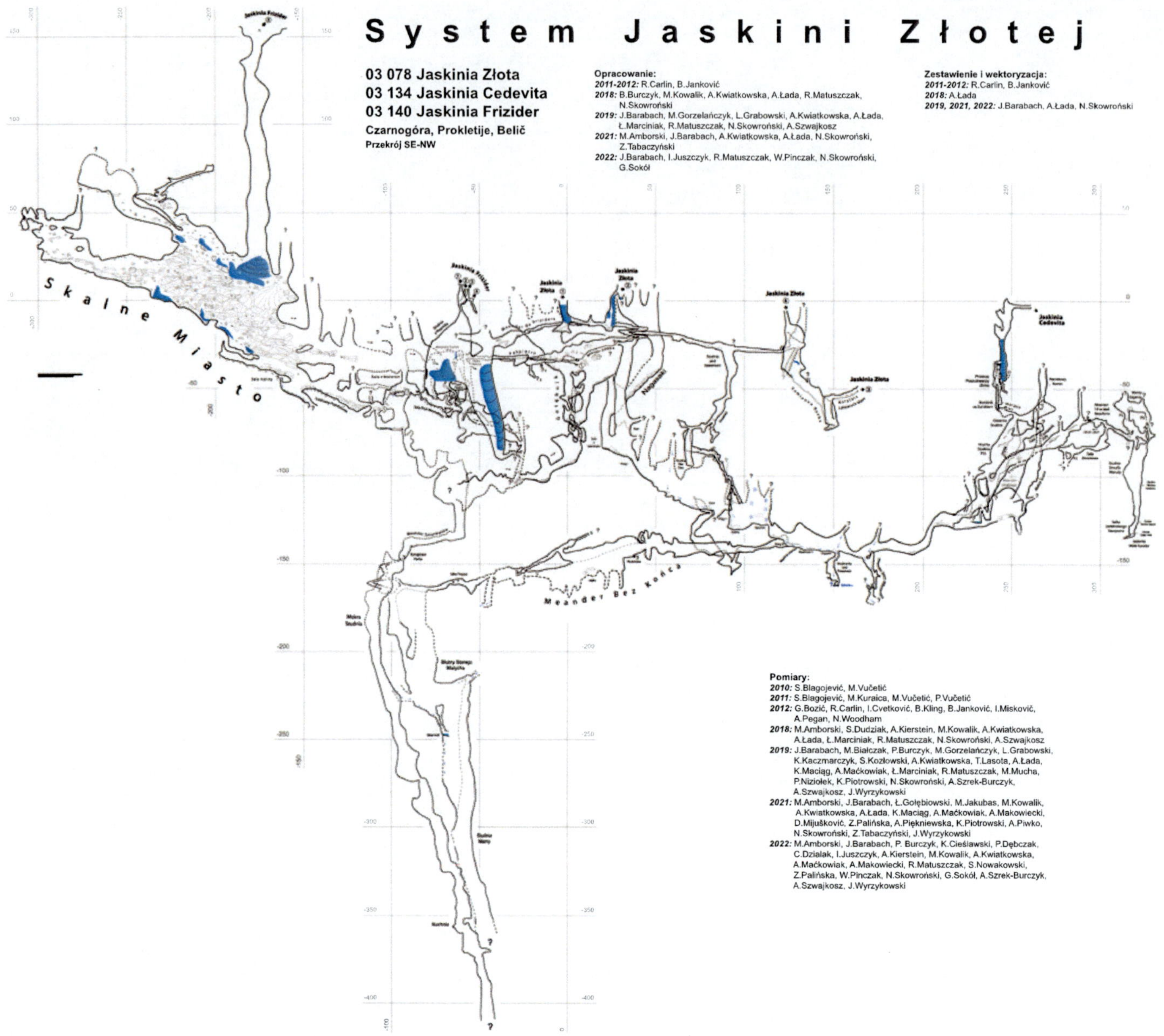

Fig. 8 The cross section of the Złota/Gold Cave System (WKTJ Archive/Prokletije)

passage called Kamienna Rzeka/Stone River (Fig. 17). Ice cover appears seasonally in this passage and its lateral parts. In 2010–2011, the icefall had a height of about 10 m and a width ranging between 7 and 8 m. There was also a frozen pond, which was formed by infiltrating water (Barovič et al., 2017). The short horizontal passages are below 400 m deep and are filled with sediment and water seasonally.

Gigant/Giant Cave (depth of -296 m, length of 1,635 m) has a big entrance located on the western part of Maja Kolata Massif (Fig. 18). From entrance the cave continues through very narrow passages several dozen centimeters high, filled with larger blocks. The passage leads to the horizontal parts of the upper level of the cave. There are two rooms here, covered with a collapse, clastic sediments and speleothems (stalagmites, dripstones, moonmilk, Fig. 19). The thickness of the moonmilk layer is 6–7 cm. The upper level of Gigant/Giant Cave is one of the oldest in the Prokletije Mts. The speleothems of this cave were dated at about 381 ka (Kicińska et al. 2021, 2023). A meander leading into the pits leads to the lower parts of the cave, where one of the largest chambers in the cave is located. There are also smaller chambers and a passage rich in speleothems (Fig. 20). Many passages of Gigant/Giant Cave are covered with collapses.

At the Red/Przy Czerwonym Cave (depth of -210 m, length of 1,867 m) is one of the bigger caves in the northwest part of Kolata Massif (Fig. 21). The cave was discovered a few years ago and then reached 19 m of height

Fig. 9 Example of phreatic passage in Złota/Gold Cave (photograph by Adam Łada)

Fig. 10 Skalne Miasto/Rock City Chamber in the Złota/Gold Cave System (photograph by Adam Łada)

Fig. 11 The ice cover in the Skalne Miasto/Rock City Chamber (photograph by Adam Łada)

difference and 140 m in length. After a break in activity, exploration of the cave has continued since 2017. Currently, the cave has 9 entrances. It is a large system with both pits and developed horizontal parts. In many places the passages are covered with ice, which in one of the larger chambers has a thickness of over 7 m (Fig. 22).

Kolektor/Collector-Łezka/Tear Cave System is one of the most interesting. Its entrance is located at an altitude of 2052 m a.s.l. in the western part of Kolata Massif. In the near-entrance part of the cave there are pits which lead to a depth of 40 m to an over 600 m long old meander. The bottom of the cave is covered with sediments of various fractions, mostly sandy. These sediments are several meters thick. There are also speleothems. The sediments from this cave were collected for research and analyzed in the laboratory.

Do Savino Oko/To Savino Oko Cave is situated in the western part of the Belić Massif. It is one of the deepest caves in this part of the massif. The cave is developed vertically. The third 50 m deep pit turns into a large chamber. A side meanders with speleothem depart from the room as well as successive pits that lead to the bottom of the cave (Najdek and Kasza 2008).

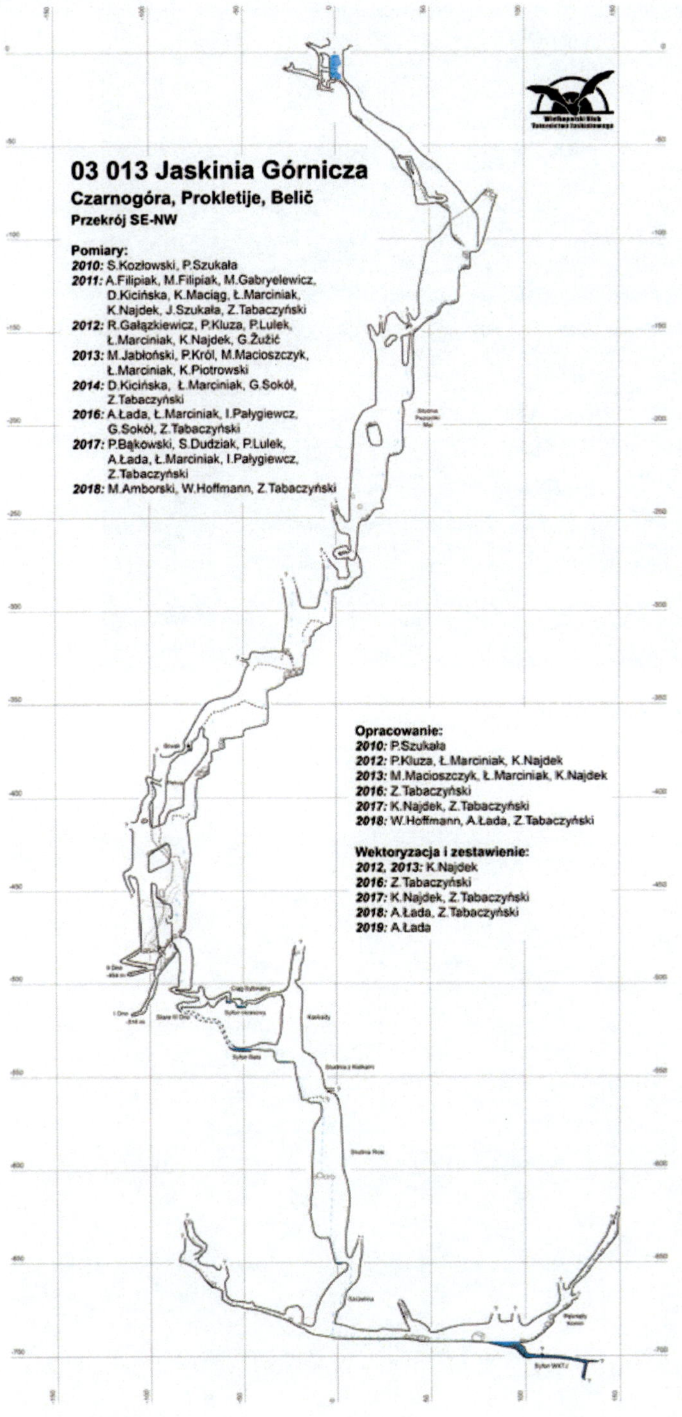

Fig. 12 The cross section of Górnicza/Mining Cave (WKTJ Archive/Prokletije)

Lodowy Smok/Ice Dragon Cave is located in the northwest of the karrenfield on the Romana slope, near the edge of a steep wall that descends into the Ropojana Valley. The greater part of the cave is developed on a meridian tectonic fissure. It is one of the deepest caves in the western part of the Belić Massif. The cave is developed vertically. The near-entrance parts of the cave are covered with snow and ice; it is a type of ice cave (Najdek and Kasza 2008).

4.1 Surroundings of the Grbaja Valley

The caves in the vicinity of the Grbaja Valley are located on the vast Ploče plateau (Fig. 23), situated in the Karanfili-Brada, Zastan Grbajski and Volučnica mountain ranges.

The Ploče plateau is located at an altitude of approx. 1,800 m above sea level. It is a vast karrens (lapiaz) with numerous pits. Most of them are shallow avens blocked with snow, collapsed or ended with narrow passages inaccessible for humans.

Of the dozens of well-proven entrances, only a few are deeper caves. These are: Lodowa/Ice Cave with a depth of −115 m, W Brekcji/In Breccia Cave (Fig. 24), Pod Karanfilem Ljuljašviča/Under Karanfili Ljuljašviča, Ukryta/Hidden Cave and Pod Szarym Kamieniem/Under the Gray Stone caves (Kicińska and Najdek 2007). The caves of this area are developed in limestone, brecciated limestone and dolomite of the Cretaceous Age.

The surroundings of the Grbaja Valley are characterized by a large number of shallow and short caves. Only a few objects were mapped.

4.2 Mount Greben

Mount Greben is part of the Visitor Massif north of Gusinje and northwest of Plav. Geologically, this area belongs to the Durmitor Unit, which includes Middle and Upper Triassic limestones and Triassic volcanic rocks. In the southern part there are Jurassic limestones and the Lower Cretaceous flysch (Djokič et al. 1976). In the part built of carbonate rocks there are surface karst forms (dolines, mainly from dissolution) and caves. However, they are not very numerous. There are several caves, the longest and deepest of which is the Čardak Cave (depth of -175 m, length of 1,054 m, Fig. 25 and 26). The entrance of Čardak Cave is located at an altitude of 1970 m above sea level. The cave is developed in Jurassic limestones (cf. Djokič et al. 1976).

In the middle part of the cave there are spacious passages and chambers (covered with blocks). In many places of the cave, there are speleothems (including small rimstone, Fig. 27). The speleothems crystallized 381 ka, in the Eemian interglacial and during the Weichselian age (Kicińska and Hercman 2014; Kicińska et al. 2021, 2023). Unfortunately, some of them are damaged. Various bones (skulls, long bones) were found on the surface of clastic deposits. The final parts were created on the tectonic fissure remodeled by the water (Kardaś 1978).

Čardak Cave is visited by tourists and local people quite rarely; probably due to the height above sea level and the long way to get there. With speleothems as well as an interesting morphology of passages that were formed under

Fig. 13 The WKTJ Siphon in Górnicza/Mining Cave (photograph by Zbigniew Tabaczyński)

Fig. 14 The Pszczółka Maja/Maya the Bee Pit in Górnicza/Mining Cave (photograph by Zbigniew Tabaczyński)

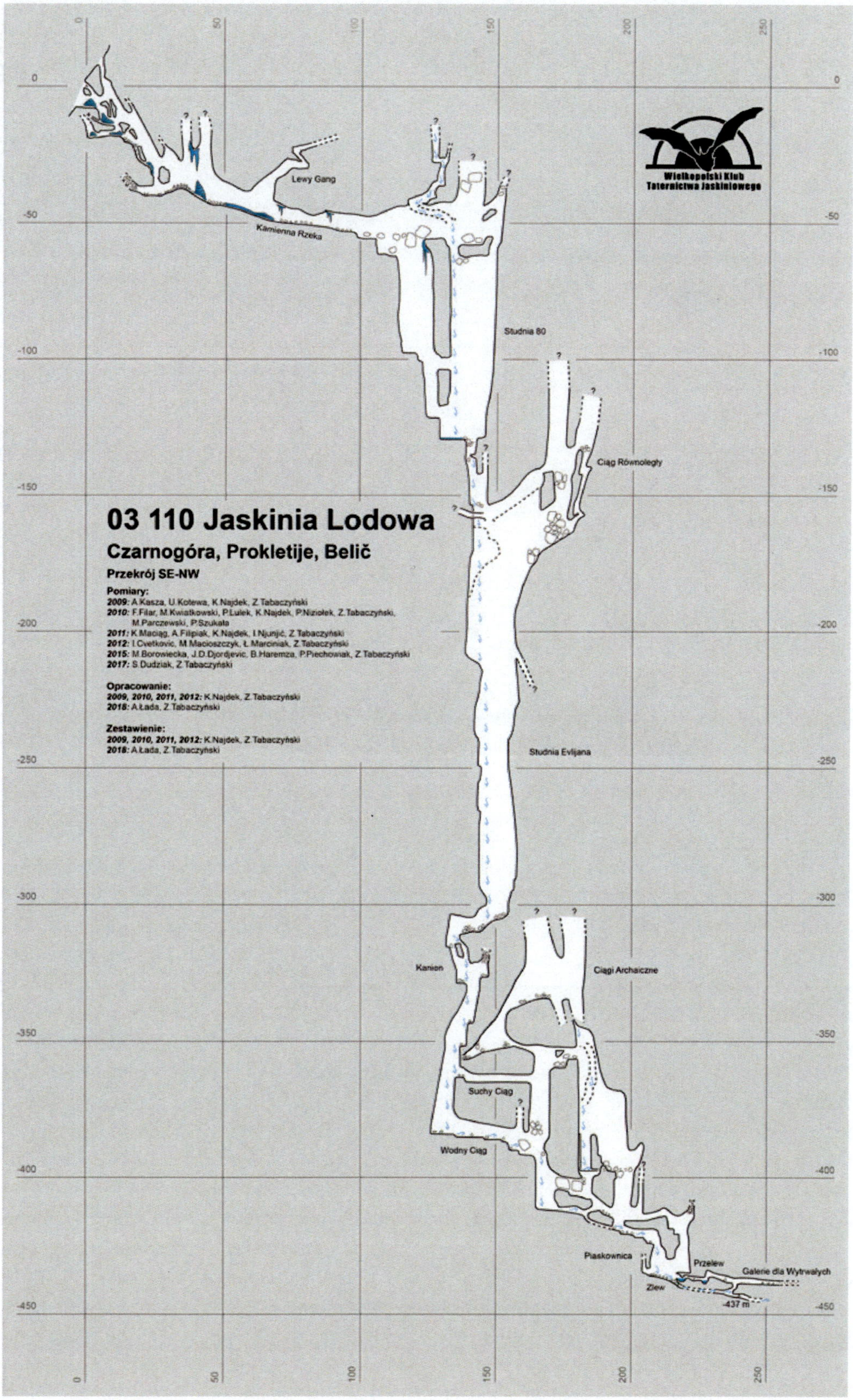

Fig. 15 The cross section of Ledena Pečina/Lodowa/ Ice Cave (WKTJ Archive/ Prokletije)

Fig. 16 Zla Kolata Peak from the entrance of Ledena pečina/Lodowa/Ice Cave (photograph by Ditta Kicińska)

various conditions (phreatic and vadose), part of the cave could be an interesting tourist object with educational and didactic values (Fig. 28).

5 Surface Karst Landforms

The Prokletije Mts are rich in forms of surface karst. Karrens (karrenfield, lapiaz), dolines, ponors, dry and blind valleys are numerous in the vicinity of the Ropojana and Grbaja valleys. In the higher parts of the mountains, extensive karrens with numerous smaller forms of surface karst are particularly noticeable. Among them there are circular (micropits, pans, kamenitza; Fig. 29), linear (grikes, solution channels and runnel, meandering runnels, Fig. 30) and polygenetic (karrenfield/lapiaz, limestone pavement, Fig. 31) forms (cf. Ford and Williams, 2007). Rillenkarren on steep slopes are common, formed by channeled flow of the rainwater (Fig. 32). Their sizes are up to several cm in width and depth (fig). These forms are found on the side walls of troughs, on roche moutonnées, blocks of moraines and many other places. Grikes are formed in many places along joints or bedding planes (Fig. 33).

In the Prokletije Mts, as in other alpine karst areas of Europe, part of the pre-Pleistocene karst forms were destroyed during the ice age. Some forms have been preserved, when the moraine of the surface was injected into karren features (cf. Ford and Williams, 2007). Karst forms such as depressions were remodeled into cirques (Milivojevič et al. 2008). Many forms were created during deglaciations. The Prokletije Mts are considered an example of glaciokarst. One of its forms in the Belić Massif area is the corrosion terraces described by Kunaver (1991). The corrosion terraces developed on the roches moutonnées, which are located near

Fig. 17 The Kamienna Rzeka/Stone River Passage in Ledena pečina/Lodowa/Ice Cave (photograph by Adam Łada)

the alpine pasture of Caf Bor (below the Caf Bor saddle). Forms such as dolines and ponors occur at different heights. Dolines are of very different sizes, from several dozen centimeters, through several meters to large forms. A large dolina with a diameter of 200 m was found on the Dobra Kolata Peak. There are also natural bridges in the study area.

6 Evolution of the Belić, Kolata and Greben Massif Caves

The caves of the Kolata, Belić and Greben massifs, located in their highest parts, at an altitude of approx. 2000 m above sea level, are horizontally developed (or their upper levels are horizontal). Like most karst systems of the Alps or the Tatra Mts, there are the oldest cave systems that were formed in a different palaeogeographic situation (the valleys were not incised as deeply as today). Some passages of the Čardak and Gigant/Giant caves have ellipsoidal cross sections, which indicates the development of these caves under phreatic conditions. These passages and larger rooms are modified by the breakdown processes and covered with collapses and speleothem (Fig. 34). The older age of these caves is evidenced not only by the height above sea level, location above the valley bottoms and the larger size of the chamber, but also by the age of the speleothems, dated at 381 ka (Kicińska et al. 2021, 2023). The age of the oldest speleothems in the cave indicates a minimum age of transition from phreatic to vadose conditions (Ford et al. 1981). Vadose-phreatic transitions have been found in these caves. These data indicate that the oldest cave systems in the Prokletije Mts began to form before 381 ka. According to Herak (1972) the development of the karst relief in the Dinaric Alps began in the Oligocene,

Fig. 18 Big entrance of Gigant/Giant Cave (photograph by Jan Barabach)

when this area has been continuously exposed to weathering. For a long time it was thought that the Dinaric karst was formed in the Middle Pliocene and most of the caves were formed in the Pleistocene. The oldest cave sediments in Slovenia are dated on 5 Ma by paleomagnetic and magnetostratigraphic methods, which proves that the caves are much older (Zupan, 2019). According to Shanov and Kostov (2014) the karst formation of the Albanian Alps "has started most likely at the end of the Eocene–Oligocene stage. By this time, the displacements of the limestone massif to the south had calmed down, and its intense uplifting started". The above data indicate that the age of the caves of the Prokletije Mountains is much older than 400 ka, it would be advisable to conduct research as the U–Pb dating of speleothem and cosmogenic radionuclides dating of clastic sediments.

On the basis of the caves discovered so far in the Kolata Massif, it is possible to conclude that there are three cave levels. Developed horizontal passages have been discovered, among others, in Złota/Gold Cave System (Fig. 8), Przy Czerwonym/At the Red Path, Gigant/Giant and Nibyczarna/Pseudoblack-03 131 caves. There are horizontal passages in the Belić Massif, for example in Entuzjastyczna/Enthusiastic (Fig. 35) and T5-03 061 cave systems, but they are less pronounced than in the Kolata Massif. The longest horizontally developed object is the Čardak Cave. These horizontal passages were formed in periods when uplift of the Prokletije Mts had ceased, reflecting the base level of erosion.

Most of the caves of the Belić and Kolata massifs developed along tectonic discontinuities (NW–SE and NE-SW) or bedding planes (i.g., the Złota/Gold, Nibyczarna/

Fig. 19 Stalagmites in the upper level of Gigant/Giant Cave (photograph by Ditta Kicińska)

Pseudoblack-03 131, Kolektor/Collector-Łezka/Tear cave systems). On the basis of structural and tectonic analyses conducted in the caves of the Albanian part of the Prokletije Mts (the Albanian Alps), Shanov and Kostov (2015) NE-SW structures related to the Pyrenean phase and NW–SE—related to neotectonic deformations. Similarly to the Albanian side, the vast majority of caves are developed vertically (cf. Shanov and Kostov 2015). According to Shanov and Kostov (2015) this situation is due to an effective infiltration which for these areas has a ratio of about 0.6–0.7. Such conditions are favorable for vast penetration into the water bearing system and do not facilitate the development of horizontal caves. However, in the Montenegrin part of Prokletije Mts, within a few years of searching, horizontal passages began to be discovered, which indicates that the rate of uplift of these mountains has varied in the past. There are episodes when it is counterbalanced by the rate of weathering and erosion.

The vertical caves of the Belić and Kolata massifs represent invasion vadose cave of proglacial type (cf. Głazek et al., 1977; Ford and Ewers, 1978; Ford and Williams, 2007). These caves were formed due to the waters flowing from melting glaciers. Pleistocene glaciations in the Prokletije Mts and in other areas of the Balkans were first observed and studied by Jovan Cvijič (1903, 1913). Intensive and systematic research conducted in the last 20 years indicates that the greatest extent of glaciers in the Balkans and its main phase took place in Marine Isotope

Fig. 20 Stalagmites and flowstone in the lower level of the Gigant/Giant Cave (photograph by Ditta Kicińska)

Stage 12, i.e. around 480–420 ka (Djurović 2009; Hughes et al. 2010, 2011; Adamson et al. 2014; Djurović and Djurović 2018). In the Prokletije Mts three different glacial events were distinguished by Milivojević et al. (2008).

In the high-mountain regions of Montenegro, it is believed that the water flowing from the melting glaciers of the oldest glacial phase (MIS 12) was the main factor in the development of caves (Djurović and Djurović 2018). The morphology of the passages and their distribution above the valley bottoms in most caves of the Belić and Kolata massifs indicate their proglacial origins (Figs. 14, 36 and 37). However, taking into account the dates of speleothem crystallization, the development of caves in the phreatic stage, the transition to the vadose stage and the modifications of the passages and chambers, the age of the upper level of the caves may be much older.

Lithology should be mentioned among the factors influencing the development of caves in the discussed massifs. The caves develop most intensively in the Upper Triassic, massive limestones of the Kolata Massif. Fewer caves formed in the dolomite and limestone breccia in the western part of the Belić Massif.

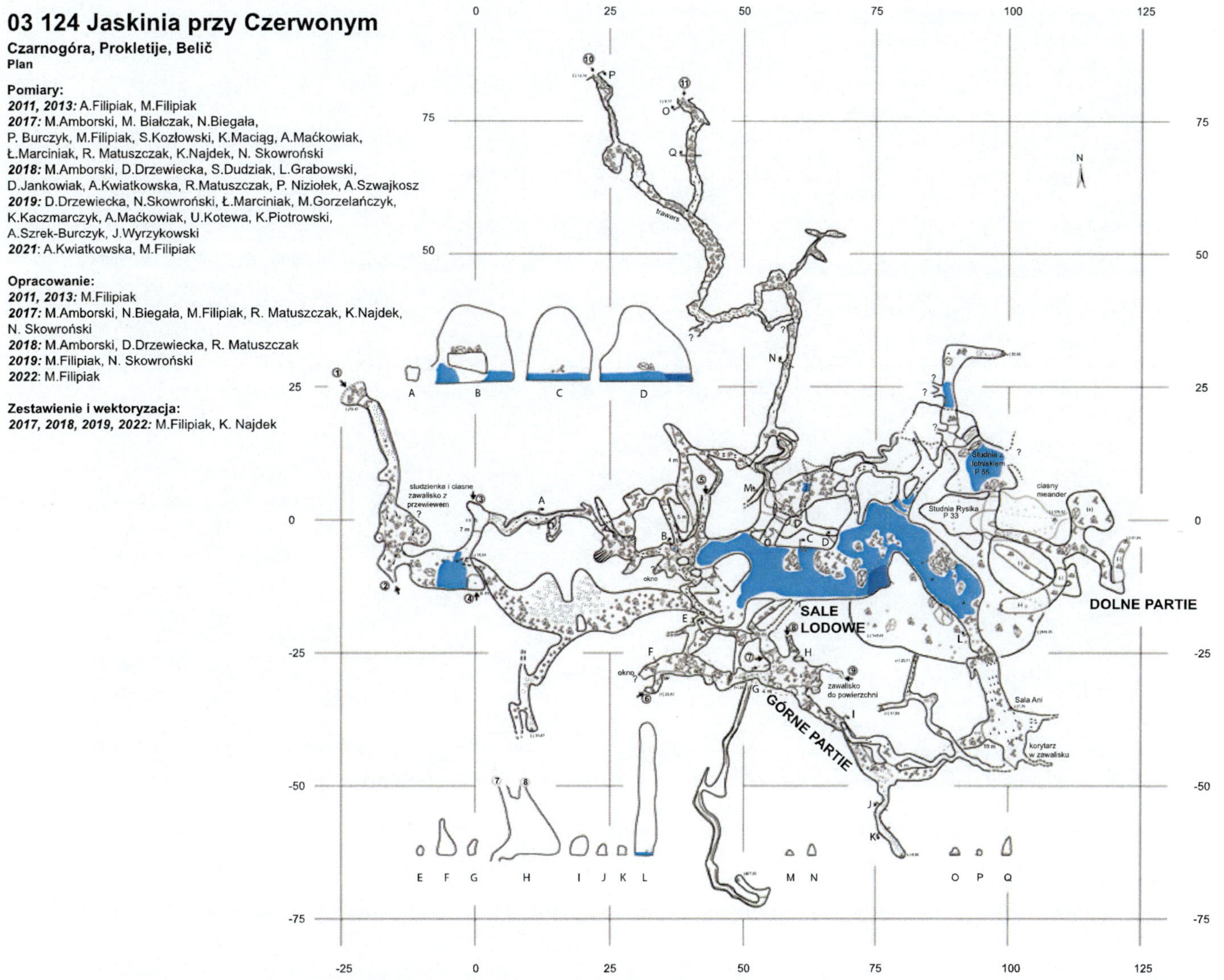

Fig. 21 The plan of Przy Czerwonym/At the Red Path Cave (WKTJ Archive/Prokletije)

Fig. 22 The ice cover in Przy Czerwonym/At the Red Path Cave (photograph by Michał Amborski)

Fig. 23 Karrenfield on the Ploče plateau (photograph by Ditta Kicińska)

As in a typical Alpine karst, clastic sediments and speleothems are not numerous in the Prokletije Mts. Passages and chambers are covered with material of various sizes, derived from the walls and ceilings. Fluvial and fluvioglacial sediments were found in a few places in the horizontal passages (Złota/Gold, Kolektor/Collector-Łezka/Tear, Nibyczarna/Pseudoblack-03 131 cave systems, Gigant/Giant and the Čardak caves).

The speleothems were found in the highest cave levels (Gigant/Giant, Čardak Cave), but also in smaller chambers or passages of other caves, such as Złota/Gold Cave System or Gigant/Giant and Kolektor/Collector caves. Some caves contain lime moonmilk, especially Gigant/Giant Cave, where it is abundant. Clastic sediments and speleothems collected for research are analyzed in laboratories.

7 Ice Caves of the Prokletije Mountains

Twelve ice caves have been identified in the explored and investigated area. Of these, 8 were described in the "Ice Caves" monograph (Barovič et al., 2017). There are Ledena pečina/Lodowa/Ice, T5-03 061, Lodowy Smok/Ice Dragon, Lodowy Gigant/Ice Giant, Górnicza/Mining, Aladinova/

Fig. 24 The entrance of W Brekcji/In Breccia Cave (photograph by Ditta Kicińska)

Aladdin, Grom Iz Vedra Neba/A bolt from the blue, Setka/Hundred-04 004 caves. In 2017–2021, another 4 caves were discovered: Kryształowa/Crystal (Fig. 38), Niebieska/Blue, Przy Czerwonym/At the Red and the Złota/Gold Cave System (Fig. 11). These caves are single-, double- and multi-entrances systems. The ice forms observed in the caves include stalactites, stalagmites, ice columns, lakes, icefalls, as well as fiber ice and ice helictites. Most caves have two entrances and are of a dynamic type. Lodowy Gigant/Ice Giant and Lodowy Smok/Ice Dragon caves are static caves with congelation ice and firn (cf. Luetscher and Jeannin 2004).

The first ice observation in Lodowy Gigant/Ice Giant was made in 2009. The marginal fissure indicated that the ice's thickness was at least 3 m (Andrzej Kasza, personal information). In 2021, the marginal fissures were so large that it was possible to get (using ropes and specialized equipment) to the base of the ice mass. Its thickness was

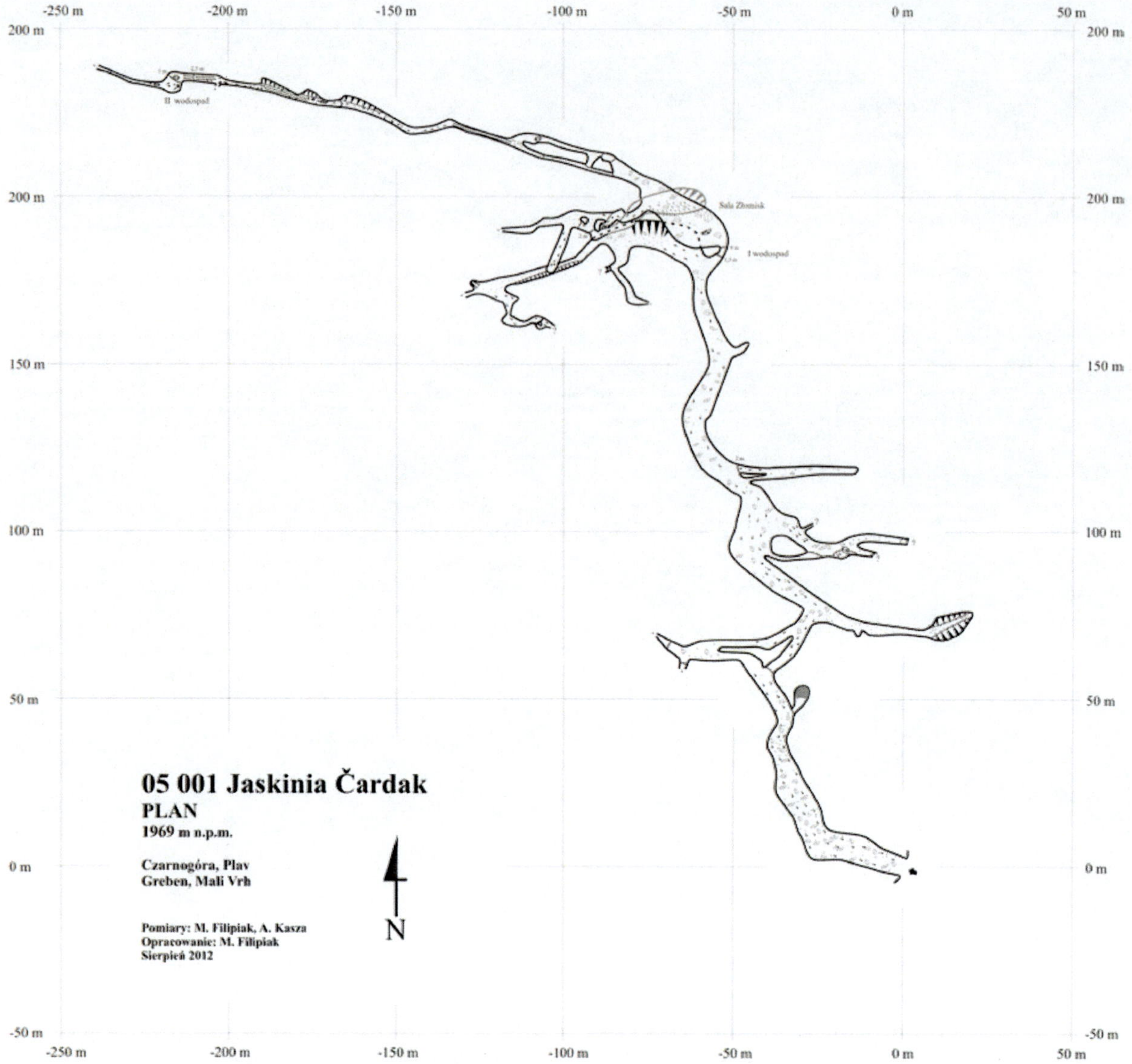

Fig. 25 The plan of Čardak Cave (WKTJ Archive/Prokletije)

estimated at 4–5 m. Moreover, the presence of fibrous ice was found (Fig. 39).

One of the most interesting objects of this type is Przy Czerwonym/At the Red Cave Path. Ice cover occurs in several places throughout the system, the greatest accumulation of it is in the Lodowe Sale/Ice Chambers. The ice thickness in this place reaches 7 m (Fig. 22). In 2019, there was a lake on its surface. In 2021, the loss of part of the ice mass was already visible. The cave is a dynamic type of ice cave.

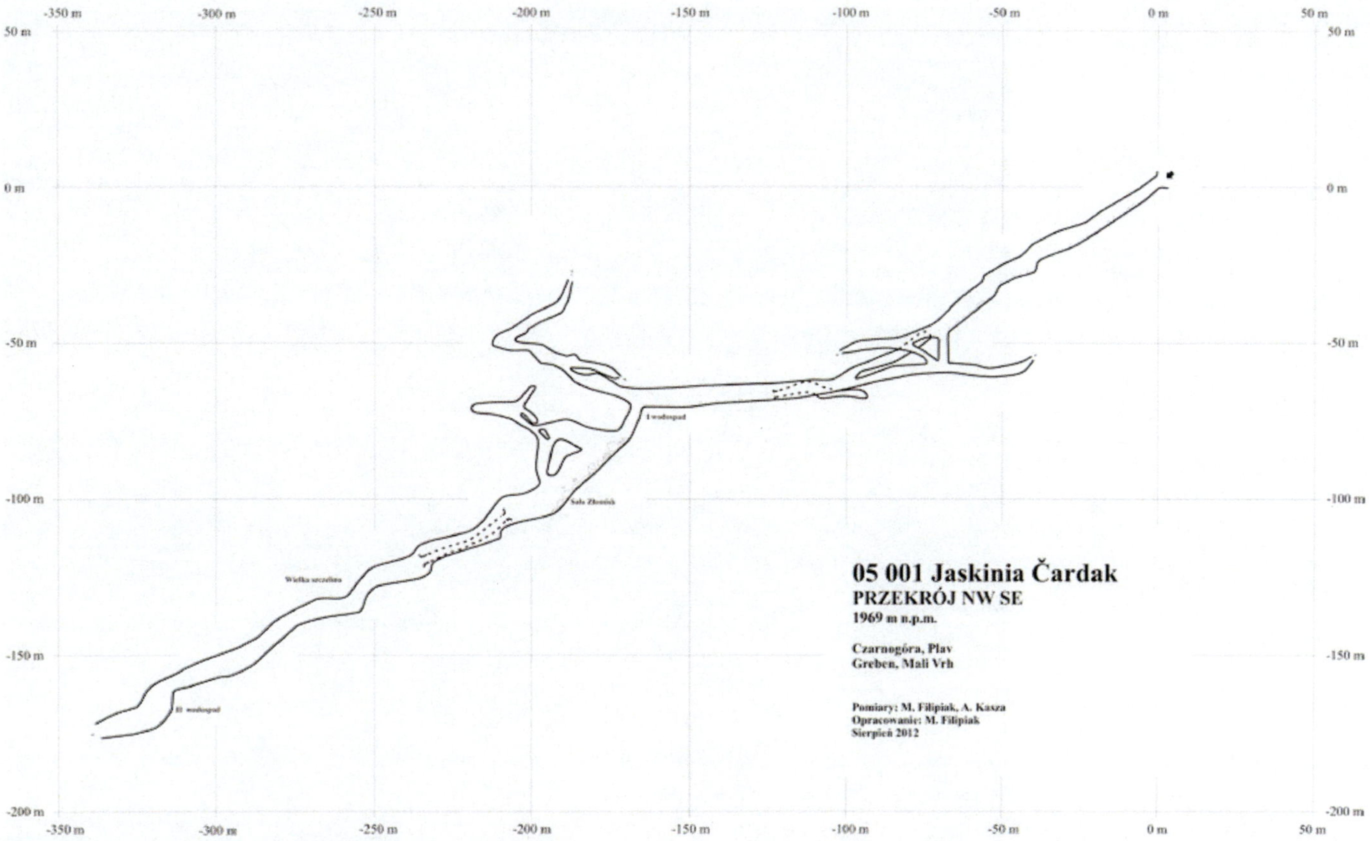

Fig. 26 The cross section of Čardak Cave (WKTJ Archive/Prokletije)

Fig. 27 Small rimstone in Čardak Cave (photograph by Mariusz Woźniak)

Fig. 28 Dripstone in the middle part of Čardak Cave (photograph by Mariusz Woźniak)

Fig. 29 Kamenitza on the Belić Massif (photograph by Ditta Kicińska)

Fig. 30 Meandering runnels on the Belić Massif (photograph by Ditta Kicińska)

Fig. 31 Glaciated limestone pavement on the Belić Massif (photograph by Ditta Kicińska)

In Kryształowa/Crystal Cave discovered in 2018, ice covered all passages and chambers from the entrance parts to the end of the cave. The cave was considered a static type. In 2021, the ice had already melted in the near-entrance part and a marginal fissure appeared in the final parts of the cave (Fig. 40). Due to the appearance of the crevice, it was possible to get to the base of the ice mass, the thickness of which was estimated at approx. 4 m. According to measurements, the cave is located in close proximity (20 m) to the Złota/Gold Cave System. Subsequent discoveries indicate that it can be a dynamic or statodynamic type.

In many caves of the Prokletije Mts, ice appears seasonally in the spring. Many entrances, especially in vertical caves, are filled with snow for years. Observations carried out during cave exploration in years 2006–2021 confirm its drastically decreasing amount (cf. Gachev and Stoyanov 2012). Since 2021, research has been carried out

Fig. 32 Small dissolution forms (rillenkarren) on the Ploče plateau (photograph by Ditta Kicińska)

Fig. 33 Grikes on the Belić Massif (photograph by Ditta Kicińska)

Fig. 34 Big chamber in Čardak Cave (photograph by Mariusz Woźniak)

Fig. 35 Phreatic passage in the Entuzjastyczna/Enthusiastic Cave (photograph by Marietta Milewska-Moult)

in two selected ice caves: Kryształowa/Crystal and Lodowy Gigant/Ice Giant. Temperature recorders were installed in these caves and measurements of the ice surface were made (Fig. 41). Continuation of research in the coming years will include, to a greater extent, topoclimatic observations, the characteristics of ice cover and three-dimensional changes in their distribution in relation to the benchmarks installed in the caves.

Fig. 36 Example of pit in Ledena Pečina/Lodowa/Ice Cave (photograph by Adam Łada)

Fig. 37 Example of pit in the Entuzjastyczna/Enthusiastic Cave (photograph by Marietta Milewska-Moult)

Fig. 38 The ice cover in Kryształowa/Cristal Cave (photograph by Adam Łada)

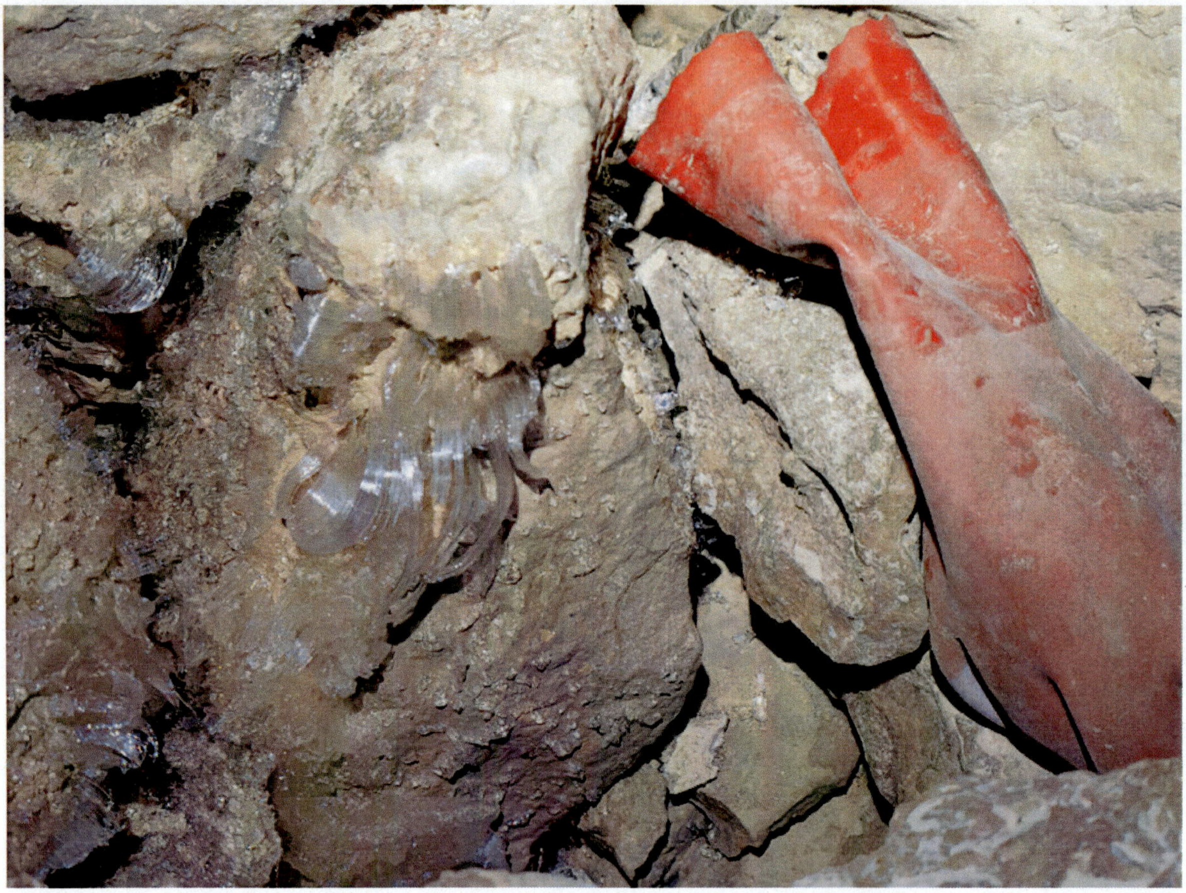

Fig. 39 Fibrous ice in Lodowy Gigant/Ice Giant (photograph by Jan Barabach)

Fig. 40 Marginal fissure in Kryształowa/Cristal Cave (photograph by Agata Piękniewska)

Fig. 41 The temperature measurements at Lodowy Gigant/Ice Giant (photograph by Jan Barabach)

Acknowledgements The authors address special thanks to Agencija za zaštitu životne sredine/Environment Protection Agency of Montenegro for permission, the Prokletije National Park for co-operation; the colleagues from caving clubs Poland, Serbia, Bosnia and Hercegovina and Croatia, Michał Amborski, Jan Barabach, Marietta Milewska-Moult, Agata Piękniewska, Mariusz Woźniak and Zbigniew Tabaczyński for photos, and Natalia Biegała for correcting the English text.

References

Adamson K R, Woodward J C, Hughes P D (2014) Glaciers and rivers: Pleistocene uncoupling in a Mediterranean mountain karst. Quaternary Science Reviews, 94: 28–43

Barović G, Kicińska D. Mandic M, Mulaomerović J (2017) Ice caves of Montenegro and Bosnia and Herzegovina. In: Persoiu A, Lauritzen S-E (eds), Ice caves, Elsevier, 263–284

Biegała N, Kicińska D, Najdek K (2009) Prokletije (Bjeskhet e Namuna) 2006–2009. In: Gradziński M, Kicińska D, Szelerewicz M (eds) Polish Caving 2005–2009 (Published on the occasion of the 15th International Speleological Congress). Caving Commision of Polish Mountaineering Association, 44. Wyd. Firma Rysunkowa Szelerewicz, Kraków, 20–22

Cvijić J (1903) Novi rezultati o glacijalnoj eposi Balkanskog poluostrva. Glas Srpske kraljevske akademije 65: 185–240

Cvijić J (1913) Ledeno doba u Prokletijama i okolnim planinama (Glacial epoch on the Prokletije and surrounding mountains). Serbian Royal Academy of Sciences, 91: 188–267

Cvijić J (1899) Glacijalne i morfološke studije Bosne, Hercegovine i Crne Gore. Glas Srpske kraljevske akademije, knj. 57

Development in Mountain Regions, Southeastern Europe. Second edition, Springer International Publishing Switzerland, 90–11

Djokič V, Živaljevič M, Petrovič Z (1976) Osnovna geološka karta i tumač za list Gusinje. Savezni geološki zavod, Beograd

Djurović P (2009) Reconstruction of the pleistocene glaciers of Mount Durmitor in Montenegro. Acta geographica Slovenica, 49(2): 263–289

Djurović P, Djurović M (2018) Review of the most significant caves in Montenegro. Acta Carsologica 50/1: 49–65

Djurović P, Djurović M. (2016) Physical Geographic Characteristics and Sustainable Development of the Mountain Area in Montenegro. In: Zhelezov G. (ed), Sustainable

Djurović P, Lješević M (1994) Speleološki objekti Djalovića klisure i njihov značaj za razumevanje cirkulacije vode u krasu Peštera. Zbornik radova Odbora za kras i speleologiju SANU, knj. 5, Beograd, 109–131

Filipiak A, Filipiak M (2013) „U nas na Göllu...", czyli Prokletije 2013. Jaskinie, 74, 19–22

Ford D C, Ewers O (1978) The development of limestone cave systems in the dimensions of length and depth. Canadian Journal of Earth Science, 15: 1783–1798

Ford D C, Wiliams P W (1989) Karst Geomorphology and Hydrology, Unwin Hyman, London, pp. 1–601

Ford D C, Schwarcz H P, Drake J J, Gascoyne M, Harmon R S, Latham A G (1981) Estimates of the age of the existing relief within the Southern Rocky Mountains of Canada. Arctic and Alpine Research, 13: 1–10

Frasheri A, Bushati S, Bare V (2009) Geophysical outlook on structure of the Albanides. Journal of the Balkan Geophysical Society, 12(1): 9–30

Gachev E, Stoyanov K (2012) Present day small perennial firn-ice patches in the mountains of the Western Balkan Peninsula. Studia Geomorphologica Carpatho-Balcanica, 51–70

Gemignani L, Mittelbach B V, Simon D, Rohrmann A, Grund M U, Bernhardt A, Hippe K, Giese J, Handy M R (2022) Response of Drainage Pattern and Basin Evolution to Tectonic and Climatic Changes Along the Dinarides-Hellenides Orogen. Frontiers in Earth Sciences, Structural Geology and Tectonics, https://doi.org/https://doi.org/10.3389/feart.2022.821707

Głazek J, Rudnicki J, Szynkiewicz A (1977) Proglacial caves - a special genetic type o f caves. In: Ford T D (ed.) Pro ceedings o f the 7th International Speleological Congress, Sheffield 1977. British Cave Research Association, pp. 215–217

Hajna N Z (2019) Dinaric karst—Geography and geology. In: White B H, Culver D C, Pipan T (eds) Encyclopedia of Caves, Elsevier, 253–362

Herak M (1972) Karst of Yugoslavia. In: Herak M, Stringfield V T (eds), Karst: Important Karst Regions of the Northern Hemisphere. Amsterdam, Elsevier. 25–83

Hughes P D, Woodward J C, Calsteren P C, Thomas L E , Adamson K R (2010) Pleistocene ice caps on the coastal mountains of the Adriatic Sea. Quaternary Science Reviews, 29, 3690–3708

Hughes P D, Woodward J C, van Calsteren P C, Thomas L E (2011) The glacial history of the Dinaric Alps, Montenegro. Quat. Sci. Rev. 30 (23–24): 3393–3412

Karaman M G (2018) Establishment of Natura 2000 network in Montenegro general report on cave habitat type. Ministry of Sustainable Development and Tourism, pp 66

Kardaś R (1978) Zjawiska krasowe w okolicach Gusinje (góry Prokletije, Jugosławia). Kras i speleologia, 2: 110–114

Kasza A, Kicińska D, Najdek K, Tabaczyński Z (2010) Prokletije Belić2009. Jaskinie, 59: 21–23

Kicińska D, Najdek K (2007) Prokletije – Góry Przeklęte. Jaskinie, 46: 20–22

Kicińska D, Najdek K, Filipiak M (2011) Prokletije 2010-2011. Taternik, 2-3, 40–45

Kicińska D, Hercman H (2014) Pierwsze datowania nacieków jaskiniowych metodą uranowo-torową w Górach Prokletije (Czarnogóra). Materiały 48. Sympozjum Speleologicznego, 16–19.10.2014, Kletno: 75

Kicińska D, Najdek K (2013b) Cave exploration of the Belićmassif in the Prokletije Mountains (Montenegro). Proceedings of 16th International Congress of Speleology, Brno, 2: 165–167

Kicińska D, Najdek K (2014) Działalność eksploracyjna w jaskiniach Gór Prokletije. Mat. 48 Sympozjum Speleologicznego, Kletno, 73–74

Kicińska D, Najdek K (2013 Prokletije 2009–2013. In: Kicińska D (ed), Polish Caving 2009–2013. Komisja Taternictwa Jaskiniowego Polskiego Związku Alpinizmu (Caving Commision of Polish Mountaineering Association). Wyd. Pracownia Kreatywna Bezliku, Kraków: 32–34

Kicińska D, Pawlak J, Stienss J, Sierpień P, Błaszczyk M, Hercman H (2021) Ewolucja i wiek systemów jaskiniowych w Górach Prokletije, Czarnogóra. Materiały 55. Sympozjum Speleologicznego Sekcji Speleologicznej PTP im. Kopernika, Bartkowa 14–17.10.2021, Sekcja Speleologiczna PTP Kraków: 59

Kicińska D, Pawlak J, Stienss J (2023) An attempt to identify source areas of clastic deposits from selected caves of the Prokletije Mountains (Montenegro): a mineralogical and U-series geochemistry approach. Int J Speleol, 52(2): 109–122

Kunaver J (1991) Corrosion terraces as geoecological response to postglacial development of glaciokarstic rock surface. Proceedings ofthe Internatlonal Conference on Envlronmental Changes in Karst Areas -I.G.U.- U.I.S.-ltaly 15–27 Sept. 1991; Quaderni del Dipartimento di Geografia n. 13. Università di Padova. pp. 325–331

Kur J (2009) Eksploracja w Jaskini Savino Oko 26–30.10.08 http://www.hotdive.com/news,117,Relacje-z-wypraw,-Eksploracja-w-Jaskini-Savino-oko-26-301008.html (update 16/01/2009)

Łada A, Kicińska D, Najdek K (2019) Eksploracja jaskiń w górach Prokletije (Czarnogóra). Materiały 53. Sympozjum Speleologicznego, Szczawnica, 56

Łada A, Kicińska D, Najdek K, Palińska Z (2021) Eksploracja jaskiń w Górach Prokletije, Czarnogóra. Materiały 55. Sympozjum Speleologicznego Sekcji Speleologicznej PTP im. Kopernika, Bartkowa 14–17.10.2021, Sekcja Speleologiczna PTP Kraków: 66

Luetscher M, Jeannin P Y (2004) A process-based classification of alpine ice caves. Theor. Appl. Karstol. 17: 5–10

Milivojevič M, Menkovič L, Čalič J (2008) Pleistocene glacial relief of the central part of Mt. Prokletije (Albanian Alps). Quaternary International, 190: 112–122

Mulič R (2009) Plavsko-Gusinjske Prokletije, četrdeset planinarskih staza. Planinarsko društvo "Karanfili", Gusinje pp. 213

Najdek K (2007) Prokletije – Bjeskhet e Namuna 2007. Taternik, 3–4: 50

Najdek K, Kasza A (2008) Prokletije - Bjeshket e Namuna 2008. Jaskinie 52: 21–24

Najdek K, Kicińska D (2015) Prokletije 2014. Jaskinie, 79–80: 27–29

Najdek K, Kicińska D (2016) Entuzjastycznie w Górach Przeklętych. Jaskinie, 83–84: 28–30

Najdek K, Łada A (2019) Prokletije 2018. Jaskinie, 95: 15–20

Najdek K, Tabaczyński Z (2013) Prokletije 2012. Jaskinie, 72: 9–12

Najdek K, Kicińska D, Tabaczyński Z (2017) Najkrótsza wyprawa i najdłuższa jaskinia. Jaskinie, 87: 8–11

Najdek K, Kicińska D (2017) Prokletije 2013–2016. In: Kicińska D (ed) Polish Caving 2013–2017 (Published on the occasion of the 17th International Speleological Congress). Caving Commision of Polish Mountaineering Association, Wyd. Firma Rysunkowa Szelerewicz, Kraków: 29–32

Najdek K, Kicińska D, Kasza A (2016) Eksploracja jaskiniowa w Górach Prokletije w latach 2015–2016. Mat. 50 Sympozjum Speleologicznego, Chęciny, 132–133

Palmentola G, Baboci K, Gruda G, Zito G (1995) A note on rock glaciers in the Albanian Alps. Permafr. Periglac. Process. 6: 251–257

Petrović A (2014) A reconstruction of the pleistocene maximum in the Žijovo range (Prokletije moutains, Montenegro). Acta Geographica Slovenica, 54 (2): 255–269

Pretner E (2011) Die Verdienste von Leo Weirather um die Biospeläologie, insbesondere Jugoslawiens, sein Höhlenkataster und seine Sammelplätze. Ber. nat.-med. Verein Innsbruck, 97: 85–234

Radojević D (2012) Protection and Sustainable Use of the Dinaric Karst Transboundary Aquifer System, Hydrogeological Report - Draft V.1, 1–8 Montenegro, pp. 1–170

Robertson A, Shallo M (2000) Mesozoic–Tertiary tectonic evolution of Albania in its regional Eastern Mediterranean context. Tectonophysics, 316: 197–254

Shanov S, Kostov K (2015) Dynamic Tectonics and Karst. Springer, Heidelberg, New York, Dordrecht, London. pp. 27–35

Szerszeń A (2008) Expedition OKO, Montenegro. http://gnj.org.pl/_php/show.php?id=119 (Update: May 31, 2019)

Vladan B, Knețević M, Laliĉić R (2012) Studija o opravdanosti osnivanja opštine Gusinje. Društveno-ekonomska opravdanost vraćanja statusa opštine Gusinju, pp 96